The Future of Wireless Networks

Architectures, Protocols, and Services

WIRELESS NETWORKS AND MOBILE COMMUNICATIONS
Dr. Yan Zhang, Series Editor
Simula Research Laboratory, Norway
E-mail: yanzhang@ieee.org

The Future of Wireless Networks

Architectures, Protocols, and Services

Edited by
Mohesen Guizani · Hsiao-Hwa Chen · Chonggang Wang

CRC Press
Taylor & Francis Group
Boca Raton London New York

CRC Press is an imprint of the
Taylor & Francis Group, an **informa** business

CRC Press
Taylor & Francis Group
6000 Broken Sound Parkway NW, Suite 300
Boca Raton, FL 33487-2742

First issued in paperback 2019

© 2016 by Taylor & Francis Group, LLC
CRC Press is an imprint of Taylor & Francis Group, an Informa business

No claim to original U.S. Government works

ISBN-13: 978-1-4822-2094-0 (hbk)
ISBN-13: 978-0-367-37741-0 (pbk)

Visit the Taylor & Francis Web site at
http://www.taylorandfrancis.com

and the CRC Press Web site at
http://www.crcpress.com

Contents

SECTION III: SERVICES AND APPLICATION IN FUTURE WIRELESS NETWORKS   305

Juan Chen and Xiaojiang Du

List of Figures

List of Tables

FUTURE WIRELESS NETWORK ARCHITECTURE

I

Chapter 1

Future Cellular Network Architecture

Ying Li

Samsung Research America

CONTENTS

1.1 Introduction

The demand of wireless data traffic is explosively increasing [26] due to the increasing popularity of smart phones and other mobile data devices such as tablets, netbooks, and eBook readers among consumers and businesses. The fourth generation (4G) cellular technologies [15] including Long Term Evolution (LTE)-Advanced and Advanced Mobile WiMAX (IEEE 802.16m) use traditional network architecture, which has limitations. In order to meet this spectacular growth in mobile data traffic, improvements in future cellular network architecture has been of paramount importance.

Today's cellular networks consist of Radio Access Network (RAN), which mainly deals with the air interface of the base stations (referred to as evolved Node Bs (eNBs)) and mobile stations (referred to as user equipment (UE)), where an eNB can consist of one or multiple cells, and Evolved Packet Core (EPC) network, which mainly deals with the packet processing after the eNB before it goes to the Internet [24]. The explosive increase in demand for wireless data has placed increasing challenges on today's RAN and EPC, which both have limitations.

In RAN, a possibility to increase the overall system capacity is to deploy a large number of smaller cells. In today's RAN, compared to traditional homogeneous networks, there are new scenarios and considerations in heterogeneous networks with base stations of diverse sizes and types [18, 27]. One consideration is traffic offloading from large cells to small cells. For example, small cell's footprint can be enlarged to offload the traffic from a macro cell to a small cell [2]. Another consideration is on the resource management considering interferences. The large cells and small cells can be deployed on multiple carriers, and carrier aggregation (CA) [24] can be used to achieve high system performance. The large cells and small cells can be deployed on a single carrier and co-channel intercell interference management can apply, such as time domain muting. Coordinated Multipoint (CoMP) transmissions [5] can also be used to coordinate the transmissions among multiple cells, such as a joint transmis- sion from multiple points to achieve higher system performance.

However, even with the techniques above for heterogeneous networks, today's RAN is with limitations and is still not optimized in several aspects.

1. First, even though CA or CoMP can be used for multiple cells with different sizes, the current RAN can only support the case that the multiple cells involved are with a same eNB. An eNB can have multiple cells, such as a large cell and a few small cells by remote radio head (RRH), transmit point (TP), etc. A CA or CoMP involving more than one eNB currently is not supported. In addition, the current CA does not much support the deployment scenario where the cells are connected via nonideal backhaul (meaning the backhaul delay is nonnegligible). These limit the deployment scenarios as well as the system performance.

2. Second, the operational expenditure (OPEX) of today's RAN is high. For example, RAN nodes constitute most of the energy consumption of a cellular network [11]. The large number of RAN nodes are usually based on proprietary platforms. A RAN node utilization is usually lower than that capacity

because the system is designed to cover the peak load; however, the average load is far lower, but today's support for resource sharing among RAN nodes is low. The high OPEX makes it difficult for operators to in- crease the revenue while the mobile traffic is explosively increasing.

3. Third, the support for self-organized networking (SON) is limited. Some SON functions are supported [24], such as plug-and-play as a home eNB. With the advances in wireless backhaul and energy harvesting, the future small cells can be with no wire (no wired backhaul or powerline), hence, they can be drop-and-play. Today's SON functions are limited and not ready for possible future drop-and-play scenarios.

The future RAN should improve to mitigate the limitations. The following can be considered for the future RAN:

1. A better support for a UE to be connected to multiple eNBs concurrently is needed, with a consideration of diverse backhaul conditions. In addition, diverse applications and diverse traffic of a UE can be assigned to cells with diverse backhaul conditions concurrently, to further improve the system performance.

2. Technologies to reduce the OPEX are needed. One of the potential technologies is cloud RAN [22], where the baseband processing of different eNBs and cells can be centralized and resources pooled and shared. It can leverage more efficient resource utilization among different eNBs and cells.

3. Advanced SON to support drop-and-play small cells can be considered. The advanced SON should support load balancing, robust routes establishment and adaptive routing, adaptive ON/OFF of small cells considering energy harvesting and traffic dynamics, and so on.

For EPC, today's architectures have some major limitations. A centralized data plane in the cellular network forces all the traffic of the UEs (including traffic between users on the same cellular network) to go through the packet gateway (P-GW) at the cellular-Internet boundary, which faces scalability challenges and makes it difficult to host popular content inside the cellular network. In addition, the network equipment has vendor-specific configuration interfaces and communicates through complex control plane protocols, with a large and growing number of tunable parameters (e.g., several thousand parameters for eNBs). As such, network operators have (at best) indirect control over the operation of their networks, with little ability to create innovative services.

Network operators are finding it difficult to introduce new revenue generating services and optimize their expensive infrastructures. Networks continue to have serious known problems with security, robustness, manageability, and mobility. Network capital costs have not been reducing fast enough, and operational costs have been growing. Even vendors and third parties are not able to provide customized cost-effective solutions to address their customers' problems.

The limitations in EPC have created opportunities for the next generation wireless network architectures incorporating Software Defined Networking (SDN) and network virtualization. SDN and virtualization are gaining momentum in wired networks [19–21] because of their advantages, such as low-cost deployments, easy management, and so on. However, SDN and virtualization have not been studied much for wireless networks, yet they have great potential to meet the challenges that today's wireless networks are facing [16, 17, 23].

SDN is a type of networking in which the control plane is physically separate from the forwarding plane. Network intelligence is (logically) centralized in a software-based SDN controller, which maintains a global view of the network. The network appears to the application and policy engines as a single, logical switch. As for network virtualization, the resources and functions in the cellular network can be virtualized. As such, SDN and network virtualization can adaptively and flexibly provide traffic offloading, revenue-adding services, and the capability to deliver services to mobile stations that require a large amount of data.

The concept and technologies of SDN and virtualization for cellular networks are still in a very early stage. They have to address cellular networks' own unique characteristics and requirements. For example, special considerations are needed for aspects such as supporting many subscribers, frequent mobility, fine-grained measurement and control, and real-time adaptation, which introduces scalability challenges.

Accordingly, the chapter is organized as follows. In the next section we describe today's cellular network architecture, including RAN and EPC. The third section discusses the future architecture for RAN, including the support for a UE to connect concurrently to multiple eNBs and associate diverse traffic to the eNBs, Cloud-RAN, and advanced SON to support drop-and-play small cells. In the fourth section, we pro- vide the future architecture for EPC, including SDN and network virtualization. The chapter is concluded with some summarizing visions.

1.2   Today's Cellular Network Architecture

Today's cellular network architecture is discussed in this section. Figure 1.1 illustrates an overview of the architecture of RAN and EPC. RAN and EPC are interfaced via S-interface, mostly S1-interface [24]. The details of the figure are presented in the following subsections.

1.2.1   Radio Access Network (RAN)

Today's RAN can deploy heterogeneous networks. Cells with different sizes can be used in a hierarchical network deployment, referred to as multitier deployment or multitier networks, where each tier can be for one type of cell of certain size. The type and location of the eNB controlling these cells will play a significant role in determining the cost and performance of the multitier deployments. For example, indoor femtocell deployments using home eNBs (HeNBs) can utilize the existing

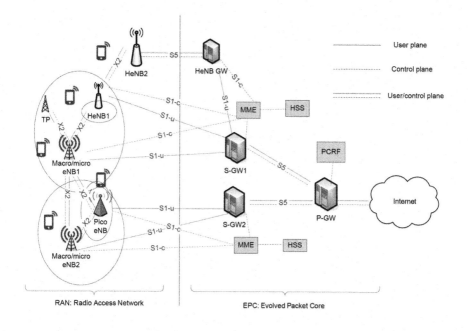

Figure 1.1: An overview of today's cellular network architecture.

backhaul, thereby significantly lowering the cost of such deployments. With outdoor picocell deployments through pico eNB, the operator will need to provide backhaul capability and manage more critical spectrum reuse challenges. Other deployment models cover indoor enterprise or outdoor campus deployments that may impose different manageability and reliability requirements.

The RAN part of Figure 1.1 illustrates an exemplary heterogeneous network with macro/micro eNB, pico eNB, and a femtocell/HeNB. For the heterogeneous network shown in Figure 1.1 the pico eNB has smaller transmission power than macro eNB— hence with smaller coverage than the macro eNB. On the other hand, the HeNB can have smaller transmission power than a pico eNB. A cell formed by a TP can belong to an eNB. Picocells typically are managed together with macro/microcells by operators. An interface of X2 can be used for the communications among the eNBs, HeNBs, and TPs. All kinds of eNBs can be connected to the servicing-gateway (S-GW) in EPC for the user plane (or the data plane), and connected to Mobility Management Entity (MME) in EPC for the control plane. For HeNB, it can also be connected to the S-GW and MME via a HeNB gateway. It is noted that Figure 1.1 does not include relay eNB for simplicity. Relay eNB can be included where an eNB that does not have wired backhaul can connect to the EPC via a relay eNB.

The coverage area of the picocell is limited not only by its transmit power, but also, to a large extent, by the intercell interference from other cells. Therefore, if the cell selection criteria are only based on downlink UE measurements such as the reference signal received power (RSRP), only UEs in close vicinity will end up being served by the pico eNB. Due to the higher deployment density of the small cells, it is beneficial to expand the footprint of the picocells, i.e., offloading UEs from macrocells to picocells, to enable more UEs to connect to the small cells to take advantage of the higher deployment density. This can be achieved through cell range expansion (RE) [2]. One of the approaches for cell range expansion is that a cell-specific bias to the UE measurement of X dB is applied for pico eNB to favor connecting to it. In this way, more UEs will be inclined to connect to pico eNBs instead of macro eNBs. Furthermore, time domain intercell interference coordination techniques can also be utilized for pico users that are served at the edge of the serving pico cell, for example, for traffic offloading from a macrocell to a picocell.

Spectrum allocation across multiple tiers is an important aspect of deployment and use of hierarchical architectures. According to the spectrum used, multitier cell deployments are possible for the following cases:

1. Multiple-carrier case: The multitier cells are deployed on multiple carriers. When multiple carriers are available, choices can be made to enable flexible cell deployment. For example, the macrocell and small cells can be deployed on distinct carriers, or on the same set of carriers while having joint carrier and power assignment/selection to better manage intercell interference.

2. Single-carrier case: The multitier cells are deployed on a single carrier. This can also be called co-channel deployment.

Techniques such as CA and CoMP can apply for the resource management. Muting in the time domain can also apply [3].

A UE can be connected to multiple cells, such as in the CA case or CoMP case. However, today's RAN does not support a UE concurrently connecting to more than one eNB. It does not support much for CA in the scenarios where cells can be connected via nonideal backhaul links.

1.2.2 *Evolved Packet Core (EPC)*

Today's cellular networks connect eNBs to the Internet using IP networking equipment. An illustration of the entities in EPC and how they connect to each other is shown in the EPC part of Figure 1.1.

For the data plane or the user plane, the traffic from an eNB goes through a serving gateWay (S-GW) over a tunnel. The S-GW serves as a local mobility anchor that enables seamless communication when the user moves from one base station to another. The S-GW must handle frequent changes in a user's location, and store a large amount of states since users retain their IP addresses when they move. The S-GW tunnels traffic to the P-GW. The P-GW enforces quality of service (QoS) policies and monitors traffic to perform billing. The P-GW also connects to the Internet and other cellular data networks, and acts as a firewall that blocks unwanted traffic. The policies at the P-GW can be very fine grained, based on whether the user is roaming, properties of the user equipment, usage caps in the service contract, parental controls, and so on.

Besides data plane functionalities, the eNB, S-GW, and P-GW also participate in several control plane protocols. In coordination with the MME, they perform hop-by-hop signaling to handle session setup, teardown, band reconfiguration, as well as mobility, e.g., location update, paging, and handoff. For example, in response to a UE's request for dedicated session setup (e.g., for VoIP call), the P-GW sends QoS and other session information (e.g., the TCP/IP 5-tuple) to the S-GW. The S-GW in turn forwards the information to the MME. The MME then asks the eNB to allocate radio resources and establish the connection to the UE. During handoff of a UE, the source eNB sends the handoff request to the target eNB. After receiving an acknowledgment, the source eNB transfers the UE state (e.g., buffered packets) to the target eNB. The target eNB also informs the MME that the UE has changed cells, and the previous eNB to release resources.

The S-GW and P-GW are also involved in routing protocols. The Policy Control and Charging Function (PCRF) manages flow-based charging in the P-GW. The PCRF is connected to the P-GW via the control interface. The PCRF also provides the QoS authorization (QoS class identifier and bit rates) that decides how to treat each traffic flow, based on the user's subscription profile. QoS policies can be dynamic, e.g., based on time of day. This must be enforced at the P-GW. The Home Sub- scriber Server (HSS) contains subscription information for each user, such as the QoS profile, any access restrictions for roaming, and the associated MME. The HSS is connected to the MME via the control interface. In times of cell congestion, a base station reduces the max rate allowed for subscribers according to their profiles, in coordination with the P-GW.

Today's EPC has limitations. Centralizing data plane functions such as monitoring, access control, and quality of service functionality at the P-GW introduces scalability challenges. This makes the equipment very expensive (e.g., more than $6 million for a Cisco P-GW). Centralizing data plane functions at the cellular- Internet boundary forces all traffic through the P-GW, including traffic between users on the same cellular network, making it difficult to host popular content inside the cellular network. In addition, the network equipment has vendor-specific configuration interfaces, and communicates through complex control plane protocols, with a large and growing number of tunable parameters (e.g., several thousand parameters for base stations). As such, carriers have (at best) indirect control over the operation of their networks, with little ability to create innovative services.

1.3   Future Radio Access Networks

Future architecture for RAN is discussed in this section, including the support for a UE to connect concurrently to multiple eNBs and associate diverse traffic to the eNBs, Cloud-RAN, and advanced SON to support drop-and-play small cells.

1.3.1 UE's Heterogeneous Traffic to Heterogeneous eNBs

As mentioned in Section 1.2.1, today's CA or CoMP technology is limited to the case that a UE is associated with multiple cells within an eNB. Such limitation has disadvantages, such as the resources among multiple eNBs cannot enjoy the advantage that CA or CoMP can bring, such as improved spectrum efficiency, enhanced radio resources sharing, etc. Hence, it is desirable to extend CA or CoMP technologies to the case that a UE can be associated to multiple eNBs. For the CA case, today's CA does not support much for the scenarios that the cells are connected via nonideal backhaul. The technology to allow in a future RAN that a UE can be connected to multiple eNBs where eNBs can be connected via ideal or nonideal backhaul is currently discussed in 3GPP [1, 7], and such technology is referred to as dual connectivity.

The dual-connectivity technology includes multistream aggregation operation. The multistream operation involves dual UE connectivity to two eNBs that are not necessarily collocated and may not have ideal backhaul connection. Enabling dual connectivity for FDD-only and TDD-only systems is discussed in [1]. All technical reasons for endorsing dual connectivity (e.g., mobility robustness and better through-put performance) for FDD-only and TDD-only systems are equally applicable for TDD-FDD multistream aggregation. Enabling dual connectivity for TDD-FDD systems discussed in [4, 10]. Examples of benefits would mainly be increased mobility robustness and increased throughput.

Figure 1.2 shows an exemplary dual-connectivity scenario. In the figure, a UE can be connected to a large eNB, and a small eNB, where the large eNB may not be connected to the small eNB via ideal backhaul; i.e., the backhaul can have nonnegligible delay.

As a further consideration, when a UE can be connected to multiple eNBs and cells where eNBs and cells can be connected via ideal or nonideal backhaul, diverse applications and diverse types of traffic of a UE can be associated to eNBs and cells with diverse backhaul conditions concurrently, to further improve the system performance. More details in this aspect are discussed below.

The future UEs may see many heterogeneous cells with different loads, wireless/wired backhaul conditions on rate, delays, etc., and the radio access links in-between a UE and cells can be diverse in rate and delay. Meanwhile, a UE can have various concurrent traffic flows with diverse QoS requirements, such as delay stringent interactive video and delay astringent best effort data.

One of the important problems in cellular networks is to properly associate UEs with serving cells. This problem usually is referred to as the user association problem. The user association would impact the radio resource allocation to UEs, and hence impact the QoS. For example, to guarantee QoS, delay stringent traffic of a UE can be associated to a first eNB via which the total delay of the UE to the eNB and the eNB to the EPC is relatively small, while delay astringent traffic of a UE can be as- sociated to a second eNB via which the total delay of the UE to the eNB and the eNB to the EPC is relatively large. It is of great interest to study in the future cellular network how to associate UEs with eNBs and cells, given the emerging new

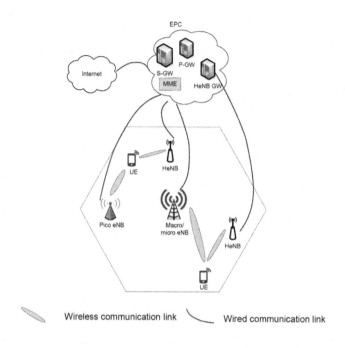

Figure 1.2: An exemplary dual-connectivity scenario.

scenarios that the UEs can have diverse application traffic flows and the cells can have diverse conditions and status.

Traditionally in cellular networks, the simplest rule is to choose the eNB that gives the strongest downlink reference signal. However, such a rule has limitations in the sense that it does not consider other factors such as cell load balancing. There have been efforts in the literature toward developing user association rules considering cell load balancing; for example, a cell can broadcast its load and a UE can be associated with a cell with lower load [13]. For a heterogeneous network, the cell association has further consideration. Small cell range expansion can expand the footprint of small cells that can coexist with large cells [2], as mentioned in Section 1.2.1. All of the above have been based on a UE being associated with only one eNB; however, concurrent connection where a UE can be connected to multiple eNBs may provide more freedom and benefits considering diverse concurrent traffic of a UE.

Moreover, diverse backhaul conditions of the eNBs and cells are playing another role for the consideration of the UE cell association rule. The impact on different back- haul conditions on CoMP performances has been studied [5], mainly in the aspect of how the backhaul delay could affect the coordinated joint scheduling for multiple transmission points. A new study item in 3GPP standardization for small cell enhance- ment also mentions the further study on backhaul impact on the performance [8]. However, it is not well studied how the backhaul conditions would impact the UE's diverse traffic associating with the diverse eNBs and cells. Not considering the eNB backhaul may not be effective in resource allocation. For instance, if a UE has a strong wireless access link to an eNB but the eNB has a backhaul with large latency, if the cell association is decided by wireless access, the UE would be associated to this eNB; however, considering the large delay at the eNB's backhaul, the time stringent traffic of the UE may not be satisfied with the total delay.

In [29], the problem of assigning UE traffic flows concurrently to mul- tiple BSs is studied, so that each flow's QoS requirement is satisfied, with the eNB backhaul condition taken into account. It shows that the dual connectivity can bring the benefits of UE-satisfied QoS for diverse traffic types. It proposes a framework to associate traffic flows with different QoS requirements of a UE to multiple eNBs, so that the corresponding QoS requirements can be satisfied, where the heterogeneous backhaul conditions of eNBs are taken into account.

For future RAN, further studies are needed to enable a UE with diverse traffic to be connected to multiple eNBs with diverse backhaul conditions.

1.3.2 Cloud-RAN

Mobile network traffic is significantly increasing by the demand generated by application of mobile devices, while the revenue is difficult to increase. To keep profit, mobile operators must reduce OPEX as well as continuously develop and provide better services to their customers.

One consideration is to reduce the operational cost by sharing resources among eNBs. A RAN node utilization is usually lower than that capacity because the system is designed to cover the peak load. The network should adequately support the peak load; however, it brings higher cost due to good QoS provisioning for the peak load. Possible pooling of eNBs may reduce the cost by sharing the resource among eNBs for more efficient resource utilization. It can also reduce power consumption to reduce the total cost of ownership (TCO), which is very important as the electricity bill takes a large portion of the TCO (about 20% according to [11]) and the cell sites are the major source of the energy consumption (about 70% according to [11]).

Cloud-RAN (C-RAN) technology [22] uses resource pooling and virtualization of eNBs. C-RAN stands for centralized processing, collaborative radio, real-time cloud computing, clean RAN system. It can leverage more efficient resource utilization among eNBs. It has great potential to help reduce OPEX. C-RAN is to realize RAN nodes onto standard IT servers, storages, and switches bringing advantages, such as lower footprint and energy consumption coming from dynamic resource alloca- tion and traffic load balancing, easier management and operation, and faster time to market. In major mobile operators' networks, multiple RAN nodes from multiple vendors are usually operated with different mobile network systems, e.g., 3G and LTE, in the same area. These multiple platforms expect to be consolidated into a physical eNB based on IT virtualization technologies, referred to as eNB virtualization.

Figure 1.3 illustrates an exemplary C-RAN architecture. In the figure, the distributed radio unit (RU) of eNBs and TPs in a same area can be connected to a digital unit (DU) or DU cloud, which is a group of DUs, via a high-bandwidth and

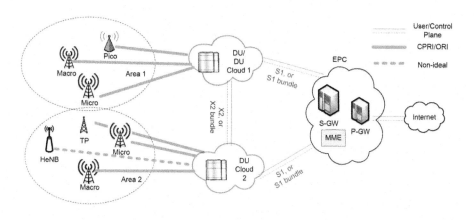

Figure 1.3: An exemplary C-RAN architecture.

low-latency transport network Common Public Radio Interface (CPRI) or Open Radio Interface (ORI). The DU can be also referred to as baseband unit (BBU). The DU and DU cloud of an area are centralized in one physical location for providing resource aggregation and pooling. The DU and DU cloud include the radio functions of the digital baseband domain. The DU is in charge of the channel coding, digital signal processing (DSP), modulation/demodulation process, and interface module. The RU is different from the traditional eNB or TP, as its baseband processing is moved to DU. The DU and DU cloud are connected to each other via an X2 or X2 bundle, and they are connected to the EPC via an S1 or S1 bundle.

ENB virtualization requires baseband radio processing using IT visualization technologies, such as high-performance general purpose processors and real-time processing virtualization to provide required signal processing capacity. eNB virtualization for C-RAN moreover requires building the processing resource, i.e., DU, pool for aggregating the resources onto a centralized virtualized environment, such as cloud infrastructure.

To support CoMP transmission/reception, UE data and channel information need to be shared among eNBs/DUs, and a high-bandwidth and low-latency interconnection for real-time cooperation among these should be supported on the virtualized environment.

To further reduce the TCO for mobile operators, SON can be used to support C-RAN. SON is especially useful as the number and structure of network parameters have become large and complex, quick evolution of wireless networks has led to parallel operation of 2G, 3G, and EPC infrastructures, and the rapidly expanding number of eNBs needs to be configured and managed with the least possible human interaction. An effective SON solution for C-RAN must be multivendor by nature and leverage the timely information from protocols such as X2 for better handling of CoMP and for such routine mobile network functions as handover and mobility optimizations.

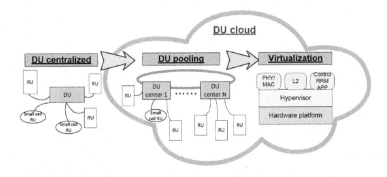

Figure 1.4: An exemplary C-RAN evolution.

Some of the high-level technical challenges for C-RAN are as follows [12]. Wireless signal processing requires strict real-time constraint in the processing. Baseband radio processing on a general purpose processor might be virtualized by Soft Defined Radio (SDR) techniques. Within a physical eNB virtualizing multiple logical RAN nodes from different mobile network systems, the processing resources must be dynamically allocated to higher-load logical RAN node keeping real-time scheduling and strict processing delay and jitter. DU pool must have a high-bandwidth and low-latency switching function with necessary data formats and protocols to interconnect among multiple DUs. I/O virtualization or API between PHY layer accelerator and standard IT platform must be addressed to access. Especially for C-RAN, higher consolidation of RRHs to a DU pool with higher I/O can benefit from a higher statistical multiplexing effect.

C-RAN can be evolved in step-by-step stages [22]. Figure 1.4 illustrates an exemplary evolution of C-RAN. In the figure, the first step is DU centralization, where DU can be in one location, and RF sites are connected to the DU using high-speed low-latency links. The second step is DU pooling, where multiple DUs are pooled and resources are not dimensioned by peak of individual DU site, but aggregated by the pool. The third step is virtualization of RAN, where the processing resources are virtualized and application independent of the hardware. The virtualization can use a hypervisor on top of the hardware platform, and different layers of the functions such as PHY/MAC, layer 2 (L2), radio resource management (RRM), and applications (APP) can be performed virtually, independent of the hardware.

1.3.3 Adaptive and Self-Organized RAN with Drop-and-Play Small Cells

Today's small cell, such as HeNB, is in a category of plug-and-play. This can reduce the cost of a planned network, where a lot of field labor, manual configuration, etc., can be minimized. With the development of SON, more and more small cells, not limited to HeNB, but also pico cells and so on, can be plug-and-play. SON can support self-configuration, self-optimization, and self-healing. For example, a cell can have self-configuration on the physical cell identifier, rather than getting a planned one as in a planned network. A cell can configure many parameters on its own, unlike in the traditional planned network. A cell can also have an automatic ON and OFF switch based on the current load. For example, if everyone is in the office campus, the HeNB can be off, while the small cells in office campus can be on, and if everyone is home, the office campus small cells can be off, while the HeNBs can be on.

The plug-and-play idea fits very well in terms of traffic load distribution, where the plug action can be related to the traffic demand, or the cluster-based UE distributions. Beyond plug-and-play, the future of small cells is place-and-play, or drop-and-play, where no wire is needed for the small cells, with the advances in wireless backhaul and energy harvesting. The SON functions should be enhanced to support the future drop-and-play deployment.

High-speed wireless backhaul is rapidly becoming a reality for small cells, which eliminates the need for wired connections. In another advance, the possibility of having a self-powered eNB is becoming realistic due to several parallel trends. First, eNBs are being deployed evermore densely and opportunistically to meet the increasing capacity demand. Small cells cover much smaller areas, and hence require significantly smaller transmit powers compared to the conventional macrocells. Second, due to the increasingly bursty nature of traffic, the loads on the eNBs will experience massive variation in space and time. In dense deployments, this means that many eNBs can, in principle, be turned off most of the time and only be requested to wake up intermittently based on the traffic demand. Third, energy harvesting techniques, such as solar power, are becoming cost-effective compared to the conventional sources. This is partly due to the technological improvements and partly due to the market forces, such as increasing taxes on conventional power sources, and subsidies and regulatory pressure for greener techniques. Therefore, being able to avoid the constraint of requiring a wired power connection or a wired backhaul is even more attractive, since it would open up entire new categories of low-cost place-and-play, or drop-and-play deployments, especially of small cells [14].

Figure 1.5 illustrates an exemplary architecture for drop-and-play deployment. In the figure, transient eNB (TeNB) is an access point or eNB without wire, which uses wireless backhaul to be connected to the core netwrok, and is self-sufficient on its energy use via energy harvester. TeNB can apply drop-and-play. TeNBs can be deployed to increase the deployment density of the wireless network. TeNB can only be turned on for a small portion of the time. In other words, the duty cycle of the TeNB is low. Due to the low duty cycle of the TeNBs, preferably only a small number of the TeNBs in the network are turned on at a time. When a TeNB is turned on, it establishes a wireless backhaul link to the core network via hubs or other eNBs such as macro/micro/pico eNBs. A TeNB can also establish multiple links with multiple eNBs or hubs. Once the backhaul link is established, a TeNB can then provide an access link to UEs. The presence of TeNBs increases the deployment density of the network and thus can increase the capacity and coverage of the access link.

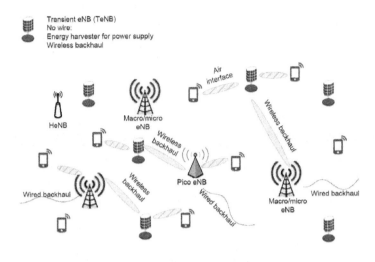

Figure 1.5: An exemplary architecture for drop-and-play deployment.

The network architecture as described above increases the robustness of the backhaul network. For example, if the communication link of one path of the wireless backhaul is congested or disrupted, TeNB can adapt the beamforming of its antenna array to establish communication via another path of the wireless backhaul. A TeNB can also have concurrent multiple paths for the wireless backhaul communication, or it can maintain multiple paths at the same time while communicating on one path at a time. SON functions can be enhanced to support route selection, establishment, and reroute.

In a TeNB, there is an energy generation module, an energy storage module (e.g., a battery), and a communication module, among others. The energy generation module can be either a solar power module, a wind power module, or power generation modules using other energy harvesting techniques. The power generated by the energy generation module can be fed either directly to the communication module or to charge the battery. The battery can then in turn power the communication module. The low duty cycle of the TeNB allows the energy generation module to be sufficiently small to ensure a small form factor of the overall device.

TeNB can also be an access point that can be turned on for a flexible portion of the time. The ON time of TeNB can be large or small. The duty cycle can be flexible. The duty cycle can be configured, indicated, updated, and sent to the other network entities, such as eNBs, UEs, backhaul hubs, etc. The network can configure or update the duty cycle based on considerations in the network such as load, distribution of the UEs, etc. This needs enhanced SON support.

The battery level of TeNB, the charging speed, etc., can be indicated and sent to the other network entities. The battery level of TeNB, the charging speed, etc., can be used as one of the factors to decide the route of the wireless backhaul, or for the UE to decide whether to access the TeNB. For example, when the battery level of a TeNB is low, a UE may not choose to connect to the TeNB; rather, the UE may choose to connect to another TeNB nearby with longer battery life. Enhanced SON functions can be used to support such.

The TeNB can follow certain algorithm and triggering conditions to turn on or turn off. The ON/OFF switch can be dependent on the battery level, the charging speed, the traffic load, the traffic distribution, the price of the energy of the power grid, and so on. As many parameters can affect the ON/OFF switch, SON functions can be enhanced to support ON/OFF switches. Different modules in a TeNB can turn on or turn off at different times. The communication module of a TeNB can be turned on (or become active in serving UEs) via a variety of mechanisms. Note that the energy generation module of a TeNB can work when a TeNB becomes idle or active.

1.4   Future Evolved Core Network

In this section, future architecture for EPC is discussed, including Mobile SDN and network virtualization in EPC.

1.4.1   Mobile SDN

Mobile SDN (referred to as MobiSDN) is a type of networking for mobile networks where the control plane of the network is physically separate from the forwarding plane (or the data plane), and the data plane uses hardware including radio hardware (such as eNBs), servers, and switches. Network intelligence is (logically) centralized in software-based controllers, which maintain a global view of the network.

MobiSDN support in the EPC has gained a lot of attention from operators and vendors. The core network may have a MobiSDN architecture in the future [17, 28]. In addition, the interface between eNB and the core network may also be impacted, introducing the need for MobiSDN-capable eNB. MobiSDN allows further development of smart edge solutions, such as content caching and local APP server hosting [28]. With MobiSDN, more and more use cases for additional revenue or value added services can be provided.

MobiSDN is comprised of smart edge and cloud EPC [28]. Figure 1.6 illustrates an example of the architecture of MobiSDN. The details of Figure 1.6 are explained as follows.

Smart edge: The smart edge includes SDN-capable eNBs. The edge controller is also SDN based, whose function can be part of the central controller. The edge server may be co-located with the eNB. The smart edge has three main functionalities: distributed computing, distributed file system, and networking controller. Distributed computing enables different processing capabilities at the edge, including computation load balancing and programming transparency. The distributed file system can support distributed storage, cache sharing, content search, etc. The network controller is based on SDN with programmable routers, flexible policy checking, and is friendly

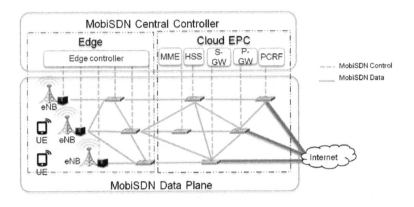

Figure 1.6: An exemplary architecture for MobiSDN.

to middle boxes. These functions are inevitable and needed, considering the volumes of mobile data and huge mobile video traffic demand.

Cloud EPC: The cloud EPC is SDN capable. It uses SDN switches and servers as the hardware. The data plane is based on SDN switches that provide data forwarding. The control plane is based on the MobiSDN central controller. The controller can take on functions including those provided by the MME, HSS, S-GW, P-GW, PCRF, etc. These functions can be applied within each of the SDN switches. The switches are also involved in routing and running transport protocols.

Although here the functions in the data and control planes are described by using the names of MME, HSS, S-GW, P-GW, PCRF, etc., there may or may not be these network entities anymore. For example, the MME, PCRF, and HSS can be absorbed in the MobiSDN central controller, while some of the functions of S-GW and P-GW will be in the data plane (e.g., the MobiSDN switches) and some will be in the control plane (absorbed in the MobiSDN central controller).

There are two essential differences with MobiSDN compared to the state-of-the-art architecture. First, MobiSDN has a clear separation of data plane and control plane. Second, MobiSDN is very flat, contrasting to the existing hierarchical architecture wherein the P-GW can be the bottleneck as the node connecting to the Internet. MobiSDN has the following advantages: more flexible routing and flows, flexible middle boxes, keeping the content in the edge, offloading traffic from the core network, handover with less overhead, flexible radio resource management and scheduling, etc.

1.4.2 *Network Virtualization in EPC*

Network virtualization in EPC is a technology by which resources of the network (for example, the hardware) can be virtualized and used transparently. It enables the creation of a competitive environment for the supply of innovative third-party network

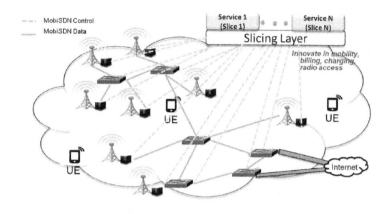

Figure 1.7: An exemplary virtualization in MobiSDN.

applications by unlocking the proprietary boundaries of mobile base station nodes. Virtualization is a good tool to achieve the paradigm of soft networking. Network function virtualization is under discussion [25]. The virtualization of eNBs in Section 1.3.2 is an example of virtualization in the RAN, and the current section focuses more on the virtualization in the EPC.

One example to achieve network virtualization in EPC can be by slicing the flow space, such as by using a hypervisor. All the network hardware can be used as shared infrastructure by all the slices. A slice may consist of part or all of different infrastructure (e.g., eNBs, switches) elements. Each slice can be used to provide a different value-added service. Since each slice can be flexibly and independently managed, it allows for the easy introduction of new revenue opportunities without additional hardware complexity costs [17].

Figure 1.7 illustrates an example of virtualization in MobiSDN. A hypervisor can be used, to support the slicing layer, on top of which different services can be supported. Each slice can be used by a different service and virtually correspond to part or all of the hardware at the physical layer.

Some examples of using virtualization for value-added services are provided as follows. One example is efficient real-time communication for enterprises with multiple-campus support. An enterprise that has multiple campuses in different geographical areas may be interested in deploying low-latency interactive wireless services, such as allowing employees to have high QoS interactive wireless video conferencing, doc collaboration query processing, etc., on top of the existing wireless network. The operator can provide this service by using MobiSDN and a slice consisting of the eNBs close to the campuses (not all the base stations are needed, which simplifies the networking) and switches involved. An advantage of this approach is that it creates new services and more revenue for operators while also reducing management complexity.

Another example is for supporting stadiums and other large venues. A slice can consist of local eNBs within the stadium and switches involved, to provide local content and services. Users can pay for premium service with better QoS since congestion is often an issue in these scenarios, or participate in certain events such as video contests, etc. Content, such as players' introduction, video replay of exciting game highlights, etc., can be stored in the local cache server. The UE can also upload its captured video or other content to the local server, to share with other local users. The venue or operator may incentivize content uploading and sharing by offering rewards for users who provide very high quality video clips that become popular.

1.5   Conclusion

This chapter provides discussions on the future cellular network architecture. For the future RAN, potential technologies, including the support for a UE to connect concurrently to multiple eNBs and associate diverse traffic to the eNBs, Cloud-RAN, and advanced SON to support drop-and-play small cells, are discussed. For the future EPC, potential technologies, including mobile SDN and network virtualization, are discussed. All these technologies provide great potential to improve today's cellular network architecture and leverage the limitations of today's architecture.

The future cellular network architecture may not be limited to what is discussed in this chapter. For example, the architecture of interworking of cellular and Wi-Fi can be further improved, as currently discussed in [9]. For another example, the architecture of supporting device-to-device communications can be consolidated to today's cellular architecture, as currently discussed in [6].

All in all, the future cellular network architecture will be supporting more use cases and be more adaptive, more optimized, more efficient, more cost effective, and easier for network management than today's architecture.

References

1. 3GPP TR 36.842 v0.2.0. Study on small cell enhancements for E-UTRA and E-UTRAN—Higher-layer aspects. 3GPP TSG RAN, 2013.

2. 3GPP R1-083813. Range expansion for efficient support of heterogeneous networks. 3GPP, TSG-RAN Work Group 1 (WG1) 54bis, Qualcomm Europe, 2008.

3. 3GPP R1-101505. Extending Rel-8/9 ICIC into Rel-10. 3GPP, TSG-RAN WG1 60, Qualcomm, February 2010.

4. 3GPP RP-130888. LTE TDD-FDD joint operation. 3GPP TSG RAN, 2013.

5. 3GPP TR 36.819 v11.1.0 (2011-2012). Coordinated multi-point operation for LTE physical layer aspects (release 11). 3GPP TSG RAN, December 2011.

6. 3GPP TR 36.843. Feasibility study on LTE device to device proximity services—radio aspects. 3GPP TSG RAN, 2013.

7. 3GPP TR 36.872. Small cell enhancements for E-UTRA and E-UTRAN—Physical layer aspects. 3GPP TSG RAN, 2013.

8. 3GPP TR 36.932 v12.0.0. Scenarios and Requirements for small cell enhancements for E-UTRA and E-UTRAN (release 12). 3GPP TSG RAN, December 2012.

9. 3GPP TR 37.834. Study on WLAN/3GPP radio interworking. 3GPP TSG RAN, 2013.

10. 3GPP TR 36.847. LTE time division duplex (TDD)—Frequency division duplex (FDD) joint operation including carrier aggregation (CA). 3GPP TSG RAN, 2013.

11. China Mobile. C-RAN: Strategy, trial and future considerations. Proceed- ings of IWPC, December 2012.

12. GS NFV 009 v015. Network function virtualization: Use cases. ETSI, June 2013.

13. H. Kim, G. de Veciana, X. Yang, and M. Venkatachalam. Distributed α-optimal user association and cell load balancing in wireless networks. *ACM Transactions on Networking*, 20(1):177–190, 2012.

14. H. S. Dhillon, Y. Li, P. Nuggehalli, Z. Pi, and J. G. Andrews. Fundamentals of base station availability in cellular networks with energy harvesting. Proceedings of the IEEE Globecom, December 2013.

15. ITU, Report M.2135. Guidelines for evaluation of radio interface technologies for IMT-Advanced. 2008.

16. J. Kempf, B. Johansson, S. Pettersson, and H. Luning. Moving the Mobile Evolved Packet Core to the cloud. Fifth International Workshop on Selected Topics in Mobile and Wireless Computing, 2012.

17. L. E. Li, M. Mao, and J. Rexford. Towards software-defined cellular networks. First European Workshop on Software Defined Networking, October 2012.

18. L. Liu, Y. Li, B. Ng, and Z. Pi. Radio resource and interference management for heterogeneous networks. *In Heterogeneous Cellular Networks*, John Wiley & Sons, 2013.

19. Open Networking Foundation. www.opennetworking.org.

20. Open Networking Summit. www.opennetsummit.org.

21. OpenDaylight. www.opendaylight.org.

22. Project Proposal. C-RAN: Centralized processing, collaborative radio, real- time cloud computing clean RAN system. NGMN Alliance, March 2011.

23. Telstra and Ericsson. Service provider SDN meets operator challenges. Open Networking Summit, April 2013.

24. TS 36.300 v11.6.0, 3rd Generation Partnership Project (3GPP); Technical Specification Group Radio Access Network (TSG RAN). Evolved Universal Terrestrial Radio Access (E-UTRA) and Evolved Universal Terrestrial Radio Access Network (E-UTRAN); overall description; stage 2. June 2013.

25. White paper. Network functions virtualization—Introductory white paper. ETSI, December 2012.

26. White paper. Cisco visual networking index: Forecast and methodology 2012–2017. May 2013.

27. Y. Li, A. Maeder, L. Fan, A. Nigam, and J. Chou. Overview of femtocell support in advanced WiMAX systems. *IEEE Communications Magazine*, July 2011.

28. Y. Li, M. Dong, D. Choe, T. Novlan, C. Zhang, and G. Morrow. MobiSDN: Vision for Mobile Software Defined Networking for Future Cellular Networks. In Proceedings of Globecom 2014, Industry Forum.

29. Y. Li, Z. Pi, and L. Liu. Distributed heterogeneous traffic delivery over heterogeneous wireless networks. Proceedings of the IEEE ICC, June 2012.

Chapter 2

Advanced Technologies in Gigabit Wireless LANs: An In-Depth Overview of 802.11ac

Michelle X. Gong

Intel Corporation, Santa Clara, California

Eldad Perahia

Intel Corporation, Santa Clara, California

Shiwen Mao

Auburn University, Auburn, Alabama

Brian Hart

Cisco Systems, San Jose, California

CONTENTS

2.1   Introduction

As the IEEE 802.11n (High Throughput) standard amendment development matured and associated products became popular in the market, IEEE 802.11 initiated a new study group in May 2007 to investigate Very High Throughput (VHT) technologies. The Wi-Fi Alliance was solicited to provide usage models to help develop requirements [8]. The general categories of the usage models included wireless display, distribution of high-definition TV, rapid upload/download, backhaul, outdoor campus, auditorium, and manufacturing floor [11]. Specific usages that will be most prevalent in the marketplace include compressed video streaming around a house, rapid sync-and-go, and wireless I/O. With streaming around the home, it is envisioned that TVs and DVRs around the home will have wireless capability and 100+ Mbps aggregate of videos from a DVR can be displayed wirelessly on TVs in different rooms. With rapid sync-and-go, users can quickly sync movies or pictures between mobile devices such as a phone, a laptop, or a tablet. With a 1 Gbps radio link, a 1 GB video file will take much less than a minute to transfer between devices. Data rates exceeding 1 Gbps will provide the capability for a wireless desktop, with wireless connections between a computer and peripherals such as monitors, printers, and storage devices.

With this input, the Very High Throughput study group developed two Project Authorization Requests (PARs), one for the 5 GHz band (802.11ac) and one for the 60 GHz band (802.11ad). This chapter will give an overview of the 802.11ac draft amendment and describe some of the advanced features. The scope for 802.11ac includes:

- Single-link throughput supporting at least 500 Mbps

- Multistation throughput of at least 1 Gbps

- Exclusion of 2.4 GHz band

- Backward compatibility and coexistence with legacy 802.11 devices in the 5 GHz band

The PAR was approved in September 2008, and the 802.11ac task group began in November 2008.

The task group initially developed a specification framework document [13], a functional requirements and evaluation methodology document [7], an amendment to 802.11n channel model document [2], and a usage model document [11]. This process was purposely different than the proposal down selection process of 802.11n. After the challenging experience in 802.11n, the group opted for the less contentious approach of developing a specification as a group based on a specification framework. Through its various revisions, the specification framework document built a list of features that would be included in the draft specification, with detail added to the features as the document evolved.

An initial draft 0.1 was developed based on the specification framework and approved by the task group in Jan. 2011. This draft went through an internal task group comment and review cycle. In May 2011, draft 1.0 was released to the 802.11 working group for the letter ballot process. After five working group letter ballots, the sponsor ballot began in May 2013. Final approval of the 802.11ac standard amendment was achieved with draft 7.0 in Dec. 2013. Whereas letter ballot includes only voting members in IEEE 802.11, the sponsor ballot pool includes members from all of the IEEE Standards Association, providing a broader review of the draft. However, as was the case with 802.11n, initial products with basic 802.11ac features based on an early draft (draft 5.0) were certified by the Wi-Fi Alliance. That certification launched in June 2013.

Figure 2.1 summarizes the building technology blocks of 802.11ac. Within the mandatory features, 20/40/80 MHz channel operation, multichannel RTS/CTS, and

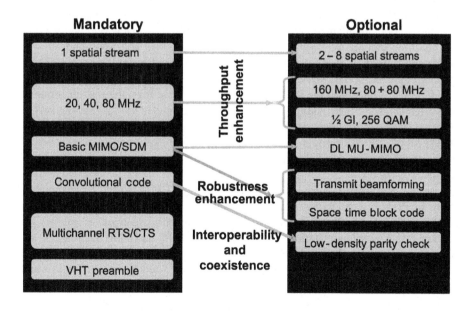

Figure 2.1: Major features defined in 802.11ac.

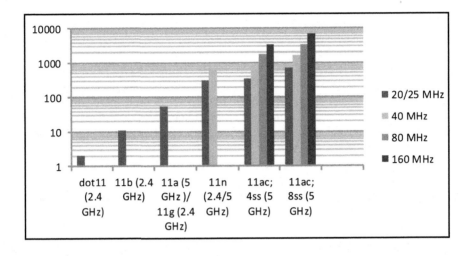

Figure 2.2: 802.11 historical PHY data rates.

VHT preambles are important new features introduced in 802.11ac. One spatial stream, 20 and 40 MHz, basic MIMO/SDM, and convolutional code are features that were originally introduced in 802.11n and modified for mandatory 802.11ac operation. Multichannel RTS/CTS and VHT preamble are new mandatory features introduced to 802.11ac. Other 802.11n features, two–four spatial streams, 1/2 guard interval (GI), transmit beamforming, space time block code, and low-density parity check code are adapted as optional features in 802.11ac. New optional features in 802.11ac include five–eight spatial streams, contiguous 160 MHz and 80 + 80 MHz noncontiguous channelization, 256 QAM, and downlink multiuser multiple input multiple output (DL MU-MIMO).

Figure 2.2 illustrates the PHY data rate improvement from the max data rate of the original 802.11 (i.e., 2 Mbps) to the max data rate defined in 802.11ac (i.e., 6,933 Mbps). The Y axis has the unit of Mbps. While the maximum data rate in 802.11ac is 6,933 Mbps (160 MHz, eight spatial streams), this will likely only ever be achieved in a multiuser scenario. The maximum data rate for four spatial streams is 3,467 Mbps (160 MHz).

In this chapter, we discuss the main features of 802.11ac, including channelization, the PHY design, channel bonding and MAC protection, and downlink multiuser MIMO (DL MU-MIMO). The remainder of this chapter is organized as follows. We review 802.11ac channelization in Section 2.2 and PHY design in Section 2.3. We examine channel bonding and MAC protection in Section 2.4, as well as a simulation study. Downlink MU MIMO is presented in Section 2.5 along with an analysis and simulations study. Section 2.6 concludes the chapter.

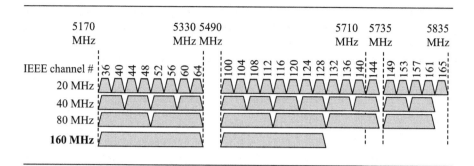

Figure 2.3: Channelization for 802.11ac.

2.2 802.11ac Channelization

The 802.11ac channelization is illustrated in Figure 2.3. It can be seen that there are no partially overlapping 20, 40, 80, or 160 MHz channels in 802.11ac channelization. For instance, channels 36 and 40 can form a valid 40 MHz channel, but channels 40 and 44 cannot form a valid 40 MHz channel, because the second 40 MHz channel would partially overlap with the first 40 MHz channel. Since partially overlapped channels introduce significant in-band interference, extremely complex coexistence schemes would have to be defined to mitigate such interference.

To avoid such an in-band interference problem and to simplify protocol design, only nonoverlapping channels are allowed in the 5 GHz band in 802.11. 802.11ac has added 80 and 160 MHz channelization. As shown in Figure 2.3, channel 144 has been added by 802.11ac, which was not included in 802.11n. The addition of channel 144 also allows a new 40 MHz channel with the combination of channels 140 and 144.

With this addition, there is a maximum of six 80 MHz channels possible, where regulatory bodies permit. However, even with the additional channel 144, there are only two 160 MHz channels available, which is the primary reason for the inclusion of the noncontiguous 160 MHz operation. Noncontiguous 160 MHz (80 + 80 MHz) channels are comprised of any two valid, nonadjacent 80 MHz channels. With the noncontiguous operation, many combinations of 80 + 80 MHz channelization become possible.

2.3 802.11ac PHY Design

The 802.11ac PHY design philosophy follows closely that of 802.11n. Interested readers are referred to [10] for more information on the 802.11n PHY. In particular, the preamble of the packet is comprised of the following fields in the order listed:

■ Legacy short training field (STF): Start of packet detection, automatic gain control (AGC) setting, initial frequency, and timing synchronization

- Legacy long training field (LTF): Channel estimation, fine frequency, and timing synchronization.

- Legacy signal field (L-SIG): Spoofs legacy devices; indicates VHT payload symbol length.

- VHT-SIG-A: Replaces 802.11n HT-SIG; contains VHT PHY single-user and some MU parameters.

- VHT-STF: Similar to 802.11n HT-STF; allows readjustment of AGC.

- VHT-LTF: Similar to 802.11n HT-LTF; used for channel estimation.

- VHT-SIG-B: New VHT field; contains additional peruser parameters.

The fields up to and including VHT-STF are comprised of a 20 MHz waveform. This is replicated in each adjacent subchannel for wider channel bandwidths.

Following the preamble is the data field. The first 16 bits of the data field is the service field. In 802.11ac, this has been modified to include a CRC for VHT-SIG-B. In addition, PHY padding comes after the data followed by tail bits. This is different from 802.11n where tail bits preceded pad bits; the change is due to adding MU and the rise in the maximum number of bytes per packet, which in turn meant that the packet length had to be signaled differently. The data are scrambled, encoded, and then interleaved. This is followed by the constellation mapper and then the spatial mapper.

The 80 MHz waveform is based on a 256-point FFT. There are 234 data subcarriers, 8 pilot subcarriers, and 14 null subcarriers, three of which are at DC. This is more than double the number of data subcarriers of the 40 MHz waveform (108 data tones), so 80 MHz data rates are more than double the 40 MHz data rates.

However, the 160 MHz subcarrier design is an exact replication of two 80 MHz segments. This allows for the same subcarrier design for contiguous 160 MHz and noncontiguous 160 MHz (80 + 80 MHz). Furthermore, the phase of the local oscillator is not required to be correlated between lower and upper portions of the signal at the transmitter for contiguous 160 MHz and noncontiguous 160 MHz (80 + 80 MHz). Again, this allows for additional commonality between contiguous 160 MHz and noncontiguous 160 MHz (80 + 80 MHz).

2.4 Channel Bonding and MAC Protection

With the numerous 20 and 40 MHz channels in the 5 GHz band in 802.11n, overlapping channels between service sets (BSSs) are easy to avoid by choosing a different channel. In the worst case, if an overlap between neighbors using 40 MHz is unavoidable, the primary 20 MHz subchannels are chosen to match to maximize coexistence capability. With much wider channels in 802.11ac, it becomes much harder to avoid overlap between neighboring BSSs. In addition, it becomes harder to choose a primary channel common to all overlapping networks. To address this problem,

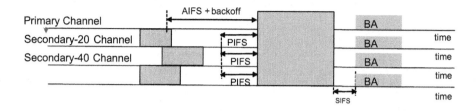

Figure 2.4: Channel bonding and PIFS medium access rule.

802.11ac improves co-channel operation with three enhancements: enhanced secondary channel Clear Channel Assessment (CCA), improved dynamic channel width operation, and a new operating mode notification frame.

In 802.11 the CCA mechanism is employed to detect other signals and defer transmission appropriately. The basic requirement for an OFDM-based device is to receive a valid 802.11 signal at a level of −82 dBm. It must also detect any other type of signal at a level of −62 dBm, termed Energy Detect (ED). When 802.11n added the 40 MHz channel comprised of a primary 20 MHz channel and a secondary 20 MHz channel, only ED was required on the secondary channel due to the added complexity of detecting a valid 802.11 signal on the secondary channel. This meant that other systems occupying the secondary channel of another 40 MHz BSS would be disadvantaged by 20 dB. In 802.11ac, valid signal detection on the secondary channels was added at a level of −72 or −69 dBm according to bandwidth, to improve CCA performance on the secondary channels. In addition, it is required that a device detect a valid packet on the secondary channels based on not just the preamble of a packet, but also the middle of the packet.

The multichannel medium access operation is illustrated in Figure 2.4. An 80 MHz-capable station (STA) performs backoff over the primary channel following the normal 802.11 backoff procedure. For the corresponding Access Category (AC), if the medium is free for Arbitrary Inter-Frame Spacing (AIFS) time, the STA starts to count down its backoff counter for every time slot that is sensed idle over the primary channel. The STA freezes its backoff counter if one time slot over the primary channel is not sensed idle. Sometime before the backoff counter counts down to zero, the STA performs CCA over all nonprimary channels. If all nonprimary channels are sensed free for the PIFS period when the backoff counter reaches zero, the STA can initiate its 80 MHz transmission. The Block ACK (BA) frames are duplicated over all 20 MHz channels using a legacy frame format to ensure backward compatibility with 20 MHz BSSs.

The basic Request-to-Send (RTS) and Clear-to-Send (CTS) mechanism of 802.11 is modified to improve multichannel operation [5]. Consider an interference scenario illustrated in Figure 2.5, whereby STA2 is transmitting to AP2 and AP1 is communicating with STA1. AP2 is occupying overlapping channels of the secondary 40 MHz channel of AP1. STA1 and STA2 can interfere with each other, but the interference is not heard by the two APs. To address this situation, bandwidth signaling is added

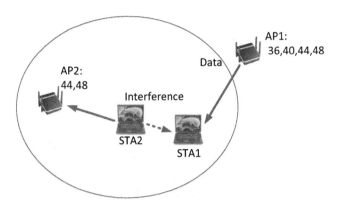

Figure 2.5: An interference scenario.

to the RTS and CTS frames. As illustrated in Figure 2.6, AP1 sends an RTS with the bandwidth of the intended transmission, which is 80 MHz comprised of channels 36, 40, 44, and 48 in this example. Before STA1 replies with a CTS frame, it senses the medium on all secondary channels for PIFS. If the secondary 40 MHz channel is not free, STA1 sends a CTS response with the bandwidth (BW) of the clear channels, i.e., 40 MHz comprised of channels 36 and 40 in this example. Then AP1 sends data to STA1 only on the clear channels and STA1 replies with Block ACK (BA) frames that are duplicated over the clear channels.

To ensure legacy compatibility, the original RTS and CTS frame formats are reused in 802.11ac's MAC protection mechanism so that legacy STAs can correctly decode RTS/CTS frames and set NAVs accordingly. The scrambler operation on RTS/CTS frames is modified to carry the extra information needed, i.e., 2 bits of BW information and 1 bit of dynamic/static BW operation mode. The 2-bit BW field, i.e., INDICATED_CH_BANDWIDTH, indicates whether the available BW is 20, 40, 80, or 160 MHz. Note that it is not necessary to include the available channel list in the RTS/CTS because all transmissions must include the primary channel and each BW mode can uniquely identify the available channels when the primary channel is known. The 1-bit dynamic/static BW operation mode field, i.e.,

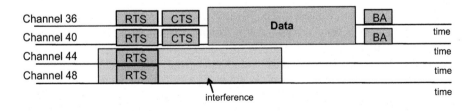

Figure 2.6: MAC protection for dynamic bandwidth operation.

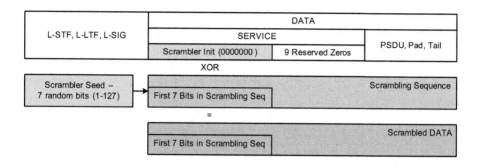

Figure 2.7: Operation of the 802.11a scrambler. (From M. Gong, B. Hart, L. Xia, and R. Want. Channel bounding and MAC protection mechanisms for 802.11ac. In *Proc. IEEE GLOBECOM 2011*, pages 1–5, Houston, TX, Dec. 2011. [5], © 2011 IEEE.)

INDICATED_DYN_BANDWIDTH, is included only in the RTS frame and indicates whether the transmitter is operating in dynamic or static BW operation mode.

Figure 2.7 illustrates an example of the 802.11a scrambler, which generates a sequence using the generator polynomial

$$S(x) = x^7 + x^4 + 1$$

and a randomly initialized nonzero 7-bit scrambler seed.

Consistent with the scrambler having a 7-bit shift register, it can be shown that the mapping between scrambler seed and the first 7 bits of the scrambling sequence (F7BOSS) is one-to-one; so defining the F7BOSS is equivalent to defining the scrambler seed. With this insight, it is possible to describe a constrained F7BOSS (or equivalently, a constrained scrambler seed) that is the straightforward concatenation of a 5- or 4-bit non-zero random field plus 2 or 3 data bits, which, respectively, are the 2-bit INDICATED_CH_BANDWIDTH field and the 1-bit INDICATED_DYN_BANDWIDTH field. The scrambling operation performed by the modified scrambler is shown in Figure 2.8.

The sequence so generated is a legitimate (self-synchronizing) scrambling sequence since the first 7 bits of Data In are all zeros, so the shift register input is the same as Scrambled Data Out, and therefore the state of the shift register after 7 bits equals the first 7 Scrambled Data Out bits. The bits inserted into the FB7OSS are not explicitly protected by a check sequence, which at first glance is a disadvantage of the scheme. However, since the scrambler is self-synchronizing, any bit errors within the FB7OSS by a receiver cause the remaining scrambling sequence to be miscalculated and then the MAC Frame Check Sequence (FCS) of any MAC Protocol Data Unit (MPDU) will almost certainly detect an error.

To notify a responder that the RTS frame contains extra information in the scrambler seed, a transmitter sets the Unicast/Multicast bit in the Transmitter Address (TA) within the RTS frame to Multicast. After receiving an RTS frame addressed to itself with the Unicast/Multicast bit set to Multicast, the responder decodes the INDICATED_CH_BANDWIDTH field and the 1-bit INDICATED_DYN_BANDWIDTH

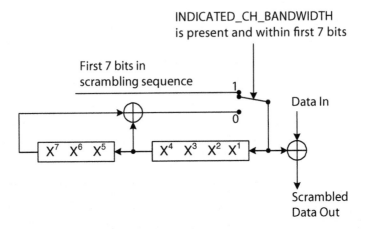

Figure 2.8: Scrambling operation performed by the modified scrambler. (From M. Gong, B. Hart, L. Xia, and R. Want. Channel bounding and MAC protection mechanisms for 802.11ac. In *Proc. IEEE GLOBECOM 2011*, pages 1–5, Houston, TX, Dec. 2011. [5]. © 2011 IEEE.)

field from the RTS frame, sets the INDICATED_CH_BANDWIDTH field to an appropriate value, and includes the field in the F7BOSS of the CTS frame.

2.4.1 Simulation Study

With OPNET simulation, we evaluate the performance of the 802.11ac MAC protection mechanism and compare the performance of dynamic bandwidth operation and static bandwidth operation. In the simulation, all STAs use MAC frame aggregation schemes, such as aggregated-MAC Protocol Data Unit (A-MPDU), and multiple transmissions in one transmit opportunity (TXOP). An STA can transmit as many A-MPDUs as the TXOP duration permits, provided that the last BA can be received within the TXOP duration.

The simulation parameters are defined as follows. The data rate over 80 MHz is 234 Mbps, aSlotTime is 9 μs, data rate over 40 MHz is 108 Mbps, aSIFSTime is 16 μs), control rate is 24 Mbps, RTS packet is 21 bytes, TXOP duration is 3 ms, MCTS is 15 bytes, max A-MPDU size is 128 KBytes, CWmin is 7, BA size is 32 bytes, and CWmax is 63. We first evaluate the 802.11ac MAC protection scheme in Section 2.4.1.1 and then compare the dynamic BW operation and static BW operation in Section 2.4.1.2.

2.4.1.1 Evaluation of the MAC Protection Scheme

We first evaluate the protection mechanism with downlink UDP traffic from each AP to a STA in each basic service set (BSS). The network topology is shown in Figure 2.9.

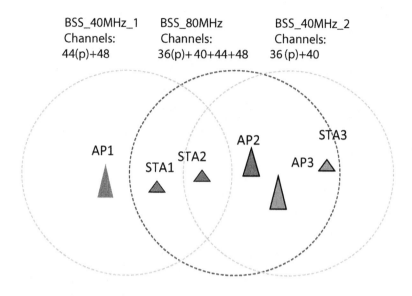

Figure 2.9: A hidden node scenario (simulation scenario 1) (From M. Gong, B. Hart, L. Xia, and R. Want. Channel bounding and MAC protection mechanisms for 802.11ac. In *Proc. IEEE GLOBECOM 2011***, pages 1–5, Houston, TX, Dec. 2011. [5]. © 2011 IEEE.)**

Fig. 2.10 presents the performance of the three BSSs with and without the new 802.11ac mechanism. The Y axis represents the saturation throughput per BSS in Mbps. When the existing 802.11 scheme is utilized, i.e., RTS/CTS without receiver CCA, neither BSS can achieve the desired throughput due to consistent collisions on non-primary channels. When the MAC protection mechanism is utilized, two 40 MHz BSSs achieve desirable throughput and BSS_80MHz achieves reasonable throughput. This is because the protection mechanism combats the hidden node problem on nonprimary channels and allows relatively fair sharing of the medium. Note that BSS_80MHz still has lower throughput than those of 40MHz BSS. This is reasonable given that BSS_80MHz shares its medium with two other BSSs while each 40 MHz BSS shares its medium with only one other BSS.

2.4.1.2 Comparison between Dynamic BW Operation and Static BW Operation

We then compare the performance of dynamic BW operation with that of static BW operation in this section. The simulation scenario is shown in Figure 2.11. There are two 40 MHz BSSs and one 80 MHz BSS. One 40 MHz BSS occupies the primary 40 MHz of the 80 MHz BSS's bandwidth, while the second 40 MHz BSS occupies the secondary 40 MHz of the 80 MHz BSS's bandwidth. Simulation scenario 2 is similar to simulation scenario 1, except that there are no hidden nodes in the network.

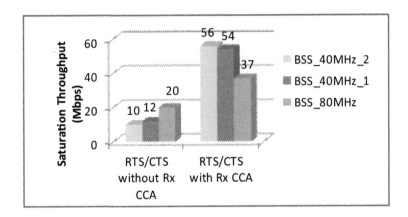

Figure 2.10: Per BSS throughput (existing scheme vs. the adopted scheme). (From M. Gong, B. Hart, L. Xia, and R. Want. Channel bounding and MAC protection mechanisms for 802.11ac. In *Proc. IEEE GLOBECOM 2011*, pages 1–5, Houston, TX, Dec. 2011. [5]. ⓒ 2011 IEEE.)

When all three BSSs are fully loaded, Figure 2.12 shows that with dynamic BW operation, BSS_80MHz has about 45 Mbps throughput, whereas with static BW operation, BSS_80MHz has extremely low throughput. The reason is that when there are two or more heavily loaded OBSSs that operate on different channels of the 80 MHz BSS, there is hardly any overlapping idle periods over the whole 80 MHz, and thus very little chance for the 80 MHz BSS to obtain full 80 MHz BW transmission opportunities.

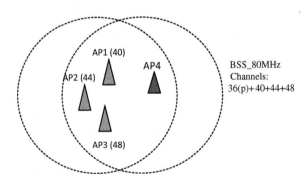

Figure 2.11: One 80 MHz BSS and two 40 MHz BSSs, no hidden node (simulation scenario 2). (From M. Gong, B. Hart, L. Xia, and R. Want. Channel bounding and MAC protection mechanisms for 802.11ac. In *Proc. IEEE GLOBECOM 2011*, pages 1–5, Houston, TX, Dec. 2011. [5]. ⓒ 2011 IEEE.)

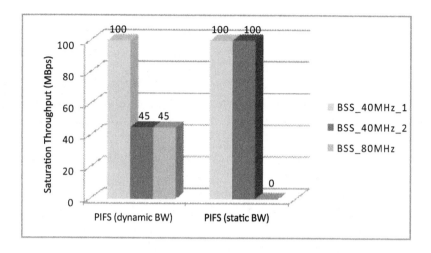

Figure 2.12: Per BSS throughput for scenario 2 (dynamic BW vs. static BW). (From M. Gong, B. Hart, L. Xia, and R. Want. Channel bounding and MAC protection mechanisms for 802.11ac. In *Proc. IEEE GLOBECOM 2011*, pages 1–5, Houston, TX, Dec. 2011. [5]. © 2011 IEEE.)

This simulation study shows that the 802.11ac MAC protection mechanism is effective in combating hidden nodes on secondary channels. The dynamic BW scheme offers better performance and is more robust than the static BW scheme, especially when there is more than one heavily loaded OBSS on different channels of the 80 MHz BSS.

If the interference on secondary channels is frequent, another new mechanism may be employed. In such a case, STA1 can send an Operating Mode Notification frame to AP1 to tell the AP that the client is changing its bandwidth on which it operates. For example, STA1 can change its operating bandwidth from 80 MHz to 40 MHz with the constraint that the client still needs to use the same primary channel as the AP. Subsequently, the AP will only send data frames at this reduced bandwidth to this client.

2.5 Downlink MU MIMO

2.5.1 *802.11ac DL MU-MIMO*

Multiple-input multiple-output (MIMO) communication techniques have been extensively studied for next generation cellular networks and have been deployed in wireless local area networks (WLANs) using the IEEE 802.11n technology. A MIMO system takes advantage of two types of gains, spatial diversity gain and spatial multiplexing gain [14].

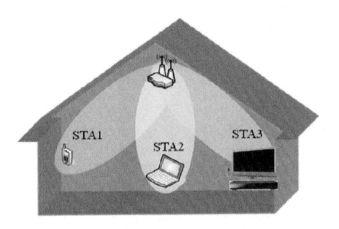

Figure 2.13: Example of downlink multiuser MIMO.

Spatial diversity is used to combat severe fading and can improve reliability of wireless links by carrying duplicate copies of the same information along multiple antennas. Spatial multiplexing creates an extra dimension in the spatial domain, which can carry independent information in multiple data streams. It has been shown that in a MIMO system with N transmit and M receive antennas, capacity grows linearly with $\min\{N,M\}$ [3]. Recent results show that similar capacity scaling applies when an N-antenna access point (AP) communicates with M users simultaneously [14]. Such a multiuser (MU) MIMO system has the potential to combine the high capacity achievable with MIMO processing with the benefits of multiuser space-division multiple access. Particularly, we're interested in downlink (DL) MU-MIMO systems, where an AP can transmit to multiple users simultaneously.

In 802.11ac DL MU-MIMO, an access point (AP) simultaneously transmits data streams to multiple client devices. For example, consider an AP with six antennas, a handheld client device with one antenna (STA1), a laptop client device with two antennas (STA2), and a TV set top box client device with two antennas (STA3). An AP can simultaneously transmit one data stream to STA1, two data streams to STA2, and two data streams to STA3. This is illustrated in Figure 2.13. The primary advantage of DL MU-MIMO is that client devices with limited capability (few or one antenna) do not degrade the network capacity by occupying too much time on air due to their lower data rates. With DL MU-MIMO, network capacity is based on the aggregate of the clients of the simultaneous transmission. However, this benefit comes with increased cost and complexity.

From a PHY perspective, the AP should have more antennas than total number of spatial streams for diversity gain. In addition, the AP requires channel state information from each of the clients participating in the DL MU-MIMO transmission in order to form the antenna weights. With DL MU-MIMO, the antenna weights are much more sensitive to changes in the channel. In the case of transmit beamforming, if the antenna weights are stale, the system performance degrades to the case without transmit beamforming. However, with DL MU-MIMO, if the antenna weights do not accurately match the channel, the streams to one client introduce interference to the other clients, leading to a negative (in dB) signal-to-interference-plus-noise ratio (SINR). Therefore, channel state information must be higher resolution and more frequently updated. To constrain the dimensions of the system to a manageable size, 802.11ac defines that the maximum number of users in a transmission is four, the maximum number of spatial streams per user is four, and the maximum total number of spatial streams (summed over the users) is eight.

Assume the AP transmits simultaneously to different stations (STAs) in the same basic service set (BSS). With N transmit antennas, the AP can transmit a total of N spatial streams. These N streams can be distributed across a maximum of N STAs. When the AP transmits different streams to multiple STAs, interference from streams intended for one STA will cause interference to the other STAs. This is represented by the following equation [4, 6]:

$$
\begin{aligned}
Y_i &= \sqrt{\frac{\rho}{M}} H_i W_1 X_1 + \cdots + \sqrt{\frac{\rho}{M}} H_i W_i X_i + \\
&\quad \cdots + \sqrt{\frac{\rho}{M}} H_i W_M X_M + Z_i \\
&= \sqrt{\frac{\rho}{M}} [W_1, \cdots, W_M] \begin{bmatrix} X_1 \\ \vdots \\ X_M \end{bmatrix} + z_i
\end{aligned} \tag{2.1}
$$

where Y_i is the received signal at the ith STA (with dimensions $N_{Rx} \times 1$), X_i is the transmitted streams to the ith STA (with dimensions $N_{ss} \times 1$), N_{ss} is the number of spatial streams for each STA, H_i is the channel between the AP and the ith STA (with dimensions $N_{Rx} \times N_{Tx}$), W_i is the weight applied at the transmitter (with dimensions $N_{Tx} \times N_{ss}$), ρ is the received power, M is the number of STAs, Z_i is addition white Gaussian noise at the ith STA (with dimensions $N_{Rx} \times 1$), N_{Rx} is the number of receiving antennas at a STA, and N_{Tx} is the number of transmitting antennas at the AP.

The signal $H_i W_j X_j$ received by Y_i causes interference when decoding its streams X_i when $i \neq j$. The AP can mitigate this interference with intelligent beamforming techniques [12]. For example, if we select weights such that $H_i W_j = 0$ when $i \neq j$, then the interference from other STAs is canceled out.

A simple linear processing approach is to precode the data with the pseudoinverse of the channel matrix [12]. To avoid the noise enhancement that accompanies zero forcing techniques, the minimum mean square error (MMSE) precoding can be used instead. To describe this approach, we first present the entire system model, including all STAs, as follows:

$$
\begin{bmatrix} Y_1 \\ \vdots \\ Y_M \end{bmatrix} = \sqrt{\frac{\rho}{M}} \begin{bmatrix} H_1 \\ \vdots \\ H_M \end{bmatrix} \begin{bmatrix} W_1 \\ \vdots \\ W_M \end{bmatrix}^T \begin{bmatrix} X_1 \\ \vdots \\ X_M \end{bmatrix} + \begin{bmatrix} Z_1 \\ \vdots \\ X_M \end{bmatrix}
$$

That is,

$$Y = \sqrt{\frac{\rho}{M}} HWX + Z \qquad (2.2)$$

The MMSE precoding weights are then given as follows:

$$W = \sqrt{\frac{\rho}{M}} H^\dagger \left(\frac{\rho}{M} HH^\dagger + \Phi_z \right)^{-1} \qquad (2.3)$$

where Φ_z is the noise covariance matrix and $H^\dagger$ is the Hermitian of H.

Interference cancellation techniques can be implemented in the receiver to further reduce degradation from multiple access interference. When the receiving STA has more receive antennas than the number of spatial streams it intends to receive, the extra antennas can be used to cancel out the spatial streams intended for other STAs. If channel state information (CSI) is known for the channel dimensions of the interference streams (i.e., $H_i W_j$), the CSI can be used to null interference in an MMSE receiver. This type of equalizer structure is given by $G_i Y_i$, where

$$G_i = \sqrt{\frac{\rho}{M}} W_i^H H_i^H \left(\sum_{k=1}^M \frac{\rho}{M} H_i W_k W_k^H H_i^H + \Phi_z \right)^{-1} \qquad (2.4)$$

To compare DL MU-MIMO to single-user 802.11n transmit beamforming (TXBF), we assume that the transmitter weights are generated using the eigenvectors from singular value decomposition (SVD). Though a specific weighting scheme is not defined in 802.11n, SVD yields maximum likelihood performance with a simple linear receiver [10]. The system equation with single-user TXBF is expressed as

$$Y = \rho HVX + Z \qquad (2.5)$$

where the SVD of H is $U\Sigma V$. When the AP has more antennas than transmitted spatial streams, the TXBF gain can be substantial even when the receiver has the same number of receive antennas as spatial streams.

As designed, an MU packet has the same preamble structure as a single-user packet. However, beginning with the VHT-STF, the remaining fields in the preamble are directionally transmitted to recipient clients, simultaneously in time and frequency. The parameter information conveyed in VHT-SIG-B and the service field is specific for each client. In addition, MAC padding is required to fill the MAC frames to the last byte to make them equal in time for each client. The PHY fills in the last few bits for each client to ensure that each has the same number of symbols.

From a MAC perspective, since with DL MU-MIMO multiple packets are transmitted simultaneously to different clients, a mechanism is needed to receive acknowledgments from these clients. The approach used for 802.11ac multiuser acknowledgments builds on the 802.11n implicit block acknowledgment feature.

As illustrated in Figure 2.14, after transmitting a DL MU-MIMO data burst, the AP uses an implicit block acknowledgment request for the first client, meaning that the first client replies immediately to the MU transmission with a Block ACK (BA). The AP subsequently polls the second client with a Block ACK Request (BAR),

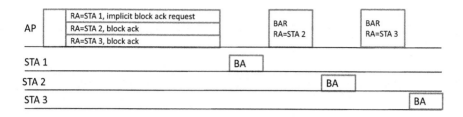

Figure 2.14: DL MU-MIMO response mechanism.

and the second client responds with a BA. This continues until all the clients in the original transmission are polled.

This procedure assumes that the clients know that they are part of the MU transmission and their order. This is achieved with the Group ID in VHT-SIG-A. Prior to the MU transmission, the group definition information is conveyed by the AP to all the DL MU-MIMO-capable clients in the basic service set (BSS). Based on the definition of the Group ID, there are 62 different possible groupings for different combinations of client devices. In addition to the Group ID, VHT-SIG-A also contains a table indicating how many data streams are being transmitted to each client in the transmission.

The sounding and feedback protocol utilizes a null data packet (NDP) frame. The sounding feedback sequence starts with the AP sending an NDP announcement (NDPA) frame immediately followed by an NDP. The NDPA identifies which client will be the first responder after the NDP and may identify other clients that will be polled subsequently. The client identified as first by the NDPA replies with a sounding feedback frame after the NDP. Then the AP polls all the remaining clients. Such a sequence requires a recovery mechanism in case a response is not received from a client. In this case a feedback poll can be re-sent to the client.

In the case of single-user transmit beamforming, the same sounding and feedback protocol is used, but the sequence stops after the single client responds with feedback.

2.5.2 Performance Analysis

In this section, we derive the saturation throughput of a DL MU-MIMO WLAN system. The system's saturation throughput is defined as the combined throughput achieved at the top of the MAC layer when all nodes in the system are fully loaded at all times.

It is assumed that the devices use MAC frame aggregation schemes, such as aggregated-MAC Protocol Data Unit (A-MPDU), and multiple transmissions in one transmit opportunity (TXOP). We follow the assumptions made in [1] and a similar 2D Markov chain model for the node behavior. In the Markov chain mode, each state is represented by $\{s(t), b(t)\}$, where $s(t)$ is defined to be the stochastic process representing the backoff stage $[0, 1, \cdots, m]$ of the station at time t and $b(t)$ is the stochastic process representing the backoff time counter for a given station. The maximum backoff stage, i.e., m, takes the value such that

$$CW_{max} = 2^m CW_{min}$$

where CW_{max} is the maximum contention window and CW_{min} is the minimum contention window.

Let S be the normalized system throughput, defined as the fraction of time when the channel is used to successfully transmit the payload bits. S can be expressed as the average payload bits transmitted in a TXOP divided by the average length of a TXOP. Based on the 2D Markov chain mode, we extend the analysis in [1] and derive the system saturation throughput as follows:

$$S = P_{AP} \frac{P_s P_{tr} \sum_{j=1}^M \sum_{i=1}^{N_j} \mathrm{E}[P_{ij}]}{(1 - P_{ij})\sigma + P_{tr}P_s T_s + P_{tr}(1 - P_s)T_c} +$$

$$P_{STA} \frac{P_s P_{tr} \sum_{j=1}^{n-1} \sum_{i=1}^{N_j} \mathrm{E}[P_{ij}]}{(1 - P_{ij})\sigma + P_{tr}P_s T_s + P_{tr}(1 - P_s)T_c} \qquad (2.6)$$

$$T_s = TXOP_{dur} \qquad (2.7)$$

$$T_c = RTS + DIFS \qquad (2.8)$$

$$P_{tr} = 1 - (1 - \tau)^n \qquad (2.9)$$

$$P_s = \frac{n\tau(1 - \tau)^{n-1}}{1 - (1 - \tau)^n} \qquad (2.10)$$

where P_{AP} is the probability that the AP wins the contention, P_{STA} is the probability that a STA wins the contention, M is the number of users to which an AP can transmit simultaneously, T_s is the average time consumed by a successful TXOP, i.e., $TXOP_{dur}$, T_c is the average medium time a collision consumes, σ is the duration of a time slot, RTS is the transmission duration of the RTS frame, n is the number of contending devices in the network, including the AP and the stations, τ is the probability that a device transmits in a randomly chosen time slot, P_s is the probability that a TXOP is successfully set up, P_{tr} is the probability that there is at least one transmission in the considered slot time, and $\sum_{i=1}^{N_j} \mathrm{E}[P_i]$ is the combined average payload size of N_j A-MPDUs that are transmitted in the TXOP.

Equation (2.6) can be rearranged as follows:

$$S = \frac{\frac{1}{n}\left(\sum_{j=1}^M \sum_{i=1}^{N_j} \mathrm{E}[P_{ij}] + \sum_{j=1}^{n-1} \sum_{i=1}^{N_j} \mathrm{E}[P_{ij}]\right)}{T_s - T_c + \frac{T_c - (1-\tau)^n(T_c - \sigma)}{n\tau(1-\tau)^{n-1}}} \qquad (2.11)$$

Under condition $\tau \ll 1$, τ can be estimated as [1]

$$\tau \approx \frac{1}{n}\sqrt{\frac{2\sigma}{T_c}}. \qquad (2.12)$$

In Figure 2.15, we plot the relationship between the optimal saturation throughput S and the number of contending devices n in the BSS. The parameter values are

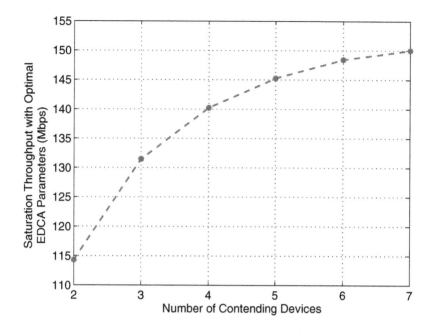

Figure 2.15: Saturation throughput *S* vs. number of contending devices *n* (optimal EDCA parameters). (From M. Gong, E. Perahia, R. Stacey, R. Want, and S. Mao. A CSMA/CA MAC protocol for multi-user MIMO wireless LANs. In *Proc. IEEE GLOBECOM 2010*, pages 1–5, Miami, FL, Dec. 2010. [6]. © 2010 IEEE.)

defined in Section 2.5.3. When all contending devices have equal transmission opportunities, the saturation throughput of the network increases with the number of contending devices due to spatial diversity gain achieved by DL MU-MIMO.

2.5.3 Simulation Study

We implement the 802.11ac DL MU-MIMO MAC protocol in OPNET Modeler [9] and compare its performance with that of the beamforming protocol. Our simulations consider a typical WLAN topology, consisting of one AP, equipped with four antennas, and multiple STAs, each of which is equipped with two antennas. Other simulation parameters are defined as follows. The DL MU-MIMO data rate is 65 Mbps, aSlotTime is 9 μs, BF data rate is 130 Mbps, aSIFSTime is 16 μs, control rate is 24 Mbps, TXOP duration is 3 ms, RTS packet is 20 bytes, MSDU size is 1,500 bytes, CTS packet is 14 Bbytes, CWmin is 7, BA size is 32 bytes, and CWmax is 63.

To support DL MU-MIMO, we assume that the STA implements interference cancellation techniques necessitating more receive antennas than received spatial streams. Therefore, the AP only transmits one spatial stream (SS) to each STA in the simulations, which has two antennas. However, when TXBF is used in the

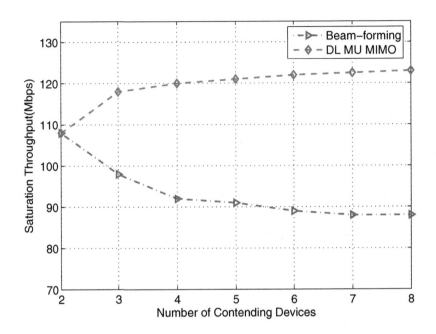

Figure 2.16: Saturation throughput S vs. number of contending devices n (bidirectional traffic). (From M. Gong, E. Perahia, R. Stacey, R. Want, and S. Mao. A CSMA/CA MAC protocol for multi-user MIMO wireless LANs. In *Proc. IEEE GLOBECOM 2010*, pages 1–5, Miami, FL, Dec. 2010. [6]. © 2010 IEEE.)

simulations, each STA can receive two spatial streams. Because STAs are placed close to the AP, on average the achievable signal-to-noise ratio (SNR) at each receiver is at least 30 dB.

We first compare the saturation throughput of DL MU-MIMO with that of TXBF with respect to the number of contending devices n. The simulation results are plotted in Figure 2.16. It can be seen that when the number of contending STAs increases, the saturation throughput achieved by DL MU-MIMO also increases, whereas the saturation throughput achieved by beamforming degrades. The reason is that DL MU-MIMO can effectively take advantage of the spatial diversity gain, which is larger when the number of contending STAs increases, while the beamforming scheme does not have this capability.

Analysis and simulation study both show that when the number of contending STAs increases, the saturation throughput achieved by DL MU-MIMO also increases. Furthermore, our simulation results show that DL MU-MIMO can achieve better performance than TXBF when there are more than two STAs in the network.

2.6 Conclusion

Wi-Fi products based on next generation gigabit per second 802.11 technologies are emerging on the market to address use cases that demand very high throughputs. 802.11ac is an evolution of 802.11n in the 5 GHz band with channels widened up to 160 MHz, modulation and coding increased to 256-QAM rate-5/6, and the maximum number of spatial streams enlarged to eight. With all these enhancements, 802.11ac enables multigigabit communications at rates up to 11 times faster than 802.11n. Moreover, 802.11ac also delivers multiuser capability to address new use cases for Wi-Fi devices, such as high-resolution video streaming to multiple screens around the home. In this chapter, we reviewed the basic concepts and specifications of 802.11ac and discussed its main features, including channelization, the PHY design, channel bonding and MAC protection, and downlink multiuser multiple input multiple output (DL MU-MIMO). We also presented simulation and analysis studies of channel bonding, MAC protection, and DL MU-MIMO, which collectively demonstrate the high potential of 802.11ac.

Acknowledgment

Shiwen Mao's work is supported in part by the U.S. NSF under grants CNS-0953513, CNS-1247955, CNS-1320664, and ECCS-0802113, and through the NSF Broadband Wireless Access and Applications Center (BWAC) at Auburn University. Any opinions, findings, and conclusions or recommendations expressed in this material are those of the author(s) and do not necessarily reflect the views of the foundation.

References

1. G. Bianchi. Performance analysis of the IEEE 802.11 Distributed Coordination Function. *IEEE J. Sel. Areas Commun.*, 18(3):535–547, 2000.

2. G. Breit, H. Sampath, S. Vermani, et al. TGac channel model addendum. IEEE 802.11-09/0308r12. March 18, 2010.

3. A. Goldsmith, S. Jafar, N. Hindal, and S. Vishwanath. Capacity limits of MIMO channels. *IEEE J. Sel. Areas Commun.*, 21(5):684–702, 2003.

4. M. Gong, D. Akhmetov, R. Want, and S. Mao. Directional CSMA/CA protocol with spatial reuse for mmWave wireless networks. In *Proceedings of the IEEE GLOBECOM 2010*, Miami, FL, December 2010, pp. 1–5.

5. M. Gong, B. Hart, L. Xia, and R. Want. Channel bounding and MAC protection mechanisms for 802.11ac. In *Proceedings of the IEEE GLOBECOM 2011*, Houston, TX, December 2011, pp. 1–5.

6. M. Gong, E. Perahia, R. Stacey, R. Want, and S. Mao. A CSMA/CA MAC protocol for multi-user MIMO wireless LANs. In *Proceedings of the IEEE GLOBECOM 2010*, Miami, FL, December 2010, pp. 1–5.

7. P. Loc and M. Cheong. TGac functional requirements and evaluation methodology. IEEE 802.11-09/0451r16. January 19, 2011.

8. A. Myles and R. de Vegt. Wi-Fi Alliance (WFA) VHT study group usage models. IEEE 802.11-07/2988r4. March 19, 2008.

9. OPNET. OPNET Modeler: http://www.opnet.com/.

10. E. Perahia and R. Stacey. *Next generation wireless LANs: Throughput, robustness, and reliability in 802.11.* Cambridge, UK: Cambridge University Press, 2008.

11. R. De Vegt. 802.11ac usage models document. IEEE 802.11-09/0161r2. March 9, 2009.

12. Q. Spencer, C. Peel, A. Swindlehurst, and M. Haardt. An introduction to the multi-user MIMO downlink. *IEEE Commun. Mag.*, 42(10):60–67, 2004.

13. R. Stacey, E. Perahia, A. Stephens, et al. Specification framework for TGac, IEEE 802.11-09/0992r21. January 19, 2011.

14. D. Tse, P. Viswanath, and L. Zheng. Diversity-multiplexing tradeoff in multiple access channels. *IEEE Trans. Inf. Theory*, 50(9):1859–1874, 2004.

Chapter 3

Future Wireless Sensor Networks for the Smart Grid

Irfan Al-Anbagi, Melike Erol-Kantarci, and Hussein T. Mouftah

University of Ottawa

CONTENTS

3.1   Introduction

Real-time and reliable data communications in the smart grid are essential for its successful automation and control processes. Wireless sensor networks (WSNs) are anticipated to be widely utilized in a broad range of smart grid applications due to their numerous advantages and unique features. Despite these advantages, the use of WSNs in such critical applications has brought forward a new challenge of fulfilling the quality of service (QoS) requirements. Providing QoS support in WSNs is a challenging issue due to the highly resource-constrained nature of sensor nodes, unreliable wireless links, and harsh operation environments. In WSN-based smart grid automation and control processes, real-time and reliable monitoring requires optimizing certain network parameters. This consequently requires an accurate and stable model that resembles the utilized network protocol. In the literature, QoS provisioning in WSNs has been considered by developing QoS-aware protocols irrespective of actual mathematical models' behavior. Analytical models that consider delay, throughput, and power consumption have not matured for QoS provisioning in smart grid applications.

The definition of QoS may vary depending on the application; e.g., QoS can indicate the capability to provide assurance that the service requirements of a specific application are met. QoS can also be defined as the ability of the network to adapt to specific classes of data such as real-time and non-real-time data. Furthermore, QoS can refer to the stability of the WSN system under varying traffic and network conditions. The International Telecommunication Union (ITU) Recommendation E.800 (09/08) has defined QoS as "totality of characteristics of a telecommunications service that bear on its ability to satisfy stated and implied needs of the user of the service" [2]. WSNs have been considered for use in various parts of the smart grid, and QoS is relevant for most of those WSN-based applications [3–5]. Choosing optimum network parameters to achieve certain QoS guarantees requires precise and realistic analytical models that take into consideration the factors that control the delay, throughput, and power consumption of the WSNs.

In the literature, several works have focused on QoS-aware protocol design in the smart grid [6–9, 11]. Furthermore, several models with different mathematical approaches have been proposed to describe the performance of WSNs under certain traffic conditions [10–15]. Some of these analytical models become too complex and impractical in the resource-constrained sensor nodes. On the other hand, some models tend to use approximations that lead to inaccuracies in the model itself by not considering certain important factors such as traffic patterns.

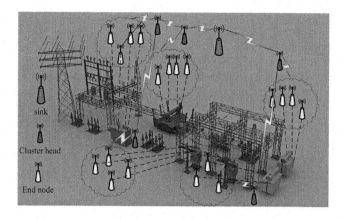

Figure 3.1: Substation monitoring via cluster-tree WSN.

In [16], we presented a realistic and stable Markov-based (RSM) model for an IEEE 802.15.4-based star topology WSN. The model addresses the inaccuracies in previous analytical models, and issues that were not covered previously in this context. Star topology is convenient for small-scale deployments due to the limited communication range of sensors within one hop reach. On the other hand, smart grid calls for WSN deployments in large-scale installations where star topology becomes inefficient due to coverage problems. WSNs can be extended for a larger coverage via cluster-tree or mesh topologies. However, it is not straightforward to use the analytical model of [16] or other available models in cluster-tree WSNs as shown in Figure 3.1. In [17] we extended the model of [16] to contention-based cluster-tree WSNs. The use of contention-based cluster-tree topology requires that all the nodes in a network hear each other to avoid collision and packet loss. However, the use of WSNs for smart grid applications may require positioning the nodes in clusters that are far apart; hence, contention-based schemes may not work well due to the hidden terminal problem.

In this chapter, we survey the state of the art in future wireless sensor networks and their applications in the smart grid. Furthermore, we present a recent Markov-based model for IEEE 802.15.4-based WSN that enhances the stability of the network. The proposed model considers WSNs with multihop topologies; these topologies are highly important in smart grid automation applications since they are suitable for large-scale deployments.We further extend the work presented in [16] and [17] to include schedule-based cluster-tree topology. Furthermore, we perform comprehensive performance evaluations of the star, contention-based, and scheduling-based cluster-tree WSN topologies under different traffic and network conditions that are suitable for smart grid applications. In addition to that, we propose a super frame (SF) structure that is suitable for the presented QoS model and the proposed application. We also perform a comprehensive performance analysis under different traffic and network conditions.

3.2 Background

WSN requirements associated with smart grid applications play a considerable role in determining how to implement the WSN technology into the power grid infrastructure. WSN QoS requirements vary depending on the criticality of the monitored power grid component. These requirements can be one or a combination of the following requirements: latency, reliability, availability, security, and spectrum availability. In this chapter we focus on the latency and the reliability requirements. Such applications vary from teleprotection systems to emergency power restoration to substation monitoring and control. Therefore, latency requirements in smart grid monitoring applications may vary from several seconds for smart metering to less than 10 ms for protection operations. Table 3.2 shows some typical latency requirements of some common smart grid monitoring applications [1]. In the literature there are many studies that discuss the use of QoS WSNs' protocol from delay and reliability critical applications. In [18], spectrum-aware and cognitive sensor networks have been proposed to overcome spatiotemporally varying spectrum characteristics and harsh environmental conditions for WSN-based smart grid applications.

In [19], the feasibility of a public LTE network in supporting worst case smart grid communications has been investigated. In [20], a distributed algorithm to minimize the data aggregation latency under the physical interference mode has been proposed in a smart grid scenario. In [21], the performance of a WSN system in the measurement of partial discharge signals and data flow optimization and management from the monitoring sensors to the base station has been optimized and evaluated.

Sun et al. [22] have proposed to use a private wireless network dedicated for power distribution system monitoring. The authors have introduced a QoS support for IEEE 802.15.4 by the differentiated service for data traffic with different priority. They have used additional queues in the MAC to store different priority traffic. Therefore, high-priority data will have higher probability of channel access, and can interrupt the service to the low-priority traffic by forcing it to backoff (BO). They have assumed N sensor nodes to monitor power distribution devices and report back to a coordinator using the IEEE 802.15.4 protocol, and that all nodes can hear each

Table 3.1   Latency Requirements for Some Smart Grid Applications

Application	Maximum Latency
Teleprotection	≤ 10 ms
Synchrophasor	~ 20 ms
Control and data acquisition	100–200 ms
Smart metering	2–3 s

Source: J. Deshpande, A. Locke, and M. Madden, "Smart choices for the smart grid: Using wireless broadband for power grid network transformation," *Alcatel-Lucent Technology White Paper*, 2010.

other. When operational data arrive at any node, they will be pushed into the queue at the MAC layer if there is a packet in service. When the emergency data arrive, they will be queued in the high-priority queue if there is a high-priority packet in service. Otherwise, they will interrupt the service of an operational data packet. They have assumed that no operational data will be serviced until the emergency data queue is empty. They have modeled the delay of QoS-MAC and the BO process using the Markov chain queue model for two classes of traffic. They have assumed that the packet arriving rate for all nodes is the same, and set the maximum number of BO stage as 5; the value of the BE for high-priority traffic ranges from 0 to 3, and for low-priority data ranges from 2 to 5. The authors have not presented the impact of the buffer or queue size on the performance of the network. The queue size of each sensor node will affect the waiting duration of the packet, and hence may affect the overall network performance.

Ruiyi et al. [23] have proposed an adaptive wireless resource allocation (AWRA) algorithm with QoS guarantee in a communication network of the smart grid. The authors have addressed adaptive wireless resource allocation, where they have assumed that if the delay of the packets is greater than the delay threshold, then the packet is discarded, while for non-real-time services, as long as the queue does not overflow, the packet will not be discarded. They have assumed that the queue is infinite and do not consider the discarded packet caused by queue overflow and the problem of retransmission and that the total transmission power of the base station in the subchannels is average distribution. The authors have proposed that the system of the smart grid contains 19 plots; each plot has 3 sectors, and each sector has N subchannels and K packets. They have defined an optimization problem based on different stages of base station tasks. The first stage is the detect stage, where the base station measures the user's SNR; the second stage is the feedback stage, where the user feeds back the channel state information; and the final stage is when the base station collects the feedback information and allocates space, time, and frequency resource for the user to transmit data based on certain scheduling criteria.

In [24], the authors have introduced a medium access scheme, delay-responsive cross-layer (DRX) data transmission, that addresses delay and service differentiation requirements of the smart grid. The DRX scheme is based on delay estimation and data prioritization procedures that are performed by the application layer for which the MAC layer responds to the delay requirements of a smart grid application and the network condition.

In addition to the above surveyed papers, some papers discuss delay-tolerant smart grid applications, which include automatic meter reading (AMR), billing, routine data measurement, and switching of appliances. These applications tolerate delays and can perform adequately with some data loss. In [25], a WSN-based intelligent light control system for indoor environments has been proposed. In [26], new field tests using open-source tools with ZigBee technologies have been proposed for monitoring photovoltaic and wind energy systems and energy management of buildings and homes. In [3], the performance of an in-home energy management application has been evaluated.

In this chapter, we initially describe our mathematical model and then use it to build our QoS schemes. We find it essential to provide an overview of available mathematical models that are used to model the MAC sublayer of the IEEE 802.15.4 standard.

Markov-based performance evaluation of the IEEE 802.15.4 MAC sublayer has been presented in several studies. The majority of these studies are based on the model derived in [27]. The work presented in [27] has assumed saturated traffic conditions; which makes this model an unrealistic WSN situation. Later studies e.g., [12, 13, 28, 29] have solved the saturation traffic model problem by modeling the IEEE 802.15.4 MAC with unsaturated traffic conditions.

Park et al. [11] have presented a generalized model for the IEEE 802.15.4 MAC sublayer. The authors proposed certain approximations in their formulation of the delay, reliability, and energy consumption. We differentiate our work by considering the idle state, traffic arrival, and buffering. The work presented in [11] has proposed an idle state duration (L_0), which is a variable that governs the model performance. In realistic WSN applications, the idle state duration cannot be quantified easily. Thus, our model does not depend on this parameter. Another major difference is that our model is designed to be less dependent on the network traffic conditions by including a finite buffer at the MAC level. This feature makes the model more stable even when the traffic changes its pattern. Furthermore, in [30], the authors have presented a model for the IEEE 802.15.4 MAC sublayer that models and optimizes the performance of a single-hop star WSN. They have also incorporated packet copying delay due to hardware limitations in to their model. In our model we also model the effect of a finite MAC buffer in our system. However, we follow a different approach in deriving and describing those buffers. We also differentiate our work by eliminating the idle state duration and presenting the model for two types of cluster-tree topologies.

Pollin et al. [12] have proposed an analytical model and performed evaluation of the slotted carrier sense multiple access with collision avoidance (CSMA/CA) algorithm in the presence of uplink and acknowledged uplink traffic, under both saturated and unsaturated conditions. They have also used a Markov-based model to evaluate the performance of the IEEE 802.15.4 standard in terms of power consumption and throughput. They have described guidelines to tune the MAC parameters to increase throughput and power savings. The authors, however, have not taken into consideration the transmission retries. In our model, similar to [11], we have considered these retries, which results in a three-dimensional Markov-based model. Furthermore, the authors in [12] have not considered the effect of MAC-level buffers on the overall performance of the network.

Zhu et al. [29] have used a Markov-based model to analyze the characteristics of the IEEE 802.15.4-based WSN in terms of packet delay, energy consumption, and throughput under unsaturated, unacknowledged traffic conditions. However, we have seen that the predetermined length of the idle state has remained in their model, and we have not seen the independence from this variable.

Busanelli et al. [31] have proposed an optimization tool that applies some classical operations research instruments to a Markov chain-based model. They have shown that their technique is suitable for the performance analysis of a generic cluster-tree IEEE 802.15.4-based network. In our model, we have specifically focused on power consumption optimization while considering the effect of both packet arrival rate and the MAC-level finite buffer. Furthermore, in [31], the power consumption metric has not been considered. In addition, the model proposed has assumed certain approximations to the traffic generation pattern.

Liu et al. [32] have studied the delay performance in a WSN with a cluster-tree topology. They have proposed a heuristic scheme to find the timeline allocations of all the cluster-heads (CHs) in a WSN in order to achieve the minimum and balanced packet drop rate for traffic originated from different levels of the cluster-tree.

Ramachandran et al. [33] have presented an analysis of the performance of the contention access period (CAP) specified in IEEE 802.15.4, in terms of throughput and energy consumption. They have modeled the CAP as nonpersistent CSMA/CA with backoff using Markov chains. Their analysis has not considered the impact of a finite MAC-level buffer size on the throughput and energy consumption of an IEEE 802.15.4-based sensing node.

The use of reliable and timely WSNs for industrial applications has been discussed in [24, 34–38]. Furthermore, QoS provisioning in smart grid monitoring applications has been discussed in [3, 22, 38, 39].

3.3 The Analytical Model

In CPSs such as the smart grid, the use of an accurate and stable model becomes an integral component for the success of such systems. In this section we develop a mathematical model for star-based and two cluster-tree-based WSN topologies. Table 10.1 shows the summary of notations used in our derivations. We initially describe the model for the star topology and then discuss the two cluster-tree topologies and study the performance of each model. All three models rely on the following assumptions:

- All of the nodes in the WSN operate in the beacon-enabled mode of the IEEE 802.15.4 MAC.

- Packets arrive at the MAC sublayer with an arrival rate of λ packets per second (pkts/s) with Poisson distribution. The arrival rate is assumed the same for all end nodes.

- We assume that the MAC-level acknowledgment (ACK) packets are used to increase the reliability.

- In the star topology, the coordinator node is the sink, while in the cluster-tree topology, the root node is the sink.

- We assume that a single packet fits into the SF period, which means a packet is assumed to be delivered to the next hop in one transmission round.

Table 3.2 Summary of Notations

$b_{i,k,j}$	The probability of being at state (i, k, j) in the Markov chain
L_s	Duration of successful transmission
L_{ack}	Duration of acknowledgment packet
L_c	Duration of packet collision
σ	Probability of packet arrival at the MAC layer
W_0	The smallest backoff window
m	macMaxCSMABackoffs
n	macMaxFrameRetries
α	Probability of finding the first clear channel assessment (CCA1) busy
β	Probability of finding the second clear channel assessment (CCA2) busy
τ	Probability of starting CCA1
P_c	Probability of collision
D_0	The size of the MAC buffer
T_i	The end-to-end delay in each network level
l	The number of levels in the cluster-tree network
BeO	Beacon order
SO	Super-frame order
B_i	Virtual buffer in a cluster-head
φ	The occupancy of B_i
η	The number of packets received from end nodes in the local cluster
ε	The number of forwarded packets from lower cluster-heads
D_{SF}	Super-frame duration
ϕ	Maximum number of packets that can be transmitted from a cluster-head in a D_{SF}
ψ	Maximum number of packets that can be received into a cluster-head in a D_{SF}
λ	Packets arrival rate at the MAC layer
δ	The number of D_{SF} a packet waits before it is transmitted to the next cluster-head

■ In both the star and cluster-tree topologies, we assume that all nodes have M/G/1/L queues and the buffer available at each node is assumed to be of a first-in first-out (FIFO) type with no flow priority. Furthermore, the packet processing time in the buffer is negligible.

3.3.1 Star Topology

In this section, we describe the modifications to the model in [11] (referred to as Park's model in the rest of this chapter), which was presented in [16] to design an accurate and realistic model for WSNs with star topology. We start by providing a brief description of Park's system model analysis and then discuss our improvement for the star topology and introduce the modifications for the two cluster-tree topologies.

Park proposed a pernode model based on Markov chain for a star topology WSN that implements the slotted CSMA/CA algorithm. The traffic used in the network is unsaturated traffic with ACK. The Markov chain is three-dimensional and described using three stochastic processes, namely, the backoff stage at time t ($s(t)$), the state of the backoff counter at time t ($c(t)$), and the state of the retransmission counter at time t ($r(t)$). A necessary assumption for the Markov chain to be applicable in the context of WSNs is that all nodes are assumed to start sensing the wireless medium independently. The resulting Markov model can be described by the tuple ($s(t), c(t), r(t)$), and the stationary distribution of the Markov chain can be written as $b_{i,k,j} = \lim_{t \to \infty} P(s(t) = i, c(t) = k, r(t) = j)$, where $i \in (-2, m), k \in (-1, max(W_i - 1, L_s - 1, L_c - 1))$, and $j \in (0, n)$. Based on all of these assumptions, we can develop the Markov chain shown in Figure 10.2. Figure 10.2 presents a single detailed transmission retry that includes up to m backoff stages. The remaining $n - 1$ retries can be realized directly from the figure. Based on the chain proposed in Figure 10.2, we can derive closed-form expressions that express our entire system. Since our Markov chain resembles Park's chain, except for the existence of the buffer and the traffic generation pattern, most of Park's derivations are applicable in this context. We outline below the final derived equations from [11] and skip their detailed derivations (the interested reader is referred to [11] for the complete derivations):

$$\sum_{i=0}^{m} \sum_{k=0}^{W_i-1} \sum_{j=0}^{n} b_{i,k,j} + \sum_{i=0}^{m} \sum_{j=0}^{n} b_{i,-1,j}$$

$$+ \sum_{j=0}^{n} \left(\sum_{k=0}^{L_S-1} b_{-1,k,j} + \sum_{k=0}^{L_C-1} b_{-2,k,j} \right) + \sum_{l=0}^{L_0-1} Q_l = 1 \quad (3.1)$$

$$\sum_{i=0}^{m} \sum_{k=0}^{W_i-1} \sum_{j=0}^{n} b_{i,k,j} \approx \frac{b_{0,0,0}}{2} \left[(1+2x)W_0 + 1 + x \right] (1+y) \quad (3.2)$$

$$\sum_{i=0}^{m} \sum_{j=0}^{n} b_{i,-1,j} \approx b_{0,0,0}(1-a)(1+x)(1+y) \quad (3.3)$$

$$\sum_{j=0}^{n} \left(\sum_{K=0}^{L_S-1} b_{-1,K,j} + \sum_{K=0}^{L_C-1} b_{-2,k,j} \right)$$

$$\approx b_{0,0,0} L_S (1 - x^{m+1})(1+y) \quad (3.4)$$

$$\sum_{l=0}^{L_0-1} Q_l \approx b_{0,0,0} \frac{1-\sigma}{\sigma} L_0 \left[1 + y + P_c(1 - x^{m+1}(y^n - y - 1)) \right] \quad (3.5)$$

$$\tau \approx (1+x)(1+y)b_{0,0,0} \quad (3.6)$$

where

$$x = \alpha + (1-\alpha)\beta \quad (3.7)$$

$$y = P_c(1 - x^{m+1}) \quad (3.8)$$

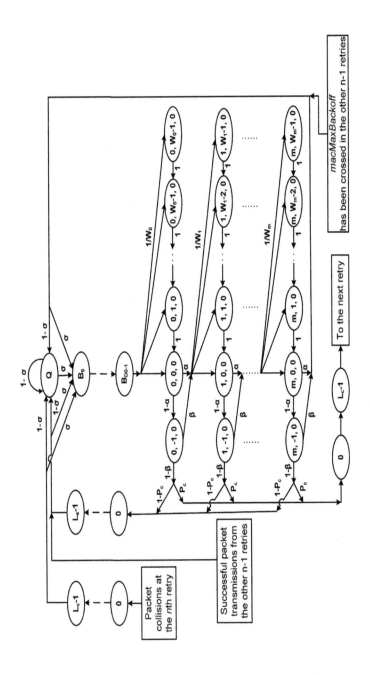

Figure 3.2: RSM transmission retry model.

$$\alpha = LP_c(1-\alpha)(1-\beta) + L_{ack}\frac{N\tau(1-\tau)^{N-1}}{1-(1-\tau)^N}P_c(1-\alpha)(1-\beta) \qquad (3.9)$$

$$\beta = \frac{P_c + N\tau(1-\tau)^{N-1}}{2-(1-\tau)^N + N\tau(1-\tau)^{N-1}} \qquad (3.10)$$

$$P_c = 1-(1-\tau)^{N-1} \qquad (3.11)$$

W_0 is defined in the standard [40] to be $2^{macMinBE}$; the probabilities α, β, and P_c are defined in [11]. The state Q is the idle state during which no packets are available for transmission. This state is modeled as Q_i (where $i = 0, 1, \ldots, L_{0-1}$) to show that it has a duration specified by L_0. Q_i models the unsaturated traffic condition. Equation (3.1) is the normalization condition of the Markov chain. The first term in this equation represents the probability of being in a backoff state. The second term refers to the probability of initiating the second clear channel assessment (CCA2). The third and fourth terms refer to the packet transmission state and packet collision state, respectively. Finally, the fifth term refers to the probability of being in the idle state when no packets are available. Equations (3.2)–(3.5) provide the mathematical expressions for all of these terms. Equations (3.2)–(3.5) can be directly used to find an expression for $b_{0,0,0}$.

A node remains in the idle state when no packets are generated based on Park's model assumption. The node remains in the idle state for a period of L_0 before checking the availability of packets. That is, even if a packet is available, a node does not leave the idle state before having the L_0 period passed. We consider sending the packets every L_0 as unrealistic, as it does not resemble real WSN scenarios. Therefore, we propose that the node should leave the idle state whenever a packet is generated. In addition to that, we introduce a MAC-level buffer in our system, which has not been done in [11]. The buffer is modeled in Figure 10.2 by the B_i $(i = 0, 1, D_{0-1})$ states. The model in this figure shows that (3.1) should be updated as follows:

$$\sum_{i=0}^{m}\sum_{K=0}^{W_i-1}\sum_{j=0}^{n}b_{i,k,j} + \sum_{i=0}^{m}\sum_{j=0}^{n}b_{i,-1,j}$$

$$+ \sum_{j=0}^{n}\left(\sum_{K=0}^{L-1}b_{-1,K,j} + \sum_{K=0}^{L-1}b_{-2,k,j}\right) + Q + \sum_{l=0}^{D_0-1}B_l \qquad (3.12)$$

We also assume $L_s = L_c = L$.

The fourth term of (3.12) is derived as follows:

By referring to Figure 10.3 and (3.5), we derive the probability of being in the idle state as follows:

By examining the first state Q_0 in Park's model we have

$$Q_0 = (1-\sigma)Q_{L_0-1} + (1-\sigma)\left[\sum_{j=0}^{n}(\sigma+(1-\sigma)\beta)b_{m,0,j}\right.$$

$$\left. + \sum_{i=0}^{m}P_c(1-\beta)b_{i,-1,n} + \sum_{i=0}^{m}\sum_{j=0}^{n}(1-P_c)(1-\beta)b_{i,-1,j}\right] \qquad (3.13)$$

From Figure 10.3 we have

$$Q_{L_0-1} = (1-\sigma)^{L_0-1}Q_0 \qquad (3.14)$$

From (3.13) and (3.14), we have

$$Q_0 = \frac{1-\sigma}{1-(1-\sigma)^{L_0}} \left[\sum_{j=0}^{n} (\alpha + (1-\alpha)\beta)b_{m,0,j} \right.$$
$$\left. + \sum_{i=0}^{m} P_c(1-\beta)b_{i,-1,n} + \sum_{i=0}^{m}\sum_{j=0}^{n} (1-P_c)(1-\beta)b_{i,-1,j} \right] \qquad (3.15)$$

From Figure 10.3 we have

$$\sum_{l=0}^{L_0-1} Q_l = Q_0 \sum_{l=0}^{L_0-1} (1-\sigma)^l \qquad (3.16)$$

and using

$$\sum_{i=0}^{x-1} y = \frac{y^x - 1}{y - 1} \qquad (3.17)$$

Using (3.17) and substituting (3.15) into (3.16) we get

$$\sum_{l=0}^{L_0-1} Q_l = \frac{1-\sigma}{\sigma} \left[\sum_{j=0}^{n} (\sigma + (1-\sigma)\beta)b_{m,0,j} \right.$$
$$\left. + \sum_{i=0}^{m} P_c(1-\beta)b_{i,-1,n} + \sum_{i=0}^{m}\sum_{j=0}^{n} (1-P_c)(1-\beta)b_{i,-1,j} \right] \qquad (3.18)$$

From (3.18) we see that the idle state duration is completely dependent on packet arrival rates and not on a predefined duration. Therefore, the L_0-1 idle states of Figure 10.3 can be augmented into a single idle state as shown in Figure 10.2.

The derivation of the last term in (3.12) is shown below (other terms remain the same as in Park's model). By simply referring to Figure 10.2 and [11], we can write the following equation for B_0:

$$B_0 = \sigma Q + \sigma \left[\sum_{j=0}^{n} x b_{m,0,j} + \sum_{i=0}^{m} P_c(1-\beta)b_{i,-1,n} + \sum_{i=0}^{m}\sum_{j=0}^{n} (1-P_c)(1-\beta)b_{i,-1,j} \right]$$
$$(3.19)$$

In (3.13), the first term is derived from Figure 10.2, the second term represents the sum of the probabilities of finding the medium busy m times in any of the n transmission retries, the third term represents the probability of experiencing a collision in any of the m backoff stages of the nth transmission retry, and the fourth term represents the probability of a successful transmission at any backoff stage in any transmission retry. By following Park's approximations, we can write (19) as follows:

$$B_0 = \sigma \left[Q + x^2(1+y) + y^{n+1} + (1-P_c)(1-x^2)(1+y) \right] b_{0,0,0} \qquad (3.20)$$

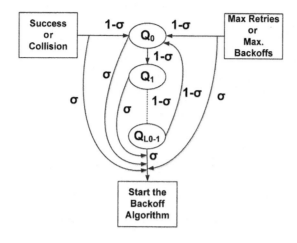

Figure 3.3: Modeling the MAC idle state.

Finally, $\sum_{l=0}^{D_0-1} B_l = D_0 B_0$. All of the terms in (3.12) can be represented in the form of $b_{0,0,0}$. Thus, this equation can be used to calculate $b_{0,0,0}$, which allows us to form a system of nonlinear equations in terms of α in (3.9), β in (3.10), and τ in (3.6) and then solve the nonlinear system to find the network operating point.

To evaluate the performance of the WSN with the modifications we introduced to Park's model, we focus on three performance metrics, namely, power consumption, reliability, and end-to-end delay.

3.3.1.1   Power Consumption

As in [11], we calculate the total average power consumed in the node (E_{tot}) by summing the average power consumed during backoff (E_{bo}), channel sensing (E_{sc}), packet transmission (E_t), packet reception (E_r), idle state (E_Q), buffering (E_B), and wake-up (E_w):

$$E_{tot} = E_{bo} + E_{sc} + E_t + E_Q + E_B + E_w + E_r \tag{3.21}$$

Each of the terms in (3.21) can be computed by knowing the probability of being at a certain state and the amount of average power consumed at that state. Since we assume that each end node only receives ACK traffic from the coordinator, the packet consumed in packet reception is negligible due to the size of the ACK packet (refer to [11] for the complete details).

3.3.1.2   Reliability

The reliability (R) is defined as the probability of successful packet reception and approximated as follows [11]:

$$R \approx 1 - x^{m+1}(1+\tilde{y}) - \tilde{y}^{n+1} \tag{3.22}$$

where $\tilde{y}$ is the approximated version of y and is given by

$$\tilde{y} = (1 - (1 - \tilde{\tau})^{N-1})(1 - x^2) \qquad (3.23)$$

and

$$\tilde{\tau} = (1 + x)(1 + \tilde{y})\tilde{b}_{0,0,0} \qquad (3.24)$$

3.3.1.3 End-to-End Delay

Similar to [11], we consider the end-to-end delay (T) to be resulting from the time spent during backoff (D_{bo}), the time wasted due to experiencing j collisions (jL_c), and the time needed to successfully transmit a packet (L_s):

$$T = L_S + jL_C + D_{bo} = (1 + j)L + D_{bo} \qquad (3.25)$$

3.3.2 Cluster-Tree Topology

In this section, we present a model for cluster-tree WSNs by extending the star topology model derived in the previous section. A cluster-tree WSN topology is widely used when an extended communication range is desired. In a large-scale CPS like the smart grid, an extended communication range is crucial. In cluster-tree topology, the traffic generated at the sources (end nodes or leafs) flows toward the sink (root) through a series of intermediate nodes called the CHs or relays. In particular, each CH receives packets coming from a specific cluster of sources. At the same time, the CHs may be grouped into higher-level clusters, which can be associated with even higher level CHs or relays or with the sink itself.

In cluster-tree topology, we consider two situations when deriving our model. The first situation is when all the CHs at the same level can hear each other (i.e., no hidden terminal problem). In this situation the CHs are assumed to use CSMA/CA to access the channel. The second situation is when the CHs cannot hear each other because they are placed far apart. This situation is more common in real WSN scenarios in smart grid deployments. Therefore, the use of contention will not be the best solution to provide channel access because of the hidden terminal problem. Hence, we use proper and careful scheduling between CHs to allow them to transmit based on specific timing and granting a minimum service guarantee all along the path through which the data are relayed (that is, using GTSs). A model for both cluster-tree situations is derived below.

3.3.2.1 Contention-Based Cluster-Tree Model

In a cluster-tree topology, the depth of the tree is obtained by grouping end nodes and CHs at various hierarchical levels [41]. For example, the network in Figure 10.4a is a particular cluster-tree network with a depth equal to 3, where CHs in each level are placed in such a way that they can hear each other and that end nodes can only communicate with their CHs. The topology described in Figure 10.4a uses multihop tree routing to transfer data from the source to the destination. CHs collect packets from sensor nodes belonging to their respective clusters, in addition to relaying packets to the higher-level CHs in the tree until reaching the sink. In this model, we assume that the communication between the nodes and their respective CHs is contention based. Furthermore, communication between the CHs is also based on contention. As we model the contention-based cluster-tree WSN, we abide by the following additional assumptions:

■ The CHs communicate with each other using contention (CSMA/CA).

■ All CHs in each hierarchical level in the cluster-tree can hear each other (i.e., no hidden terminal problem between CHs at the same level).

■ Each cluster is having a finite number of end devices contending to send data to its CH.

■ The packet arrival rate to the MAC sublayer (λ pkts/s) is the same for all end devices.

■ The traffic received by a CH in an upper level is equal to the aggregate of traffic from CHs at lower levels.

■ All CHs have M/G/1/L queues; the difference between CHs is in the packet arrival rate.

■ Each cluster is modeled with the same Markov model described above.

In our study of the CSMA/CA cluster-tree-based WSN, we concentrate on the following metrics to evaluate the performance:

Total end-to-end delay. The total end-to-end delay to transmit a packet in the contention-based cluster-tree topology is assumed to be equal to the sum of the end-to-end delays along the path from the source node to the sink node. The total end-to-end delay (T_{CCT}) is dependent on the number of nodes and the packet arrival rate in each level, and its value is given by the following equation:

$$T_{CCT} = \sum_{i=0}^{l-1} Ti \qquad (3.26)$$

T_i is measured from an end node to a CH, from a CH, to the next CH or from a CH to the sink, and l is equal to the number of levels starting from the end node (for which the delay is to be calculated) to the sink.

End-to-end reliability. The end-to-end reliability is defined as the probability of successful packet reception from any end node in the network to the sink node. The end-to-end reliability (R_{e2e}) is equal to the product of the reliabilities along the path to the sink node:

$$R_{e2e} = \prod_{i=0}^{l-1} R_i \qquad (3.27)$$

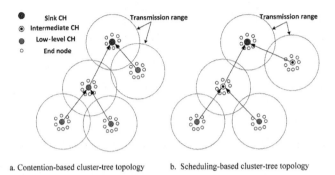

a. Contention-based cluster-tree topology   b. Scheduling-based cluster-tree topology

---

Figure 3.4: Proposed cluster-tree topology.

where R_i is the reliability from an end node to a CH, a CH, to a CH or a CH to the sink.

Power consumption. The average power consumed ($E_{tot_{(CT)}}$) in transmitting a packet from an end node to the sink in a cluster-tree topology is equal to the sum of all power consumed along the path from an end node to the sink. The total power consumed includes the power consumed by each CH in receiving a packet from a lower-level CH or end node. The total power consumption in a cluster-tree topology is given by the following equation:

$$E_{tot_{(CT)}} = \sum_{i=0}^{l-1} E_{tot}i \qquad (3.28)$$

where E_{tot} is the average power consumed in transmitting a packet from an end node to a CH, a CH to another CH and a CH to the sink (including the power consumed in receiving the packet at each CH).

3.3.2.2   Scheduling-Based Cluster-Tree Model

The cluster-tree model considered here is assumed to have one sink and CHs that form a multilevel wireless backbone with the cluster-tree topology. We assume that there are l levels in each cluster-tree topology. We classify the CHs into three categories based on their location in the network. These CHs are sink CH, which is the root of the tree and at the highest level; intermediate CHs, which have one or multiple child CHs (at a lower level); and low-level CHs, which are not connected to any CHs at the lower levels.

Figure 10.4b shows the cluster-tree topology with the three CH categories. We assume that each CH communicates with its sensors using the CAP (i.e., using CSMA/CA). This is a practical situation because generally end devices connected to a single CH are located geographically close to each other within the same personal area network (PAN). We refer to the CH collecting data from end devices as a parent CH. Each parent CH forwards the data from end devices to upper-level CHs until the sink CH is reached. In a typical WSN, most traffic is from the sensors to the sink. We assume that traffic in the opposite direction is mostly for control signaling and ACK transmission. Therefore, their effect on the data traffic transmission delay is not significant and hence can be neglected. Hence, we only consider traffic transmissions from the sensors to the sink.

We identify the traffic from the end nodes to their parent CH as local traffic and the traffic between CHs as the forwarded traffic. We assume that there is no cochannel interference between the transmissions in neighboring clusters. We assume that there is time synchronization between communicating CHs so that when one CH is transmitting at a frequency channel, the intended receiving CH should tune its radio to receive at the same frequency channel. With the cluster-tree topology, communicating CHs have a strict parent-child relationship, which makes time synchronization between them much simpler than in the mesh topology. All the children CHs listen to the beacons from their parent CHs and synchronize with them. Furthermore, the time synchronization between CHs does not have to be reperformed for each individual packet transmission, but only when the CHs should switch from the receiving mode to the transmitting mode. We assume that the effect of time synchronization on the packet transmission delay is negligible.

Figure 3.5 shows the proposed SF structure for the three types of CHs. Each beacon interval (BI) specifies the CAP for contention-based channel access, the contention-free period (CFP) for contention-free channel access, and the inactive period, where the node is in either idle state or sleep mode to save power [40]. The CAP of each SF is allocated for intracluster transmission (i.e., from end devices to their parent CH). We assume that the CAP of all the low-level CHs is longer than the CAP of intermediate CHs and the sink CH. This assumption is made to allow a longer CFP to the intermediate and sink CHs, since these devices are expected to handle higher intercluster traffic rates than the low-level CHs. Another reason for allocating a longer CAP to the low-level CHs is that we assume that these clusters have a higher number of nodes than the upper-level clusters, and thus having a longer CAP will reduce the delay from the end devices to the parent CH. However, according to the IEEE 802.15.4 standard [40], the maximum number of GTSs that can be allocated during a single SF is equal to seven; therefore, we follow the standard and allocate a maximum of seven GTSs to intermediate CH, and the sink CH as shown in Figure 3.5. Furthermore, we divide the CFP of all intermediate-level CHs into two periods: one is GTS for transmitting to upper CHs (GTS_{TX}), and the other is for receiving from lower-level CHs (GTS_{RX}). In doing so, we guarantee that intermediate CHs do not transmit and receive at the same time, and hence avoid collision [42–44]. We assume that the CFP period of the low-level CH is completely allocated for transmitting, and the CFP period of the sink CH is completely allocated for receiving from intermediate CHs. To avoid beacon frame collisions between neighboring CHs, we use the beacon frame collision avoidance approach described in [45]. In this approach, the time is divided such that beacon frames and the SF duration of a given coordinator are scheduled in the inactive period of its neighbor coordinators. We implement this approach by carefully selecting the duty cycle of each CH in the net-

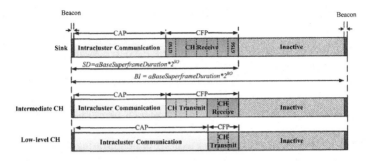

Figure 3.5: Proposed SF structure.

work. This is done by selecting a specific *BeO* and *SO* [45]. We assume that each CH maintains a buffer (B_i) to store its received packets, which can be either from its own end devices or from child CHs at lower levels. These buffers can accommodate all of the incoming traffic. We also abide by the following additional assumptions:

■ Each cluster has a finite number of end devices contending to send data to its CH.

■ The traffic received by a CH in an upper level $(l+1)$ is equal to the aggregate of traffic from CHs at lower levels (l).

■ Each cluster is modeled with the same Markov model described above.

In our study of the scheduling-based cluster-tree WSN, we focus on the following metrics to evaluate the performance:

Total end-to-end delay. The total end-to-end delay to transmit a packet in the scheduling-based cluster-tree topology is assumed to be equal to the sum of the end-to-end delays along the path from the source node to the sink node. The total end-to-end delay (T_{SCT}) is dependent on the number of nodes and packet arrival rate in each level (l), and its value is given by the following equation:

$$T_{SCT} = T + \sum_{i=0}^{l-1} D_i \qquad (3.29)$$

where T is the end-to-end delay from the end device to its parent CH and is given by (3.25). D_i is the inter-CH delay (D_H) and is calculated as follows:

Let φ_i be the occupancy of B_i at the ith level CH. φ_i can be obtained through the following relation:

$$\varphi_i = \eta_i + \varepsilon_i \qquad (3.30)$$

η_i can be obtained directly from λ and R in (3.22).

The values of ϕ_i and ψ_i depend on the packet length and the GTS_{TX} and GTS_{RX}.

Similar to [32], we let $\pi - 1$ be the number of packets that are in B_i at the time the tagged packet arrives in B_i. For simplicity, we assume that all the packets in B_i experience the same one-hop D_H. The one hop D_H can be given by the following relation:

$$D_H = \delta D_{SF} + \phi \tag{3.31}$$

D_H is measured in time slots; similar to [32], δ can be given as follows:

$$\delta = \left\lceil \frac{\pi}{\phi} \right\rceil - 1 \tag{3.32}$$

Similar to [32], we consider a special case where the CH is the sink node; in this case, the one-hop delay of the packet transmission can be given by

$$\tilde{D}_H = \delta D_{SF} + \tilde{\delta} \tag{3.33}$$

where $\tilde{\delta}$ represents the number of packets served in the same SF as the tagged packet and can be given by the following relation [32]:

$$\tilde{\delta} = \pi - \delta\phi \tag{3.34}$$

End-to-end reliability and power consumption. To calculate the end-to-end reliability and power consumption for scheduling-based cluster-tree topology, we assume that there are no packets lost during buffering at the CHs and that the power consumption during the buffering stage is negligible. Based on these assumptions, we can use (3.22) to calculate the end-to-end reliability in transmitting a packet from an end node to the sink and (3.28) to calculate the power consumption in transmitting a packet from an end node to the sink.

3.4   Priority and Delay-Aware Medium Access in WSNs

The delay-responsive cross-layer (DRX) scheme and the fair and delay-responsive cross-layer (FDRX) scheme aim to address data prioritization and delay-sensitive data transmission for WSNs in smart grid environments. The DRX scheme [38] uses the application layer data prioritization to control medium access of sensor nodes. DRX first performs delay estimation; if the estimated delay cannot meet the delay requirements of the smart grid application, then channel access of the node is fast-tracked by reducing CCA duration. FDRX [39] incorporates fairness into DRX. Similar to DRX, FDRX initially executes delay assessment; if the estimated delay is higher than the delay requirements of the application, then the node is given higher priority to access the channel. To provide fairness, a node periodically yields to other nodes in the PAN. Hence, FDRX provides fairness by periodically allowing other nodes in the PAN to contend fairly to access the channel. In this section, the DRX and FDRX schemes are presented. We include comparisons with previously proposed QoS supporting mechanisms [22, 46, 47]. We evaluate the yielding factor (α_y) of FDRX and different CCA durations of DRX.

3.4.1   DRX and FDRX Schemes

As an example to implement the DRX and the FDRX schemes, we propose a WSN with star topology to monitor delay-critical data in a smart grid environment. The data collected by certain sensors are assumed to have high priority and should be delivered with minimum end-to-end delay.

Both the DRX and FDRX schemes include an adaptation module that facilitates the interaction of the application layer with the MAC and physical layers. They aim to reduce the end-to-end delay by estimating the delay of critical data, and then insuring that these data are delivered to the destination with minimum delay. Each node in the PAN initially implements the delay estimation algorithm that estimates the expected delay based on the model described in Section 3.3.1. Thus, a node makes a decision based on the delay estimation algorithm, by making the MAC layer respond to a specific delay requirement of the application.

If a node finds out that the estimated delay is higher than a predefined threshold (τ_{TH}), then the application layer places a flag in the application layer header indicating that lower layers should treat the packet accordingly. Thus, upon the arrival of those packets to the MAC layer, it requests the physical layer to make changes in its parameters.

In DRX, the MAC sublayer requests the physical layer to reduce the CCA duration from eight symbol periods to four symbol periods (i.e., from 128 μs to 64 μs). In doing so, the physical layer senses the channel in half of the regular CCA duration and reports the results to the MAC layer. Thus, this node can acquire the channel and transmit its data with higher probability than other contending nodes. If the node finds the channel busy, it invokes the BO algorithm as described in [40]. To avoid any possible coexistence problems, no devices are assumed to be transmitting at the same frequency band other than the IEEE 802.15.4 nodes. Algorithm 3.1 describes the DRX scheme; initially the application layer evaluates the captured data and decides if the level of the monitored parameter value (Φ_M) is beyond an acceptable threshold (Φ_{TH}) (i.e., higher or lower than normal limit values). These values can vary from one application to an other (in the smart grid, values of Φ_{TH} are taken from [48]). The algorithm invokes the delay estimation process $E[D]$. If the estimated delay is found to be higher than the threshold τ_{TH} value (different delay thresholds for different smart grid applications are obtained from [48] and used later in the performance evaluation section), then the CCA duration is divided by two; otherwise, the algorithm does not make any changes to the physical layer parameters and transmits the data using a regular CCA duration process.

FDRX includes an improvement to the DRX scheme. The DRX scheme aims to reduce the end-to-end delay without taking other nodes in the PAN into consideration. The proposed FDRX scheme can achieve the delay reduction and additionally allow other nodes to transmit fairly. Similar to the DRX scheme, the FDRX scheme initially implements the delay estimation algorithm described in Section 3.3.1. Based on the resulting values of the delay estimation, the MAC layer responds to the delay requirement of the application. The main difference between the DRX and the FDRX

Algorithm 3.1 DRX Algorithm

//Measure the data//
if $\Phi_M \geq \Phi_{TH}$ **then**
    // Invoke delay estimation algorithm //
    $E[D]$
    if $E[D] \geq \tau_{TH}$ **then**
        // Insert a flag in the application layer header//
        $APP_{Header} = APP_{Header^*}$
        $CCA_{duration} = CCA_{duration/2}$
        $MAC - CSMA/CA()$
    else
        $CCA_{duration} = 8symboldurations$
        $MAC - CSMA/CA()$
    end if
else
    $CCA_{duration} = 8symboldurations$
    $MAC - CSMA/CA()$
end if
(Execute IEEE 802.15.4 CSMA/CA Algorithm)

---

schemes is that the latter yields to other nodes in the PAN periodically to allow them to transmit. Thus, FDRX is fairer to other nodes.

In the FDRX scheme, the MAC sublayer requests the physical layer to reduce the CCA duration from eight symbol periods to four symbol periods. This request is done based on a predefined duration (yielding factor α_y). The value α_y varies from 0 to 1; 0 means the node is not yielding to other nodes (corresponds to DRX) and 1 means that the node uses the default IEEE 802.15.4 MAC settings. The fairness property is added to ensure that a node only utilizes this scheme for a short period of time and then inverts back to default to allow other nodes to transmit.

3.5   Simulation and Analysis

To validate the analytical results of the RSM model proposed in Section 3.3.1, we use QualNet [51] network simulator to simulate a beacon-enabled WSN of star and cluster-tree topologies. We set all the simulation parameters similar to the mathematical model environment. We use Poisson traffic arrival, and as an example of a typical WSN monitoring application in a CPS system, we assume that end nodes are distributed in a 20×20 m area to form a local cluster; a cluster-tree topology is formed by the combination of a number of these clusters. All of the sensor nodes are operating in the 2.4 GHz band with a maximum bit rate of 250 Kbps. We run each simulation for 300 s and repeat each simulation 10 times. All nodes transmit with sufficient power, which means that all nodes in a single PAN or cluster can hear

Table 3.3 Initial Simulation Parameters

Parameter	Value
Transmission power (dBm)	3.5
Noise factor (dB)	10.0
Contention window	2
Packet size (byte)	120
Beacon order	1
Super-frame order	1

each other. We also assume that the noise level is constant throughout the entire simulation (i.e., constant noise factor). The ACK mechanism is activated to improve the reliability of the system. We assume that the power consumed during the buffering state as well as the backoff state is equal to the power consumed during the idle state. Table 10.2 shows some of our simulation parameters. We acquire the rest of the parameters from the IEEE 802.15.4 standard document [40] and the actual specification document of the MicaZ platform.

Note that in simulation, results for the star topology show a strong match with the behavior predicted by our analytical model. We run simulations of a star topology where each node is having an arrival rate (λ) of 10 pkts/s and compare the results with the analytical results of RSM. We do not show simulations of other arrival rates (i.e., 50 and 90 pkts/s) due to space limitation in the figures. We study the following performance metrics: power consumption, end-to-end delay, and reliability. In all of the presented results and similar to [16], we refer to our proposed model as the RSM model.

3.5.1 Star Topology

Figure 10.6 shows the end-to-end delay of packet transmission from the tagged node (i.e., a node where we perform our measurements) to the PAN coordinator against the number of nodes for different λ values and a buffer size of 512 B. We show that in Figure 10.6, when λ changes in the default IEEE 802.15.4 model, the end-to-end delay fluctuates, whereas the RSM model shows a more stable behavior with different λ values. This is a more realistic QoS model, which represents a stable WSN system because the designer and the operator of WSN want a system that is independent of the traffic arrival rates. Simulation results of $\lambda = 10$ pkts/s agree very much with our analytical results.

Figure 3.7 shows the reliability of packet transmission from the tagged node against the number of nodes for different λ values and a buffer size of 512 B. We show that there is a fluctuation in the reliability when λ changes in Park's model. On the other hand, the RSM model results show very stable performance with varying λ. This happens because the nodes in the RSM model tend to shortly buffer the packets before contending for the channel access, and this reduces the collision rate

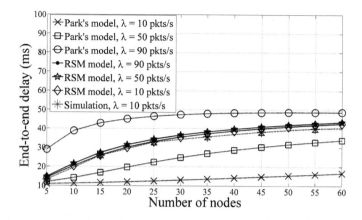

Figure 3.6: End-to-end delay in a star topology.

drastically and makes it almost independent of λ. A simulation of $\lambda=10$ pkts/s agrees with the RSM model.

Figure 3.8 shows the total power consumed in transmitting a packet from the tagged node against the number of nodes for different λ values and buffer size equal to 512 B. We show that there is an obvious fluctuation in the total consumed power when λ changes from 10 to 90 pkts/s in Park's model. On the other hand, the RSM model results show a clear stability in terms of the total power consumption and that the power consumed is independent of the probability of traffic arrival. This indeed is what is required from a model to be realistic and stable. We also show that the simulation results for $\lambda=10$ pkts/s agree with the RSM model.

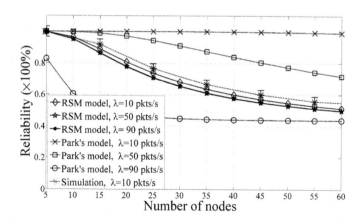

Figure 3.7: Reliability in a star topology.

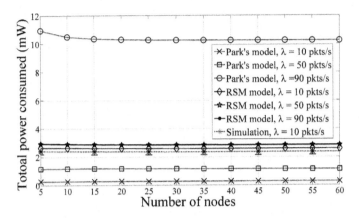

Figure 3.8: Total power consumption in a star topology.

3.5.2 Cluster-Tree Topology

3.5.2.1 Contention-Based Cluster-Tree

For the contention-based cluster-tree topology, we follow the same scenario described in Section 3.3.2.

Figure 3.9 shows the end-to-end delay of packet transmission from the tagged node as a function of λ for different MAC buffer sizes. This delay is measured from the end device (leaf node) to the sink node. We show that as λ increases, the end-

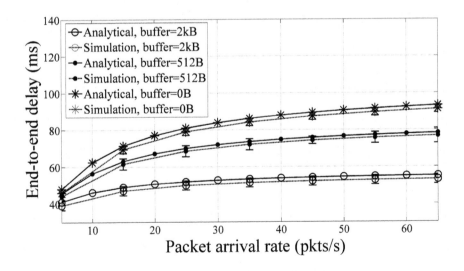

Figure 3.9: End-to-end delay in a contention-based cluster-tree topology.

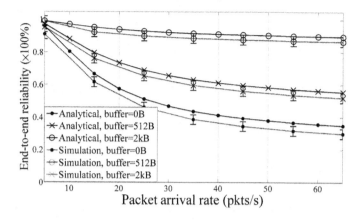

Figure 3.10: End-to-end reliability in a contention-based cluster-tree topology.

to-end delay slightly increases; then the delay reaches saturation. For a buffer size of 512 B the increase of the end-to-end delay with λ is more significant than the cases where the buffer is equal to 1 and 2 kB. This happens because with higher MAC buffers, the nodes tend to buffer the arriving packets before transmitting them, and hence decrease the gradient of the delay with λ. This takes place because as the buffer size increases, the contention among nodes decreases, which leads to less collisions, and hence delay is decreased. There is a good agreement between simulation and analytical results for all buffer sizes.

Figure 3.10 shows the end-to-end reliability of packet transmission from the tagged node to the sink (i.e., the probability of successful packet reception by the sink from an end node) as a function of packet arrival rate for different buffer sizes. We show that as the MAC buffer increases, the total end-to-end reliability increases. This is attributed to the buffers that reduce the contention, which leads to fewer collisions. As a result, the total reliability is increased. The simulation and analytical results agree for all buffer sizes.

Figure 3.11 shows the total power consumed in transmitting a packet from the tagged node in the cluster-tree network. The power consumption is dependent on the number of relays between the end node and the sink. We show that the total power consumption decreases as the MAC buffer sizes increases. This reduction in power consumption takes place because the nodes experience fewer retransmissions as they experience fewer collisions.

3.5.2.2 Scheduling-Based Cluster-Tree

For the scheduling-based cluster-tree topology we assume that the tagged node is at the lowest level and that $l = 3$ (refer to Figure 10.4). We assume that the number of nodes in the low-level clusters is 20 and the number of nodes in the intermediate clusters is 10.

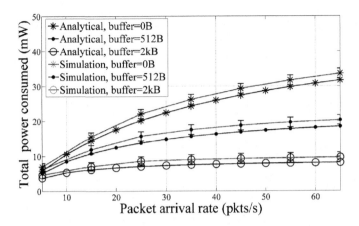

Figure 3.11: Total power consumption in a contention-based cluster-tree topology.

Figure 3.12 shows the end-to-end delay of packet transmission in the tagged node against the packet arrival rate for different end nodes' MAC buffer sizes for a tagged node located in the lower-level cluster two hops away from the sink. We show that when the packets are transmitted within the same SF, the end-to-end delay is lower when the MAC buffer sizes of local nodes are higher. However, when the packet arrival rate increases, we show that increasing the MAC buffer size of a local node will increase the end-to-end delay. This happens because at certain packet arrival rates and higher MAC buffer sizes, the packets wait longer, and hence miss the current SF. Therefore, a careful optimization between the packet arrival rate and the local nodes' MAC buffer sizes needs to be put in place to minimize the end-to-end delay.

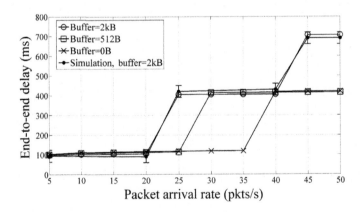

Figure 3.12: End-to-end delay in a cluster-tree topology for different MAC buffer sizes.

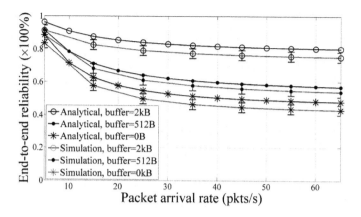

Figure 3.13: End-to-end reliability in a scheduling-based cluster-tree topology.

Simulation results for a MAC buffer size of 2 kB agree with the theoretical results presented in Figure 3.12.

Figure 3.13 shows the end-to-end reliability of the tagged node against the packet arrival rate for different local nodes' MAC buffer sizes. The tagged node is assumed to be in a low-level cluster and two hops away from the sink. We assume that there are no packets lost during the inter-CH communication due to the synchronization between CHs. We show that the reliability increases as the end nodes' MAC buffer increases for all packet arrival rates. We also show that the reliability levels are higher than the contention-based cluster-tree scenario.

Figure 3.14 shows the power consumed in packet transmission from the tagged node as a function of packet arrival rate for different end nodes' MAC buffer sizes.

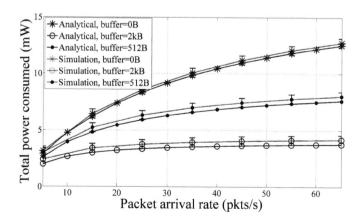

Figure 3.14: Power consumption in a scheduling-based cluster-tree topology.

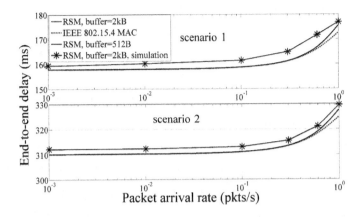

Figure 3.15: End-to-end delay for low packet arrival rates.

We assume that the power consumed during buffering of data packets is negligible. We show that as the local nodes' buffer sizes increase, the power consumption decreases for all packet arrival rates. We also show that the power consumption is lower than the contention-based cluster-tree scenario. This is because the CHs do not experience retransmissions and back-offs as they communicate with each other.

In the previous results we use packet arrival rates ranging from 5 to 50 pkts/s. These arrival rates are considered normal for applications with high data rates, such as PD monitoring [49, 50]. Figure 3.15 shows the end-to-end delay for low arrival rates (i.e., from 0.001 to 10 pkts/s). We consider two scenarios. Scenario 1, is when the tagged node is two hops away from the sink CH. In this scenario we assume that the packet arrival rate for all the upper-level CHs varies from 0.001 to 10 pkts/s. Scenario 2, is when the tagged node is three hops away from the sink CH and the packet arrival rate of all the upper-level CHs varies from 0.001 to 1 pkts/s. We show that for scenario 1, the end-to-end delay remains within acceptable ranges until the packet arrival rate is approximately more than 2 pkts/s, and then it increases for both the RSM model and the IEEE 802.15.4. In scenario 2, the end-to-end delay remains within acceptable ranges until the packet arrival rate exceeds 0.2 pkts/s. We also show that the simulation results agree with the analytical model.

3.5.3 DRX and FDRX Performance Evaluation

To evaluate the performance of the DRX and FDRX schemes, we also use the QualNet [51] network simulator platform. We test the two schemes with different numbers of nodes and traffic conditions. Furthermore, to investigate the performance of the DRX and FDRX schemes in realistic smart grid environments, we investigate smart grid-specific shadowing deviation and path loss properties. In addition to that, we select simulation parameters similar to those of the analytical model described in Section 3.3.1. We use a beacon-enabled star topology having N nodes and a coor-

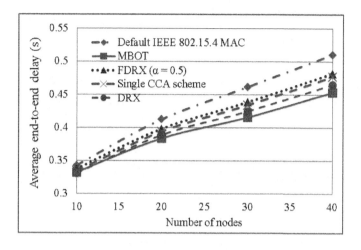

Figure 3.16: Average end-to-end delay of DRX and FDRX.

dinator. In this setup, we assume that all nodes have constant bit rate (CBR) traffic. Initially, we let one node receive high-priority packets during the simulation time. The transmission range is set to 20 m, and all the nodes are in the same PAN. Each simulation is run for 300 s, and each result represents an average of 10 runs. In the initial simulations, the delay threshold τ_{TH} is set to 0.400 s (following actual delay bound requirements presented in [48]). We assume that all nodes are transmitting with sufficient power (i.e., all nodes in the PAN can hear each other). The noise factor is assumed to be constant throughout the entire simulation. Table 10.2 shows the default parameters used in the simulations; the remaining parameters are taken from [40]. We compare the performance of the presented schemes with an existing QoS supporting scheme [46] in terms of end-to-end delay and the packet delivery ratio. The scheme presented in [46] reduces the BO duration of a contending node to make it BO for a shorter period than the rest of the nodes. The authors reduce the BO time by reducing the value of the BE. We also compare the results with [47], where the authors reduce the number of CCAs performed in high-priority nodes from two to one and perform frame tailoring to avoid collision. Throughout this section, the choice of the number of nodes is selected based on actual smart grid scenarios (i.e., the number of nodes vary from 10 to 50 in most cases).

Figure 3.16 shows the relation between the average end-to-end delay and the number of nodes in the default IEEE 802.15.4 MAC settings, the modified backoff time (MBOT) scheme of [46], the single CCA scheme [47], the FDRX scheme, and the DRX scheme. An obvious reduction in the end-to-end delay in the DRX scheme against the default IEEE 802.15.4 MAC settings and MBOT scheme is observed. Furthermore, there is a slight improvement in the delay when using DRX compared to the single CCA scheme [47]. The significance of this delay reduction is illustrated more clearly later in this section in a smart grid case study. We show that the DRX

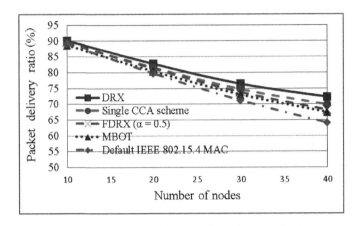

Figure 3.17: Packet delivery ratio of DRX and FDRX.

scheme has higher impact on delay reduction than the FDRX ($\alpha_y = 0.5$) scheme; $\alpha_y = 0.5$ implies that FDRX is yielding 50% of the time. This higher delay reduction takes place because the DRX scheme allows the node to utilize the channel more often and does not share the resources with other nodes in the PAN. On the other hand, FDRX is considered to be fair because it yields to other nodes to allow them to transmit their data; hence, it is observed that the delay reduction is less than that the in DRX scheme. The FDRX scheme performs slightly better than the MBOT scheme.

Figure 3.17 shows the percentage of data packets received by the PAN coordinator (packet delivery ratio) from an individual node versus the number of nodes for the default IEEE 802.15.4 MAC settings, the MBOT scheme, the single CCA scheme, the FDRX scheme, and the DRX scheme. The packet delivery ratio drops as the number of nodes increases since the number of collisions is proportional to the number of nodes in the PAN. As seen in the figure, the DRX scheme performs better than the default IEEE 802.15.4 MAC settings, the single CCA scheme, and the MBOT scheme. The FDRX scheme has a slightly higher percentage of delivery ratio than the default IEEE 802.15.4 and the MBOT scheme. Again, DRX performs better in terms of packet delivery ratio, because that node transmits at a higher rate than other nodes in the PAN. The FDRX scheme comes next in terms of the packet delivery ratio because it yields to other nodes.

To show the effect of different CCA durations and why they are divided by two in the proposed schemes, we investigate the effect of reducing the CCA symbol duration on the average end-to-end delay. Figure 3.18 shows the effect of changing the CCA symbol duration from the default value to the DRX value. The average end-to-end delay starts to increase as the symbol duration increases. The results presented in Figure 3.18 assist in selecting an optimum value for the CCA symbol period that will minimize the end-to-end delay and maintain an acceptable packet collision rate in the entire PAN. We further investigate the effects of α_y of the FDRX scheme on

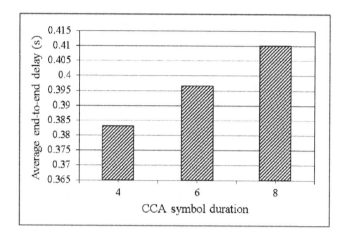

Figure 3.18: Effect of the CCA symbol duration on the average end-to-end delay.

the performance of the WSN. This investigation assists in optimizing the value of α_y to obtain certain delay bounds, packet delivery ratios, and packet collision rates. Figure 3.19 shows the effect of yielding factor, α_y, on the average end-to-end delay of a particular node implementing the FDRX scheme. As α_y increases, the average end-to-end delay also increases for all numbers of nodes. This is because when α_y approaches 1, the scheme converges to the default setting, and as it approaches zero, it converges to DRX. Hence, based on the application and the delay bound requirements, certain values of α_y that guarantee delay reduction and fairness at the same time are selected.

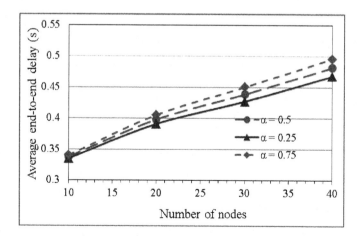

Figure 3.19: Effect of α_y on the average end-to-end delay.

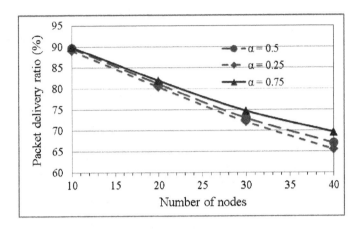

Figure 3.20: Effect of α_y on the packet delivery ratio.

Figure 3.20 shows the effect of α_y on the packet delivery ratio of a particular node implementing the FDRX scheme. The results presented in this figure agree with general behavior of the FDRX protocol; i.e., as α_y increases, the packet delivery ratio decreases. This is because the node implementing FDRX at lower α_y values will acquire the channel more often than the rest of the nodes, and hence have a higher packet delivery ratio.

We study the effect of the DRX and FDRX schemes on the energy consumption of sensor nodes. We use the energy model of the MicaZ nodes. In Figure 3.21, the energy consumed in the transmit mode is slightly higher for DRX and FDRX schemes than the default settings since nodes implementing these schemes will have the opportunity to transmit more often than their neighboring nodes. However, the increase in energy consumption is not significant (only 0.9%) compared to the increase in the packet delivery ratio and the reduction in the end-to-end delay.

We investigate the effect of different α_y values on the energy consumed in a node implementing the FDRX scheme. The value of α_y can be adjusted according to the power requirements of individual nodes. Figure 3.22 shows the effect of α_y on the energy consumed in the transmission mode. Again, as α_y approaches 1, a performance close to the default settings is obtained. In the previous set of results, we investigated the effect of the DRX and FDRX schemes on the performance of the node implementing these schemes. In the next set of results, we test the effect of implementing these two schemes on the overall WSN performance in terms of the number of packets lost at the sink due to collision. We use the same assumptions made previously. Furthermore, to have a wider perspective, we compare the impact of two CCA methods; on the network performance: CCA with energy detection and CCA with carrier sensing and energy detection.

Figure 3.23 shows the number of packets lost due to collision seen by the PAN coordinator in the entire WSN. In this set of simulations, we, use carrier sensing with energy detection method. As the number of nodes increase, the number of packets

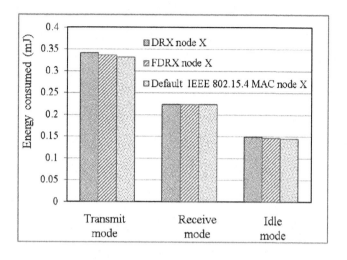

Figure 3.21: Energy consumption of DRX and FDRX.

lost due to collisions also increases, as expected. Furthermore, as the number of nodes increases, the number of packets lost in the DRX scheme becomes higher than the default IEEE 802.15.4 MAC settings. This slight increase of packets lost due to collisions is experienced by nodes do not implement the DRX scheme since they fail to have their data transmitted to the PAN coordinator due to packet collisions. In the worst case scenario, when the number of nodes is 40, the difference in the number of packets lost due to collision at the PAN coordinator is approximately 6%. However, for a lower number of nodes (10–20 nodes), the difference between the

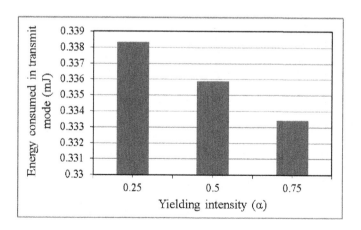

Figure 3.22: Effect of α_y on the energy consumed.

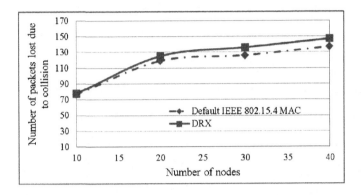

Figure 3.23: Packets lost due to collision in DRX using carrier sensing with energy detection.

packets lost due to collision is negligible. Figure 3.24 shows the number of packets lost due to collision at the PAN coordinator in the entire WSN. In this simulation, we use the energy detection method. As the number of nodes increases, the packet loss also increases and the difference is negligible at a lower number of nodes. However, the number of packets lost is very much higher than that of the carrier sensing with energy detection method (Figure 3.23). This agrees with [40], and the results are presented in [52].

Figures 3.25 and 3.26 show the number of packets lost due to collision at the PAN coordinator in the entire WSN in the FDRX scheme for different yielding intensities with energy detection and carrier sensing methods, respectively. The trend of the results presented in these figures agrees with the general results presented previously.

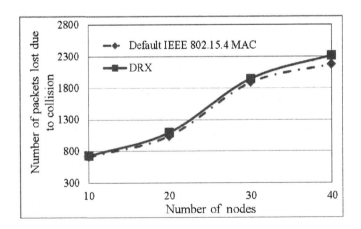

Figure 3.24: Packets lost due to collision in DRX using energy detection method.

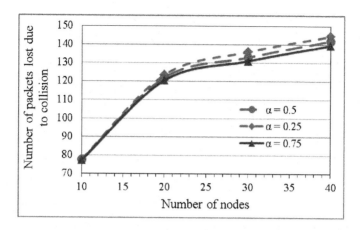

Figure 3.25: Packets lost due to collision in FDRX using energy detection method.

It is worth noting that if the application requires certain bounds on the data delivery from the entire WSN, certain values of α_y can be chosen to maintain certain levels of packet collisions and end-to-end delay at the same time.

We test the performance of the DRX and FDRX schemes in real smart grid environment by taking the effect of the path loss models into consideration. The path loss is defined as the difference (in dB) between the transmitted power and the received power, which represents the signal level attenuation caused by free space propagation, reflection, diffraction, and scattering.

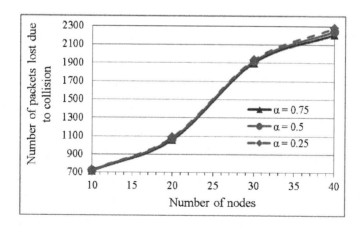

Figure 3.26: Packets lost due to collision in FDRX using carrier sensing with energy detection method.

Table 3.4   Data Packets Lost due to Collision in Different Electric Power Environments

Path Loss Model	Propagation Environment	IEEE 802.15.4	DRX DRX	FDRX $\alpha_y = 0.5$
Empirical model	Outdoor 500 kV substation	169	163	165
Empirical model	Indoor main power room	165	159	162
Empirical model	Underground transformer vault	162	160	160
JTC model	Indoor commercial	163	158	159
Two-ray model	Outdoor	172	166	167

There are three types of path loss models: empirical models, which are based on data measurement; deterministic models, which depend on the geometry of the site; and semideterministic models, which are based on empirical models in addition to deterministic models. We investigate the performance of the DRX and FDRX schemes in both empirical and deterministic path loss models. For the deterministic path loss model, we consider the two-ray path loss model in an outdoor environment (i.e., transformers in a substation) where there are normally two signal paths; one is direct from the sensor node to the sink, and the other is reflected through a metal object or through the ground. The other deterministic path loss model that we consider is the Telecommunications Industry Association/American National Standards Institute (TIA/ANSI) Joint Technical Committee (JTC) path loss model for the Personal Communication Service (PCS) bands for indoor areas, recommended by a technical working group for 1900 MHz PCS bands [53]. The parameters for path loss calculations in the indoor as well as the outdoor environments are taken from [53].

For the empirical path loss model, we follow the work presented in [54], where the authors have conducted experiments with actual sensor nodes operating in the 2.4 GHz industrial, scientific, and medical band with an effective data rate of 250 Kbps. The work in [54] is performed to measure the link quality indicator (LQI) and the received signal strength indicator (RSSI) with certain radio propagation parameters for different electric power system environments. Their experimental studies showed that their results provided more accurate multipath channel models than the Nakagami and Rayleigh models. We simulate the DRX and FDRX schemes in similar environments to those of [54], namely, outdoor 500 kV substation environment, indoor main power room, and underground transformer vault. We use the following values for channel propagation parameters: outdoor substation (path loss = 3.51, shadowing deviation = 2.95), indoor main power room (path loss = 2.38, shadowing deviation = 2.25), and underground transformer vault (path loss = 3.15, shadowing deviation = 3.19). We assume that the channel is assumed to have a lognormal shadowing model with a shadowing mean of 2.25 dB, and all sensor nodes are operating in non-line-of-sight (NLOS) mode.

Table 10.2 shows the number of data packets lost due to collision at the sink considering three different electrical power environments for the default settings, the DRX and FDRX schemes. In this scenario, we used 15 nodes and overloaded the nodes with CBR traffic to test the scheme in extreme traffic conditions. The DRX

Table 3.5   End-to-End Delay Values for Critical Smart Grid Applications

Application	Delay Requirements	IEEE 802.15.4	DRX DRX	FDRX $\alpha_y = 0.5$
Capacitor bank control	500 ms	510 ms	450 ms	475 ms
Fault current indicator	500 ms	510 ms	450 ms	475 ms
Transformer monitoring	500 ms	510 ms	450 ms	475 ms

and FDRX schemes outperform the default IEEE 802.15.4 settings. The results for the JTC path loss model are close to those for the indoor empirical path loss model, and those two for the ray model are somehow close to those for the outdoor 500 kV substation model for this simulation scenario.

3.5.4   Case Study

As a case study, we consider three critical smart grid monitoring applications that have strict end-to-end delay requirements. These applications are capacitor bank control, fault current indicator and transformer monitoring [48]. We obtain the functional requirements for these applications from [48]. We evaluate the performance of a WSN with priority and delay-aware medium access schemes for those smart grid applications. For this set of results, we consider a WSN with 40 nodes in a star topology where sensor nodes monitor certain parameters, such as current or voltage, and transmit their data to a PAN coordinator. We assume that the PAN coordinator is connected to a high-speed network, e.g., Ethernet Passive Optical Network (EPON); hence, the delay from the PAN coordinator to the user is negligible. We simulate the WSNs using the default IEEE 802.15.4 MAC settings and the DRX and FDRX schemes. In Table 3.5, the default IEEE 802.15.4 MAC setting has higher latency than the functional requirements of all applications, while both DRX and FDRX schemes succeed in reducing the latency below the functional requirements. The DRX and FDRX schemes are able to reduce the end-to-end delay by 60 and 35 ms, respectively.

3.6   Concluding Remarks

CPSs are promising technologies for next generation power grid systems, particularly applications that target real-time and resilient monitoring of critical assets. Furthermore, CPSs are undergoing rapid development and inspiring numerous application domains.

In this chapter, we presented an analytical model for the MAC sublayer of the IEEE 802.15.4 standard that can provide QoS to certain smart grid applications such as PD detection. The model can provide QoS by reducing the fluctuations of the WSN parameters as the traffic rates and number of nodes vary. The model considers a star topology and two cluster-tree-based WSN topologies. We included the actual traffic generation rates rather than a predefined idle state length to study the overall performance in terms of the end-to-end delay, reliability, and power consumption. We also studied the impact of utilizing MAC-level finite buffers on the performance of these WSNs. We validated our model and performed performance analysis using extensive simulations. Our analytical results agree with the simulation results for different numbers of nodes operating at different traffic generation rates. We show through analytical and simulation studies that introducing a finite MAC buffer has a significant impact on improving the network performance in terms of end-to-end delay, reliability, and power consumption. We also showed the impact of implementing our model on the performance of contention-based and scheduling-based cluster-tree topologies. We showed that the contention-based cluster-tree topology can achieve lower end-to-end delays than the scheduling-based cluster-tree topology at high packet generation rates. We illustrated that the scheduling-based cluster-tree topology (which is a more realistic scenario for distributed condition monitoring applications) can achieve higher reliability and lower power consumption for all traffic generation rates.

As a future work, we plan to investigate the performance of our analytical model with downlink traffic flows.

References

1. J. Deshpande, A. Locke, and M. Madden, Smart choices for the smart grid: Using wireless broadband for power grid network transformation, *Alcatel-Lucent Technology White Paper*, 2010.

2. International Telecommunication Union (ITU), http://www.itu.int/en/Pages/default.aspx.

3. M. Erol-Kantarci and H. Mouftah, Wireless sensor networks for cost-efficient residential energy management in the smart grid, *IEEE Transactions on Smart Grid*, vol. 2, no. 2, pp. 314–325, 2011.

4. Y. Liu, Wireless sensor network applications in smart grid: Recent trends and challenges, *International Journal of Distributed Sensor Networks*, vol. 2012, 2012.

5. Z. M. Fadlullah, M. M. Fouda, N. Kato, A. Takeuchi, N. Iwasaki, and Y. Nozaki, Toward intelligent machine-to-machine communications in smart grid, *IEEE Communications Magazine*, vol. 49, no. 4, pp. 60–65, 2011.

6. P. Han, J. Wang, Y. Han, and Q. Zhao, Novel WSN-based residential energy management scheme in smart grid, in *International Conference on Information Science and Technology (ICIST 2012)*, 2012, pp. 393–396.

7. H. Li and W. Zhang, QOS routing in smart grid, in *IEEE Global Telecommunications Conference (GLOBECOM 2010)*, 2010, pp. 1–6.

8. A. R. Alkhawaja, L. L. Ferreira, and M. Albano, Message oriented middleware with QoS support for smart grids, HURRAY-TR-120709, 2012.

9. S. C. Lu, Q. Wu, and W. K. Seah, *Quality of service provisioning for smart meter networks using stream control transport protocol,* School of Engineering and Computer Science, Victoria University of Wellington, 2012.

10. T. R. Park, T. Kim, J. Choi, S. Choi, and W. Kwon, Throughput and energy consumption analysis of IEEE 802.15.4 slotted CSMA/CA, *Electronics Letters,* vol. 41, no. 18, pp. 1017–1019, 2005.

11. P. Park, P. Di Marco, P. Soldati, C. Fischione, and K. Johansson, A generalized Markov chain model for effective analysis of slotted IEEE 802.15.4, in *6th IEEE International Conference on Mobile Adhoc and Sensor Systems, 2009 (MASS '09).,* October 2009, pp. 130–139.

12. S. Pollin, M. Ergen, S. Ergen, B. Bougard, L. Der Perre, I. Moerman, A. Bahai, P. Varaiya, and F. Catthoor, Performance analysis of slotted carrier sense IEEE 802.15.4 medium access layer, *IEEE Transactions on Wireless Communications,* vol. 7, no. 9, pp. 3359–3371, 2008.

13. J. Misic, S. Shafi, and V. Misic, Performance of a beacon enabled IEEE 802.15.4 cluster with downlink and uplink traffic, *IEEE Transactions on Parallel and Distributed Systems,* vol. 17, no. 4, pp. 361–376, 2006.

14. C. Buratti, Performance analysis of IEEE 802.15.4 beacon-enabled mode, *IEEE Transactions on Vehicular Technology,* vol. 59, no. 4, pp. 2031–2045, May 2010.

15. S. Guo, Z. Qian, and S. Lu, A general energy optimization model for wireless networks using configurable antennas, in *ACM Symposium on Applied Computing,* 2010, pp. 246–250.

16. I. Al-Anbagi, M. Khanafer, and H. T. Mouftah, A realistic and stable Markov-based model for WSNs, in *IEEE International Conference on Computing, Networking and Communications (ICNC 2013),* pp. 802–807, January 2013.

17. I. Al-Anbagi, M. Khanafer, and H. T. Mouftah, MAC finite buffer impact on the performance of cluster-tree based WSNs, in *IEEE International Conference on Communications (ICC 2013),* pp. 1485–1490, June 2013.

18. A. Bicen, O. Akan, and V. Gungor, Spectrum-aware and cognitive sensor networks for smart grid applications, *IEEE Communications Magazine,* vol. 50, no. 5, pp. 158–165, 2012.

19. J. Markkula and J. Haapola, Impact of smart grid traffic peak loads on shared LTE network performance, in *Proceedings of the IEEE International Conference on Communications (ICC),* 2013, pp. 4046–4051.

20. B. Wang and J. S. Baras, Minimizing aggregation latency under the physical interference model in wireless sensor networks, in *Proceedings of the IEEE Third International Conference on Smart Grid Communications (SmartGridComm)*, 2012, pp. 19–24.

21. I. S. Hammoodi, B. Stewart, A. Kocian, S. McMeekin, and A. Nesbit, Wireless sensor networks for partial discharge condition monitoring, in *Proceedings of the 44th IEEE Universities Power Engineering Conf. (UPEC 2009)*, September 2009, pp. 1–5.

22. W. Sun, X. Yuan, J. Wang, D. Han, and C. Zhang, Quality of service networking for smart grid distribution monitoring, in *First IEEE International Conference on Smart Grid Communications (SmartGridComm 2010)*, October 2010, pp. 373–378.

23. Z. Ruiyi, T. Xiaobin, Y. Jian, C. Shi, W. Haifeng, Y. Kai, and B. Zhiyong, An adaptive wireless resource allocation scheme with QoS guaranteed in smart grid, in *Proceedings of the IEEE Innovative Smart Grid Technologies (ISGT'13)*, Washington, DC, February 2013, pp. 1–6.

24. I. Al-Anbagi, M. Erol-Kantarci, and H. Mouftah, Priority and delay-aware medium access for wireless sensor networks in the smart grid, *IEEE Systems Journal*, vol. PP, no. 99, pp. 1–11, 2013.

25. M. S. Pan, L. W. Yeh, Y. A. Chen, Y. H. Lin, and Y. C. Tseng, A WSN-based intelligent light control system considering user activities and profiles, *IEEE Sensors Journal*, vol. 8, no. 10, pp. 1710–1721, 2008.

26. N. Batista, R. Melício, J. Matias, and J. Catalão, Photovoltaic and wind energy systems monitoring and building/home energy management using ZigBee devices within a smart grid, *Elsevier Journal of Energy*, pp. 306–315, 2012.

27. G. Bianchi, Performance analysis of the IEEE 802.11 distributed coordination function, *IEEE Journal on Selected Areas in Communications*, vol. 18, no. 3, pp. 535–547, March 2000.

28. J. Misic and V. Misic, *Wireless personal area networks: Performance, interconnection, and security with IEEE 802.15.4*, The Atrium Southern Gate, Chichester West Sussex PO19 8SQ England, Wiley, 2008.

29. J. Zhu, Z. Tao, and C. Lv, Performance evaluation of IEEE 802.15. 4 CSMA/CA scheme adopting a modified lib model, *Wireless Personal Communications*, vol. 65, no. 1, pp. 25–51, 2012.

30. P. Park, P. Di Marco, C. Fischione, and K. Johansson, Modeling and optimization of the IEEE 802.15.4 protocol for reliable and timely communications, *IEEE Transactions on Parallel and Distributed Systems*, vol. 24, no. 3, pp. 550–564, March 2013.

31. S. Busanelli, M. Martalò, and G. Ferrari, Markov chain-based optimization of multihop IEEE 802.15.4 wireless sensor networks, in *4th International ICST Conference on Performance Evaluation Methodologies and Tools*, 2009, p. 78.

32. W. Liu, D. Zhao, and G. Zhu, End-to-end delay and packet drop rate performance for a wireless sensor network with a cluster-tree topology, *Wireless Communications and Mobile Computing*, vol. 14, no. 7, pp. 729–744, 2014.

33. I. Ramachandran, A. K. Das, and S. Roy, Analysis of the contention access period of IEEE 802.15.4 MAC, *ACM Transactions on Sensor Networks (TOSN)*, vol. 3, no. 1, p. 4, 2007.

34. M. Jonsson and K. Kunert, Towards reliable wireless industrial communication with real-time guarantees, *IEEE Transactions on Industrial Informatics*, vol. 5, no. 4, pp. 429–442, 2009.

35. H.-J. Korber, H. Wattar, and G. Scholl, Modular wireless real-time sensor/actuator network for factory automation applications, *IEEE Transactions on Industrial Informatics*, vol. 3, no. 2, pp. 111–119, 2007.

36. J. R. Moyne and D. Tilbury, The emergence of industrial control networks for manufacturing control, diagnostics, and safety data, *Proceedings of the IEEE*, vol. 95, no. 1, pp. 29–47, 2007.

37. P. Soldati, H. Zhang, and M. Johansson, Deadline-constrained transmission scheduling and data evacuation in wirelessHART networks, in *European Control Conference 2009 (ECC09)*, pp. 751–758, 2009.

38. I. Al-Anbagi, M. Erol-Kantarci, and H. T. Mouftah, A low latency data transmission scheme for smart grid condition monitoring applications, in *IEEE Electrical Power and Energy Conference (EPEC)*, 2012, pp. 20–25.

39. I. Al-Anbagi, M. Erol-Kantarci, and H. Mouftah, Fairness in delay-aware cross-layer data transmission scheme for wireless sensor networks, in *26th Biennial Symposium on Communications (QBSC)*, 2012, pp. 146–149.

40. IEEE std 802.15.4-2006 (revision of IEEE std 802.15.4-2003), online. http://standards.ieee.org/findstds/standard/802.15.4-2006.html

41. A. Koubaa, A. Cunha, and M. Alves, A time division beacon scheduling mechanism for IEEE 802.15.4/ZigBee cluster-tree wireless sensor networks, in *19th Euromicro Conference on Real-Time Systems 2007. (ECRTS '07).*, July 2007, pp. 125–135.

42. E. Toscano and L. Lo Bello, A multichannel approach to avoid beacon collisions in IEEE 802.15.4 cluster-tree industrial networks, in *IEEE Conference on Emerging Technologies & Factory Automation 2009. (ETFA 2009)*, 2009, pp. 1–9.

43. E. Toscano and L. Lo Bello, Multichannel superframe scheduling for IEEE 802.15.4 industrial wireless sensor networks, *IEEE Transactions on Industrial Informatics*, vol. 8, no. 2, pp. 337–350, 2012.

44. S. Takagawa, M. Shirazi, B. Zhang, J. Cheng, and R. Miura, A reliable and energy-efficient MAC protocol for cluster-tree wireless sensor networks, in *International Conference on Computing, Networking and Communications (ICNC)*, 2012, pp. 159–163.

45. IEEE 802.15 WPAN Task Group 4b (TG4B), http://www.{ieee}802.org/15/pub/TG4b.html.

46. M. Youn, Y.-Y. Oh, J. Lee, and Y. Kim, IEEE 802.15.4 based QoS support slotted CSMA/CA MAC for wireless sensor networks, in *Proceedings of the IEEE International Conference on Sensor Technologies and Applications (SensorComm'07)*, Valencia, Spain, October 2007, pp. 113–117.

47. T. H. Kim and S. Choi, Priority-based delay mitigation for event-monitoring IEEE 802.15. 4 LR-WPANs, *IEEE Communications Letters*, vol. 10, no. 3, pp. 213–215, 2006.

48. M. Oldak and B. Kilbourne, Communications requirements: Comments of Utilities Telecom Council, http://www.energy.gov.

49. R. Sarathi, A. Reid, and M. Judd, Partial discharge study in transformer oil due to particle movement under dc voltage using the UHF technique, *Electric Power Systems Research*, vol. 78, no. 11, pp. 1819–1825, 2008.

50. C. Hudon and M. Belec, Partial discharge signal interpretation for generator diagnostics, *IEEE Transactions on Dielectrics and Electrical Insulation*, vol. 12, no. 2, pp. 297–319, 2005.

51. Qualnet network simulator, Available: http://www.scalable-networks.com/content/.

52. I. Ramachandran and S. Roy, Clear channel assessment in energy-constrained wideband wireless networks, *IEEE Wireless Communications Journal*, vol. 14, no. 3, pp. 70–78, 2007.

53. K. Pahlavan and A. H. Levesque, *Wireless information networks*, vol. 93, Wiley, Hoboken, New Jersey, 2005.

54. V. C. Gungor, B. Lu, and G. P. Hancke, Opportunities and challenges of wireless sensor networks in smart grid, *IEEE Transactions on Industrial Electronics*, vol. 57, no. 10, pp. 3557–3564, 2010.

PROTOCOLS AND ENABLING TECHNOLOGIES FOR FUTURE WIRELESS NETWORKS

II

Chapter 4

Cooperative Multiuser Networks

Zhiguo Ding, Kanapathippillai Cumanan, Bayan Sharif, and Gui Yun Tian

School of Electrical and Electronic Engineering, Newcastle University, UK

CONTENTS

In the last decade, the exponential growth of mobile users and newly emerging high-data-rate wireless applications have created a huge demand for the available wireless resources and opened up new challenges in terms of throughput and quality of services. In order to circumvent this huge demand and challenges, cooperative communications have been recently proposed for future generation wireless communications which include relay technology as evidenced by the development of standards in 3GPP LTE-Advanced systems. This relay approach replaces a long wireless link by shorter hops with relay nodes and increases the throughput dramatically at the cell edges by improving the received signal-to-interference-plus-noise ratio (SINR). In addition, incorporation of these relay nodes significantly reduces power consumption and makes future wireless networks more environmentally friendly, while introducing multiplexing and diversity gains in the network.

4.1   Introduction

In wireless communications, the transmitted signal experiences different types of fading due to multipath propagation between the source and destination. This multipath fading significantly influences the performance of the system in terms of SINR and bit error rate. Multiple-input multiple-output (MIMO) technology has become more attractive in wireless communications due to the diversity gain introduced by employing multiple antennas at the transmitter and the receiver [14]. Moreover, this approach has the potential to mitigate the multipath fading effects by introducing another degree of freedom while enhancing the reception reliability without scarifying bandwidth. However, this MIMO scheme is not always feasible due to the size of the terminals.

Recently, cooperative communications has been recognized as a new design paradigm to provide spatial diversity in the network, where multiple single antenna nodes cooperate to form a virtual MIMO system and help each other for data transmission by using relays. These wireless relays are considered to be an essential enabling technology for achieving energy and spectral efficiencies in the design of wireless networks. In addition, incorporation of relays has been proposed to increase the data rate at the cell edges by improving the received SINR, as discussed in Long Term Evolution (LTE)-Advanced systems [22, 28]. In addition, these relays enhance the quality of the wireless links influenced by multipath fading, shadowing, and path losses. Hence, the relays have the potential to support the required quality of services (QoSs) at the destination by mitigating the co-channel interference and improving the reliability of the links between the sources and destinations, while facilitating a better frequency reusage and lower energy consumption [24, 26].

Cooperative diversity was first investigated in [4] by developing capacity theorems for relay channels, whereas efficient protocols and outage behavior were studied in [20] based on amplify-and-forward, decode-and-forward, and relay selection strategies for one source-destination pair with a relay node. These protocols were then extended by incorporating space-time coding in [19] and for an *N*-user scenario in [31] with the explicit expression of the outage probability. The requirement of extra channel use for relaying introduces bandwidth inefficiency in these protocols. To overcome this issue, spectrally efficient schemes were developed in [23] by exploiting multiple access techniques where multiple users transmit their information simultaneously based on nonorthogonal cooperative schemes. However, these multiple access techniques introduce more complexity in the design of medium access control [1, 23]. As an alternative approach, a group of relays can be employed to replace unreliable links between source and destination and to relay information. This approach was first explored in [2, 30] based on the available a priori information at the source nodes.

In this chapter, different cooperative strategies are investigated for a source-destination pair with multiple relays based on the available a priori channel information at the source node in Section 4.2. The performance of these schemes is evaluated by deriving explicit expression of the outage probability and the diversity-multiplexing trade-off [27, 32]. In Section 4.3, two cooperative protocols are proposed for multiple access networks based on source cooperation and relay assistance, whereas Section 4.4 proposes cooperative transmission schemes for broadcast networks with and without direct links between source and destination. In Section 4.5, network coding techniques are developed for multiple access networks and two-way relay networks. Finally, Section 4.6 concludes this chapter.

4.1.1 Notation

The uppercase and lowercase boldface letters are used for matrices and vectors, respectively. $(\cdot)^T$, $(\cdot)^*$, and $(\cdot)^H$ denote the transpose, conjugate, and conjugate transpose, respectively. $\text{Tr}(\cdot)$ and $\mathscr{E}\{\cdot\}$ stand for trace of a matrix and the statistical expectation for random variables. $\mathbf{I}$ and $(\cdot)^{-1}$ denote the identity matrix with appropriate size and the inverse of a matrix, respectively. The notation $diag\{\cdot\}$ represents a vector consisting the diagonal elements of a matrix or a diagonal matrix where the diagonal elements are from a vector. $|\cdot|$ and $\det\{\mathbf{A}\}$ stand for absolute value of a complex number and determinant of matrix $\mathbf{A}$. $(x)^+$ denotes $\max\{x,0\}$, whereas $\lceil x \rceil$ and $\lfloor y \rfloor$ represent the smallest integer equal to or greater than x and the largest integer equal to or less than than y, respectively.

4.1.2 Preliminaries

In this subsection, the preliminaries used to derive the diversity-multiplexing trade-off are provided as discussed in [27, 32]. The multiplexing gain r is defined as

$$r \triangleq \lim_{\rho \to \infty} \frac{R(\rho)}{\log \rho} \tag{4.1}$$

where ρ is SNR and $R(\rho)$ is the rate in bits per channel use (BPCU). Similarly, the diversity gain d is expressed as follows:

$$d \triangleq -\lim_{\rho \to \infty} \frac{\log[P_e(\rho)]}{\log \rho} \tag{4.2}$$

where $P_e(\rho)$ represents the maximum-likelihood error probability.

4.2 Single Source-Destination with Multiple Relays

In this section, the effect of a priori channel information at the transmitter is explored in a relay network, where a source communicates its destination with the help of multiple relays, as shown in Figure 4.1 [7]. In a conventional cooperative network, relaying nodes are randomly selected, which significantly degrades the performance of the network. This performance loss can be improved by exploiting a priori channel information at the transmitter. In this section, various numbers of opportunistic relaying schemes are investigated with different types of a priori channel information.

To be specific, five cooperative schemes are considered, ranging from the case that no a priori channel information is known to the source node, to the ideal case that full channel knowledge is available. Except for the first case with no a priori channel information, as it has been studied in [31], in this section, the explicit expressions of the outage probability at arbitrary SNRs, as well as the the diversity-multiplexing trade-off [27, 32] at high SNR for all other cases, are developed. The analytical results, which are shown to fit well with the Monte-Carlo simulations, prove the intuition that the more a priori channel information available, the higher spectral efficiency the cooperative system can achieve. Moreover, except for the ideal case that the full diversity can be achieved with only one relay link, for all other cases, the more relay nodes the system uses, the higher diversity it reaches. It is also shown that although a good performance index at the high SNR, the diversity-multiplexing trade-off is not an approximate one at the low SNR. Hence, the same high SNR behavior does not always promise the same performance at low SNR.

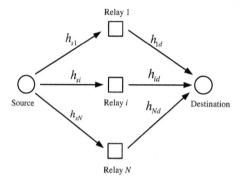

Figure 4.1: A relay network consists of a single source-destination pair with multiple relays.

4.2.1 Cooperative Protocols with Partial CSI

To simplify the presentation, protocols studied in this section are based on time division schemes as in [20], where each user is assigned a unique time slot that is further divided into several sub-time slots. As an illustration, Figure 4.2 shows the details of the first time slot, where there are one source and m relays. During the sub-time slot 0, only the source node transmits, while the other users keep silence by "listening." During the following sub-time slots, the relays then forward the overheard signals to the destination. In this section, the decode-and-forward type of schemes are considered; i.e., the relay nodes decode the information before they are forwarded, if necessary, to the destination.

In order to enhance the spectral efficiency, the source node exploits a priori channel information. In this section, it is assumed that the channel state information (CSI) is available at the transmitter as in [2]. Moreover, it is not always possible to have

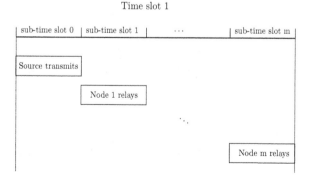

Figure 4.2: The sub-time slot assignment for the time slot 1.

Table 4.1   Type of A Priori Channel Information

Type	Order Information Channel S→R	Exact CSI Channel S→R	Exact CSI Channel S→D	Exact CSI Channel R→D
I	No	No	No	No
II	Yes	No	No	No
III	Yes	Yes	No	No
IV	Yes	Yes	Yes	No
V	Yes	Yes	Yes	Yes

the perfect CSI at a transmitter due to the time-varying nature of the wireless channels. Hence, a transmitter could have either the perfect CSI or the order information. Depending on the limit of the overhead introduced by the feedback, the available a priori CSI at the source can be classified into five categories, as shown in Table 4.1. By exploiting the available CSI at the transmitter, the following cooperative schemes are proposed:

1. *Random n-relay scheme with type I CSI:* With no a priori CSI available at all, a random n number of relays are chosen.

2. *Opportunistic n-relay scheme with type II CSI:* With the order information of the source-relay channels, the best n number of the relays are chosen.

3. *Opportunistic relay scheme with type III CSI:* With the exact CSI from the source to relay nodes, only the relay nodes that correctly decode the source information participate in the communications.

4. *Opportunistic incremental scheme with type IV CSI:* Besides the source to relay, the exact CSI from source to destination is also available. The source first decides whether it needs cooperative channels by examining the CSI from the source to destination. If it does, the scheme then operates in the same manner as that for the type III CSI.

5. *Opportunistic best-relay scheme with type V CSI:* With full knowledge of all channels among the source, relays, and destination, only the relay nodes having good links when both the source and destination are chosen.

The random n-relay scheme has been studied in [31], and the other four schemes are focused on in this section. It will be shown that when only the order information of the source-relaying channels is available, the cooperative system still suffers from performance loss at low SNR. Only when the exact CSI of the source-relaying channels becomes available can the performance loss at low SNR be effectively reduced, and the full diversity is reached with all of the possible relay nodes participating in the communication. On the other hand, for the ideal case that the source node has full information of all channels, full diversity is achieved with only one relay. These observations match well with the original intuition that the more a priori channel information available at the source, the better the performance that can be achieved.

For simplicity, in this section it is assumed that the channels are flat fading with slow-varying values, all of the channels have the same SNR, ρ, and the receivers know the channel information.

4.2.1.1 Opportunistic n-Relay Scheme with Type II CSI

The opportunistic n-relay scheme assumes the source node has the order information of the source-to-relay channels. As it is natural to use only the best n-relay links, an interesting question is: What is the relation between the value of n and the system performance, where n is no larger than N and N is the number of all possible relays? Denoting $x_{(i)} = |h_{si}|^2$ as the channel value square from the source to the ith relay node, it is assumed, without loss of generality, that

$$x_{(1)} \le x_{(2)} \le \cdots \le x_{(N-1)} \le x_{(N)} \tag{4.3}$$

In this scheme, only the $(N - n + 1)$th to the Nth relay nodes are involved in the transmission, and only when the overheard information can be successfully decoded does the relay node forward the information to the destination; otherwise, it keeps silent. Because the receiver is assumed to have the channel information, the meaning of "successful decode" translates into the expression of $R < I_{(i)}$, where R is source transmission rate and $I_{(i)} = \frac{1}{n+1} \log(1 + \rho x_{(i)})$, which is the mutual information between the source and the ith relay node. Since the source has no CSI, the channel between the source and a chosen relay node may still suffer from deep fading. Theorem 4.1 provides the outage probability of this opportunistic scheme:

Theorem 4.1

The outage probability of the opportunistic n-relay scheme with type II CSI can be expressed as

$$P_{out,I} = \sum_{k=0}^{n-1} \left[1 - e^{-\lambda \gamma_n} \sum_{i=0}^{k} \frac{(\lambda \gamma_n)^i}{i!} \right] \frac{N!}{(N-k)!(k)!} [1 - e^{-\lambda \gamma_n}]^{N-k} [e^{-\lambda \gamma_n}]^k + \tag{4.4}$$
$$\left[1 - e^{-\lambda \gamma_n} \sum_{i=0}^{n} \frac{(\lambda \gamma_n)^i}{i!} \right] \sum_{m=n}^{N} \frac{N!}{(N-m)!(m)!} [1 - e^{-\lambda \gamma_n}]^{N-m} [e^{-\lambda \gamma_n}]^m$$

and the outage probability at high SNR can be approximated as

$$P_{out,I} \approx \frac{[\lambda \gamma_n]^{n+1}}{(n+1)!} \sum_{m=n}^{N} \frac{N!}{(N-m)!(m)!} \tag{4.5}$$

where $\gamma_n = \frac{2^{(n+1)R} - 1}{\rho}$.

Proof: *Please refer to [7].* ■

From Theorem 4.1, the diversity-multiplexing trade-off can be immediately derived, as shown in Theorem 4.2.

Theorem 4.2

The diversity-multiplexing trade-off of the opportunistic n-relay scheme with type II CSI can be expressed as

$$d(r) = [n+1][1 - (n+1)r] \tag{4.6}$$

where $r = R/\log\rho$, which is the normalized transmission rate.

Proof: *The diversity-multiplexing trade-off of a system is defined as (see [32])*

$$d(r) = -\lim_{\rho \to \infty} \frac{\log P_{out}}{\log \rho} \tag{4.7}$$

Substituting (4.5) into (4.7) and letting $R = r\log\rho$ gives

$$
\begin{aligned}
d(r) &= -\lim_{\rho \to \infty} \frac{\log[\lambda \frac{\rho^{(n+1)r-1}}{\rho}]^{n+1} + \log C}{\log \rho} \\
&= [n+1][1 - (n+1)r]
\end{aligned}
\tag{4.8}
$$

where $C = \sum_{m=n}^{N} \frac{N!}{(N-m)!(m)!}$, which is independent of ρ.

Theorems 4.1 and 4.2 reveal that this cooperative scheme has to use all of the possible relaying nodes to achieve the full diversity of $N+1$, but increasing the number of the relay nodes also decreases the multiplexing gain. These observations will be verified through simulations.

4.2.1.2 Opportunistic Relay Scheme with Type III CSI

With type III CSI, the source node has access to not only the order information, but also the exact CSI between the source and the relay nodes. Thus, the source node invites only the *qualified* relay nodes that can successfully decode the information to join in the communication, which is realized by investigating the expression of $R < I_{(i)}$ at the source node. For all existing N number of relay nodes, it is assumed that n of them satisfy the above expression and become *qualified*. A question is then whether all or part of the n nodes should be chosen.

4.2.1.3 Opportunistic n-Relay Scheme with Type III CSI

First, assuming all of the n *qualified* nodes are used for the cooperation, the following theorems are provided:

Theorem 4.3

Given that type III CSI is available at the source node and all n qualified relaying nodes will participate in communication, the outage probability of such an oppor-

tunistic relay scheme can be expressed as

$$P_{out} = \sum_{n=1}^{N-1} \left[1 - e^{-\lambda \gamma_n} \sum_{i=0}^{n} \frac{[\lambda \gamma_n]^i}{i!}\right] \frac{N!}{(N-n)!n!}[1 - e^{-\lambda \gamma_{n+1}}]^{N-n}[e^{-\lambda \gamma_n}]^n \quad (4.9)$$

$$+ \left[1 - e^{-\lambda \gamma_0}\right]\left[1 - e^{-\lambda \gamma_1}\right]^N + \left[1 - e^{-\lambda \gamma_N} \sum_{i=0}^{N} \frac{[\lambda \gamma_N]^i}{i!}\right][e^{-\lambda \gamma_N}]^N$$

and the outage probability at high SNR can be approximated as

$$P_{out} \approx \sum_{n=1}^{N-1} \frac{N!}{(N-n)!n!} \frac{[\lambda \gamma_n]^{n+1}}{(n+1)!}[\lambda \gamma_{n+1}]^{N-n} + \lambda \gamma_0[\lambda \gamma_1]^N + \frac{[\lambda \gamma_N]^{N+1}}{(N+1)!}. \quad (4.10)$$

Proof: *Please refer to [7].* ■

Theorem 4.4

The diversity-multiplexing trade-off of the opportunistic relay scheme can be expressed as

$$d(r) = [N+1][1 - (N+1)r] \quad (4.11)$$

Proof: *Substituting $R = r \log \rho$ into (4.10) gives*

$$P_{out}(r) \approx \sum_{n=1}^{N-1} \frac{N!}{(N-n)!n!} \frac{[\lambda(\rho^{(n+1)r} - 1)]^{n+1}}{(n+1)!\rho^{N+1}}[\lambda(\rho^{(n+2)r} - 1)]^{N-n} \quad (4.12)$$

$$+ \lambda \frac{(\rho^r - 1)}{\rho^{N+1}}[\lambda(\rho^{2r} - 1)]^N + \frac{[\lambda(\rho^{(N+1)r} - 1)]^{N+1}}{(N+1)!\rho^{N+1}}$$

Substituting (4.12) into (4.7), and further noting that when $\rho \to \infty$, the last term of the right side of (4.12) dominates the equation, it can be easily obtained (4.11).

4.2.1.4 Opportunistic m-Relay Scheme with Type III CSI

In this scheme, the total n number of the qualified relay nodes are ranked in an ascending order similar to (4.3), and only the best m of them are chosen for the cooperation. While m is a prechosen value, n is a variable according to the channel condition. Thus, when $n < m$ for some channel realization, all of the n nodes are chosen and the scheme reduces to that described in the previous subsection. The following theorem gives the outage probability for this scheme.

Theorem 4.5

Given that type III CSI is available at the source node and only m of the n (m < n) qualified relaying nodes participate in the communications, the outage probability of *the opportunistic relay scheme can be expressed as*

$$
P_{out} = \left[1 - e^{-\lambda \gamma_m} \sum_{i=0}^{m} \frac{[\lambda \gamma_m]^i}{i!} \right] \left[\sum_{k=m}^{N-1} \frac{N!}{(N-k)!k!} [1 - e^{-\lambda \gamma_{k+1}}]^{N-k} [e^{-\lambda \gamma_k}]^k + [e^{-\lambda \gamma_N}]^N \right]
$$
$$
+ \sum_{n=1}^{m-1} \left[1 - e^{-\lambda \gamma_n} \sum_{i=0}^{n} \frac{[\lambda \gamma_n]^i}{i!} \right] \frac{N!}{(N-n)!n!} [1 - e^{-\lambda \gamma_{n+1}}]^{N-n} [e^{-\lambda \gamma_n}]^n
$$
$$
+ \left[1 - e^{-\lambda \gamma_0} \right] \left[1 - e^{-\lambda \gamma_1} \right]^N \tag{4.13}
$$

and the outage probability can be approximated at high SNR as

$$
P_{out} \approx \frac{[\lambda \gamma_m]^{m+1}}{(m+1)!} \sum_{k=m}^{N-1} \frac{N!}{(N-k)!k!} [\lambda \gamma_{k+1}]^{N-k} + \sum_{n=1}^{m-1} \frac{N!}{(N-n)!n!} \frac{[\lambda \gamma_n]^{n+1}}{(n+1)!} [\lambda \gamma_{n+1}]^{N-n}
$$
$$
+ [\lambda \gamma_0][\lambda \gamma_1]^N \tag{4.14}
$$

Proof: *Please refer to [7].* ■

Similar to those for the above analyzed schemes, the diversity-multiplexing trade-off can be easily obtained as shown in Theorem 4.6.

Theorem 4.6
Given that type III CSI is available at the source node and the m of the n (m < n) qualified relaying nodes participate in the communication, the diversity-multiplexing trade-off of the opportunistic relay scheme can be expressed as

$$
d(r) = [m+1][1 - (m+1)r] \tag{4.15}
$$

Comparing Theorems 4.4 and 4.6 clearly shows that full diversity $N + 1$ is reached with all of the qualified relays being involved in the cooperation. This is not surprising since at high SNR, all of the N possible relay nodes receive the source information correctly, making $n = N$. It is also interesting to compare Theorems 4.2 and 4.6, which shows that the diversity-multiplexing trade-off of the m-relay scheme with type III CSI and that with type II CSI are the same, and so are their performances at high SNR. But this does not promise the same performance at low SNR. On the contrary, with more channel information available at the transmitter, the m-relay scheme with type III CSI has significantly better performance than that with type II CSI. This is because exploiting the CSI at the transmitter is similar to applying the *precoding* technique. The coding gain can be easily observed through the simulation results.

4.2.1.5 Opportunistic Relay Incremental Scheme with Type IV CSI

Compared to type III CSI, type IV CSI provides additional CSI information between the source and destination. As in [20], the source node will first decide whether it needs cooperation or not, which is done by examining $R < I_{sd}$, where I_{sd} is the mutual information between the source and destination. Only when the direct transmission becomes unreliable do the relay nodes start transmission with the same schemes as those for the type III CSI. Below, we will study two schemes similar to opportunistic n-relay scheme with type III CSI and opportunistic m-relay scheme with type III CSI, respectively.

4.2.1.6 Opportunistic n-Relay Incremental Scheme with Type IV CSI

As in opportunistic n-relay scheme with type III CSI, here it is assumed that all of the n relay nodes that can decode the source information successfully are used for the cooperation. The following theorem provides the outage probability:

Theorem 4.7
Given that type IV CSI is available at the source node and all of the n qualified relaying nodes will participate in communications, the outage probability of the opportunistic relay scheme is given by

$$
P_{out} = \sum_{n=1}^{N-1} \left\{ [1 - e^{-\lambda \gamma_0}] \Psi(n-1, \gamma_n - \gamma_0) + \Psi(n, \gamma_n) e^{-\lambda [\gamma_n - \gamma_0]} \sum_{i=0}^{n-1} \frac{[\lambda(\gamma_n - \gamma_0)]^i}{i!} \right\}
$$
$$(4.16)$$
$$
\times \frac{N!}{(N-n)!n!} [1 - e^{-\lambda \gamma_{n+1}}]^{N-n} [e^{-\lambda \gamma_n}]^n
$$
$$
+ \left\{ [1 - e^{-\lambda \gamma_0}] \Psi(N-1, \gamma_N - \gamma_0) + \Psi(N, \gamma_n) e^{-\lambda [\gamma_N - \gamma_0]} \sum_{i=0}^{N-1} \frac{[\lambda(\gamma_N - \gamma_0)]^i}{i!} \right\}
$$
$$
\left[e^{-\lambda \gamma_N} \right]^N + \left[1 - e^{-\lambda \gamma_0} \right] \left[1 - e^{-\lambda \gamma_1} \right]^N
$$

where $\Psi(n, z) = \left[1 - e^{-\lambda z} \sum_{i=0}^n \frac{[\lambda z]^i}{i!} \right]$ and the outage probability can be approximated at high SNR as

$$
P_{out} \approx \sum_{n=1}^{N} \frac{N!}{(N-n)!n!} [\lambda \gamma_{n+1}]^{N-n} \left[\frac{[\lambda(\gamma_n - \gamma_0)]^n}{n!} \lambda \gamma_0 + \frac{[\lambda \gamma_n]^{n+1}}{(n+1)!} \right]
$$
$$
+ \lambda \gamma_0 [\lambda \gamma_1]^N.
$$
$$(4.17)$$

Proof: *Please refer to [7].* ■

From Theorem 4.7, the diversity-multiplexing trade-off is given by the following theorem:

Theorem 4.8
Given that type IV CSI is available at the source node and all of the n qualified

relaying nodes will participate in communications, the diversity-multiplexing trade-off of the opportunistic relay scheme can be expressed as

$$d(r) = [N+1][1-(N+1)r] \tag{4.18}$$

4.2.1.7 Opportunistic m-Relay Scheme with Type IV CSI

As in opportunistic m-relay scheme with type III CSI, here it is assumed m of the n number of the qualified relay nodes participate in the communication. Following similar procedures as those in deriving Theorems 4.7 and 4.8, the outage probability and the diversity-multiplexing trade-off can be easily obtained, which are given by the following two theorems, respectively:

Theorem 4.9
Given that type IV CSI is available at the source node and only m of the n qualified relaying nodes will participate in communications, the outage probability of the opportunistic relay scheme CSI can be expressed as

$$P_{out} = \left\{ [1 - e^{-\lambda \gamma_0}]\Psi(m-1, \gamma_m - \gamma_0) + \Psi(m, \gamma_m)e^{-\lambda[\gamma_m - \gamma_0]} \sum_{i=0}^{m-1} \frac{[\lambda(\gamma_m - \gamma_0)]^i}{i!} \right\} \tag{4.19}$$

$$\times \left[\sum_{k=m}^{N-1} \frac{N!}{(N-k)!k!}[1 - e^{-\lambda \gamma_{k+1}}]^{N-k}[e^{-\lambda \gamma_k}]^k + \left[e^{-\lambda \gamma_N} \right]^N \right]$$

$$+ \sum_{n=1}^{m-1} \left\{ [1 - e^{-\lambda \gamma_0}]\Psi(n-1, \gamma_n - \gamma_0) + \Psi(n, \gamma_n)e^{-\lambda[\gamma_n - \gamma_0]} \sum_{i=0}^{n-1} \frac{[\lambda(\gamma_n - \gamma_0)]^i}{i!} \right\}$$

$$\times \frac{N!}{(N-n)!n!}[1 - e^{-\lambda \gamma_{n+1}}]^{N-n}[e^{-\lambda \gamma_n}]^n + \left[1 - e^{-\lambda \gamma_0} \right] \left[1 - e^{-\lambda \gamma_1} \right]^N$$

where $\Psi(n,z) = \left[1 - e^{-\lambda z} \sum_{i=0}^{n} \frac{[\lambda z]^i}{i!} \right]$ *and the outage probability can be approximated at high SNR as*

$$P_{out} \approx \left[\lambda \gamma_0 \frac{[\lambda(\gamma_m - \gamma_0)]^m}{m!} + \frac{[\lambda \gamma_m]^{m+1}}{(m+1)!} \right] \sum_{k=m}^{N-1} \frac{N!}{(N-k)!k!}[\lambda \gamma_{k+1}]^{N-k} \tag{4.20}$$

$$+ \sum_{n=1}^{m-1} \frac{N!}{(N-n)!n!}[\lambda \gamma_{n+1}]^{N-n} \left[\lambda \gamma_0 \frac{[\lambda(\gamma_n - \gamma_0)]^n}{n!} + \frac{[\lambda \gamma_n^{n+1}]}{(n+1)!} \right] + \lambda \gamma_0 [\lambda \gamma_1]^N$$

Theorem 4.10
Consider that type IV CSI is available at the source node and only m of the n qualified relaying nodes will participate in communication. The diversity-multiplexing trade-off of the opportunistic relay scheme can be expressed as

$$d(r) = [m+1][1-(m+1)r] \tag{4.21}$$

In general, due to the higher, but not too much, *coding gain* from the extra channel information between the source and the destination, the schemes associated with the type IV CSI have slightly better performance than, if not the same as, those with the type II CSI. These observations will be verified through the simulations.

4.2.2 Cooperative Protocol with Full CSI

With type V CSI where both the source-to-relay and relay-to-destination CSI are available, the source node first specifies the qualified relay nodes that can successfully decode the information. Then it arranges the relay to destination channels for all of the n specified relay nodes, without loss of generality, in an ascending order as

$$z_{(1)} \leq z_{(2)} \leq \cdots \leq z_{(n-1)} \leq z_{(n)} \tag{4.22}$$

where $z_{(i)} = |\bar{h}_i|^2$ is the channel value square from the ith chosen relay node to the destination. Finally, the source chooses the best link, i.e., *the source* to *the nth qualified relay node* to *the destination*, for cooperation. The outage probability and the diversity-multiplexing trade-off are given by the following two theorems, respectively:

Theorem 4.11
Given that type V CSI is available at the source node, the outage probability of the opportunistic best relay scheme can be expressed as

$$P_{out} = \sum_{n=0}^{N} \frac{N!\lambda}{(N-n)!n!}[1 - e^{-\lambda\gamma(1)}]^{N-n}[e^{-\lambda\gamma(1)}]^{n+1} \tag{4.23}$$

$$\times \left\{ \sum_{\substack{i=0 \\ i\neq 1}}^{n} C_n^i(-1)^i \frac{1}{(i-1)\lambda}[1 - e^{-(i-1)\lambda\gamma_1}] - n\lambda\gamma_1 \right\}$$

and the outage probability at high SNR can be approximated as

$$P_{out} \approx [\lambda\gamma(1)]^{N+1} \sum_{n=1}^{N-1} \frac{N!}{(N-n)!(n+1)!} \tag{4.24}$$

Proof: *Please refer to [7].* ■

Theorem 4.12
With the type V CSI available at the source node, the diversity-multiplexing trade-off of the opportunistic bestrelay scheme can be expressed as

$$d(r) = [N+1][1 - 2r] \tag{4.25}$$

Theorems 4.11 and 4.12 provide an interesting result that the best relay scheme

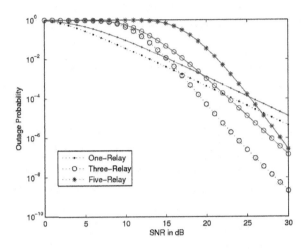

Figure 4.3: The outage probability vs. SNR. The dotted lines are for the scheme with no a priori information of CSI, and the solid lines are for the proposed relaying scheme with type II CSI.

can achieve full diversity with only one relay link. This can be explained as follows. If all of the n qualified relay nodes are used for the cooperation, the system has the best reliability of reception, or the full diversity, since the outage event occurs when all of the n relay-destination links fail to convey reliable communication. Such an outage event is obviously equal to the event that the best relay-destination link becomes unreliable. Thus, the best relay scheme achieves the same reliability of reception as the scheme that uses all of the n nodes. But the best relay scheme has the highest spectral efficiency since it only uses one relay node.

To verify the analytical results, numerical simulations are provided. The number of the possible relaying nodes is assumed to be $N = 5$ and the data rate at the source node is set as $R = 1$ bit/s/Hz. As a first example, the proposed strategy with type II CSI is compared with the scheme studied in [31] with no a priori channel information. Note that the scheme in [31] is a multiple-node extension of the classical scheme in [20]. It is shown in Figure 4.3 that the proposed scheme with the order information achieves better performance than that without. It is interesting to observe that the fewer the relay nodes chosen, the bigger the performance gap between the two strategies. For example, if only one node is used (i.e., $n = 1$), there is around a 5 dB performance difference between the two approaches. On the contrary, in the extreme case that all 5 nodes are chosen, the two schemes have identical performance, since then there is no difference between having knowledge of the order information or not.

Figure 4.4 depicts the outage probabilities of the scheme with type II CSI, where the curves obtained by the Monte Carlo simulation and our analytical results are almost overlapped, which verifies the accuracy of the corresponding explicit ex-

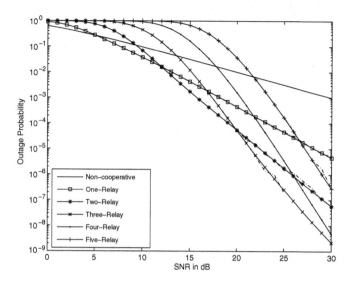

Figure 4.4: The outage probability of the relaying scheme with type II CSI vs. SNR. The dashed lines are for the Monte Carlo simulation, and the solid lines are for the analytical results.

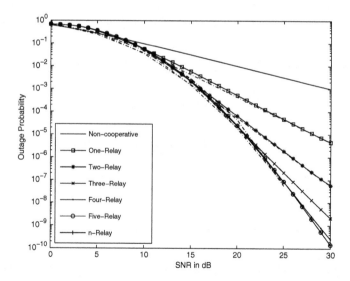

Figure 4.5: The outage probability of the relaying scheme with type III CSI vs. SNR. The dotted lines are for the Monte Carlo simulation, and the solid lines are for the analytical results.

pression. It is also clear from this result that the proposed scheme achieves better performance than the direct transmission at high SNR, but still suffers from performance loss at low SNR. Specifically, with more relay nodes being involved in the cooperation, while the superiority of the proposed schemes to the direct transmission is higher, the performance loss at low SNR also becomes worse. Similar observations were also reported in [31]. This is because an extra channel resource has to be assigned with every relay node joining the cooperation. Such unbalanced performance behavior may be difficult in practical cooperative system design, as the number of relay nodes has to be carefully chosen to balance the performance at low and high SNR.

In the second example, Figure 4.5 shows the outage probabilities of the scheme with type III CSI. Again, the curves obtained by the analytical and Monte Carlo simulations match very well. As expected, with the exact CSI at the source, the performance loss at low SNR is effectively reduced. To be specific, it can be clearly observed that with more relay nodes being chosen, the outage performance at both low and high SNR is improved.

The outage probabilities of the incremental scheme with type IV CSI are depicted in Figure 4.6. With more a priori information used by the source node, the performance of this scheme is shown to be close to, or slightly better than, that in Figure 4.5. This observation matches well with our previous statement.

In the last example, the best relay scheme with type V CSI is investigated and the outage performance of this scheme is compared with those for direct transmission as well as other schemes studied in this section. It is obviously shown in Figure

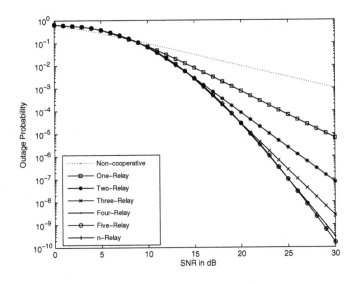

Figure 4.6: The outage probability vs. SNR. The dotted line is for the noncooperative scheme, and the solid lines are for the relaying scheme with type IV CSI.

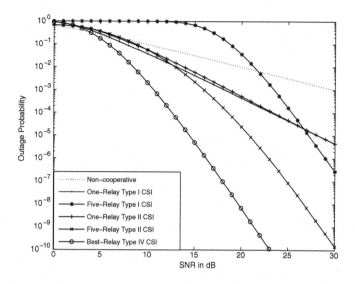

Figure 4.7: The outage probability vs. SNR. The dotted line is for the noncooperative scheme, and the solid lines are for the proposed relaying schemes.

4.7 that the best relay scheme has the best performance among all of the schemes. Specifically, at high SNR, the curves for the best relay scheme, the five-relay scheme with type II CSI, and the five-relay scheme with type III CSI have the same slope, since all of them can achieve the full diversity order $d = 5$. At the low SNR, on the other hand, the best relay scheme and those with type III/IV CSI can effectively suppress the performance loss, whereas the scheme with type II CSI cannot.

4.3   Cooperative Multiple Access Networks

In this section, two cooperative schemes are proposed for multiple access networks based on source cooperation and with relay assistance. The performance of these schemes is evaluated by studying the diversity-multiplexing trade-off. In addition, simulation results are provided to validate the analytical results.

4.3.1   *Protocol Based on Sources Cooperation*

In this subsection, a spectrally efficient scheme is proposed for cooperative multiple access channels. This scheme is developed based on superposition modulation, where each user transmits a mixture of its own signal and the signals received from other users. This cooperative scheme can be viewed as a precoded point-to-point multiple antenna systems; hence, the performance of this scheme is accessed through multiplexing trade-off analysis. These analytical results reveal that the proposed cooper-

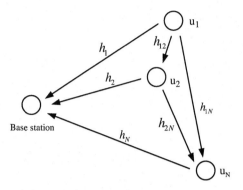

Figure 4.8: A multiple access network with source cooperation.

ative scheme outperforms the compared transmission schemes in terms of diversity-multiplexing gain.

In this subsection, a centralized communication system is considered with N single antenna sources communicating a central node, as shown in Figure 4.8. This network setup could be seen in a cellular system with multiple users transmitting their information to a base station or in a sensor network with scattered sensors sending their data to a fusion center. Each user communicates the central node through an orthogonal channel use in a time or frequency or code domain by sending its own information as well as the received signal during the previous $N - 1$ channel uses. For this scheme, the received signal at the base station during the nth channel use of the kth frame can be written as

$$y_n(k) = h_n \left[\gamma_{n,n} s_n(k) + \sum_{i=1, i \neq n}^{N} \hat{s}_i(k) \right] + w_n(k) \tag{4.26}$$

where h_n is the channel coefficient between the base station and the nth user and $w_n(k)$ is zero mean circularly symmetric additive white Gaussian noise with variance σ^2. In addition, $\gamma_{n,i}$ are the coefficients to allocate the transmission power between different signals. As in the selection relaying protocol [20], $\hat{s}_i(k)$ is defined as follows:

$$\hat{s}_i(k) = \begin{cases} \gamma_{i,n} s_i(k) & \text{if} \quad |h_{ij}|^2 > g(\rho) \quad \& \quad i < n \\ \gamma_{i,n} s_i(k-1) & \text{if} \quad |h_{ij}|^2 > g(\rho) \quad \& \quad i > n \\ 0 & \text{if} \quad |h_{ij}|^2 \leq g(\rho) \end{cases} \tag{4.27}$$

where ρ represents the SNR and h_{xy} is the channel coefficient between node x and node y. The threshold $g(\rho)$ can be chosen as $(2^{2R} - 1)/\rho$ for repetition coding as in [20]. However, the performance analysis presented for the proposed scheme is partly influenced by the choice of the threshold as shown in the following. In order to simplify the performance analysis, it is assumed that $s_n(k)$ contains only one symbol. The signal model for the proposed cooperative scheme can be represented by considering all N users as follows:

$$y(k) = \mathbf{H}\mathbf{\Gamma}\mathbf{s}(k) + \mathbf{w}(k) \tag{4.28}$$

where $\mathbf{y}(k) = [y_N(k) \cdots y_1(k)]^T$, $\mathbf{H} = \text{diag}\{h_N, \ldots, h_1\}$, $\mathbf{w}(k) = [w_N(k) \cdots w_1(k)]^T$, and $\mathbf{s}(k) = [s_N(k) \cdots s_1(k)\hat{s}_N(k-1) \cdots \hat{s}_2(k-1)]^T$. The matrix $\mathbf{\Gamma}$ could take different form depending on the choice of $\hat{s}_i$. At high SNR, the following structure of $\mathbf{\Gamma}$ will dominate due to $P[|h_{ij}|^2 > g(\rho)] \to 1$:

$$\mathbf{\Gamma} \approx \begin{bmatrix} \gamma_1 & \cdots & \gamma_N & 0 & \cdots & 0 \\ 0 & \gamma_1 & \cdots & \gamma_N & \cdots & 0 \\ \vdots & \ddots & \ddots & \ddots & \ddots & \vdots \\ 0 & \cdots & 0 & \gamma_1 & \cdots & \gamma_N \end{bmatrix}_{N \times (2N-1)}$$

Since a symmetric system is assumed, all users will have the same power coefficients and $\sum_{n=1}^{N} \gamma_n^2 = 1$. For the two-user scenario, the matrix $\mathbf{\Gamma}$ can be provided as follows:

$$\mathbf{\Gamma} = \begin{cases} \mathbf{\Gamma}_0 & \text{if} \quad |h_{12}|^2 > g(\rho) \\ \mathbf{\Gamma}_1 & \text{if} \quad |h_{12}|^2 \le g(\rho) \end{cases}$$

where $\mathbf{\Gamma}_0 = \begin{bmatrix} \sqrt{1-\gamma^2} & \gamma & 0 \\ 0 & \sqrt{1-\gamma^2} & \gamma \end{bmatrix}$ and $\blacksquare_1 = \begin{bmatrix} 1 & 0 & 0 \\ 0 & \sqrt{1-\gamma^2} & \gamma \end{bmatrix}$ and at high SNR, $P(\mathbf{\Gamma} = \mathbf{\Gamma}) \to 1$ as $P(|h_{12}|^2 \le g(\rho)) \to 0$ [21].

The proposed cooperative multiple access system in (4.28) can be considered a special case of a MIMO system, where $\mathbf{\Gamma}$ consists of transmit filters. The performance analysis of this scheme can be evaluated by studying its diversity-multiplexing trade-off, which is a fundamental concept to understand and design a multiantenna system. The following theorem provides the multiplexing-diversity trade-off for the proposed cooperative multiple access scheme:

Theorem 4.13
Let us assume that the Rayleigh fading channels from N source nodes to the destination node are independent and identically distributed. Then the diversity-multiplexing trade-off for the superposition cooperative access system is given by

$$d_{SP}^*(r) = N(1 - r) \tag{4.29}$$

Proof: *Please refer to [12].* ■

In order to understand the performance of the proposed cooperative scheme, two multiple access schemes developed based on direct transmission and relaying schemes are considered in the following [20]:

Theorem 4.14
Let us assume that the Rayleigh fading channels from N source nodes to the destination node are independent and identically distributed. Then the diversity-multiplexing

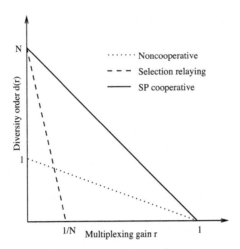

Figure 4.9: The multiplexing-diversity trade-off for the superposition (SP) and selection cooperative schemes and noncooperative scheme.

trade-off for the multiple access system with direct (or noncooperative) transmission is given by

$$d_D^*(r) = 1 - r \tag{4.30}$$

Proof: *Please refer to [12].* ■

Theorem 4.15
Let us assume that the Rayleigh fading channels from N source nodes to the destination node are independent and identically distributed. Then the diversity-multiplexing trade-off for the multiple access system with selective relaying is given by

$$d_S^*(r) = N(1 - Nr) \tag{4.31}$$

Proof: *Please refer to [12].* ■

The performance of the proposed cooperative multiple access scheme is evaluated by analyzing the multiplexing trade-off. Figure 4.9 depicts the multiplexing trade-off for the three transmission schemes. In the noncooperative transmission scheme, the each user's data are transmitted through an orthogonal channel by one user, whereas in the selective relaying scheme each user's information is received at the central node through N different channels. Hence, the noncooperative transmission and selective relaying schemes achieve the multiplexing gains 1 and N, respectively. In the proposed cooperative multiple access scheme, the same data is transmitted N times through N independent channels, which results in a better

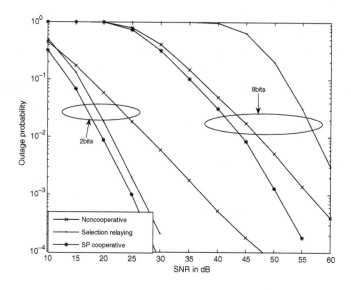

Figure 4.10: The outage probability for the superposition (SP) and selection cooperative schemes and noncooperative scheme.

performance in terms of diversity-multiplexing trade-off compared to the other two schemes. This performance gain is achieved without any loss in the resource of frequency/time/code. By exploiting the superposition modulation, each user shares its transmission power to transmit other users' information, which yields the better spectral efficiency while significantly enhancing the reception reliability.

In order to evaluate the performance of the proposed scheme, a multiple access network is considered with two sources (i.e., $N = 2$) and $\gamma_1^2 = 0.13$. The fixed data rates have been set to 2 bits/s/Hz and 8 bits/s/Hz. Figure 4.10 depicts the outage probability of the three transmission schemes. From this result, the proposed cooperative multiple access scheme outperforms the other two schemes in terms of outage probability. In addition, as reported in [21], this scheme achieves 2 dB performance gain at low data rates compared to the other schemes. On the other hand, the performance of the relaying scheme significantly degrades at high data rate and becomes inferior to the direct transmission scheme. This performance loss is due to the requirement of extra $N - 1$ channel uses, which results in data rate loss and severely affects the system performance at high data rates. However, the proposed scheme does not require any extra use of a frequency or temporal resource, and the performance loss is effectively controlled as evidence from the simulation shows.

4.3.2 A Relay-Assisted Cooperative Multiple Access Protocol

In this subsection, a spectrally efficient cooperative multiple access (CMA) protocol is proposed with the help of relays. This protocol is developed by carefully

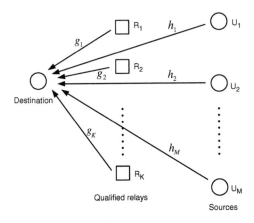

Figure 4.11: A relay-assisted multiple access network.

scheduling multiple sources and relays' transmissions according to the channel conditions between the sources and relays as well as the destination. In this scheme, it is with high probability that a relay will be scheduled to a source having poor connection with the destination. To evaluate the performance of the proposed protocol, an achievable diversity-multiplexing trade-off is developed. Based on the outage event discussed in [27], the closed-form expression of outage probability is derived by exploiting the order statistics [5]. From this analytical result, it is shown that the proposed scheme can approximately achieve the optimal multiple-input single-output (MISO) upper bound with a large number of relays. In most conditions, the simulation results show that the proposed scheme outperforms the other alternative schemes [1, 12].

To propose this relay-assisted cooperative multiple access protocol, a multiple access network with M source nodes transmitting information to a common destination with the help of L relay nodes is considered, as shown in Figure 4.11. It is assumed that all the nodes are half duplex and all the channel coefficients between the relay and sources as well as the destination are quasi-static identical and independent Rayleigh fading. Moreover, the available transmission power at sources and relays is assumed to be the same and time division multiple access is considered. Each relay employs the decode-and-forward strategy and has the CSI between the relay and sources as well as the destination.

As proposed in [20], it is assumed that a set of relays (i.e., qualified relays) is capable of decoding the mth source information satisfying $\log(1 + \rho|h_{mr_i}|^2) \geq R$, where h_{mr_i} is the channel coefficient between the mth source and the relay r_i, R denotes the target data rate, and ρ denotes the average received signal-to-noise ratio (SNR) at each relay and the destination. The number of these qualified relays is informed to the destination through an error-free signaling channel, and these relays will be considered for whole frame transmission due to the quasi-static fading channels. Based on the number of qualified relays, the destination determines the size of each data frame as $N = QM$, where $Q = \lceil \frac{K+1}{M} \rceil$.

This scenario could arise in a classical sensor network with a large number of sensorss or in a cellular network with a large number of mobile users, where only a small set of sensors or mobile users will be active and the rest of them will be idle. This large number of idle mobiles and sensors could be considered as relays for the proposed relay-assisted protocol. Different from the strategies discussed in [1, 12], the source nodes are scheduled in the proposed scheme as follows:

$$|h_1|^2 \leq |h_2|^2 \leq \cdots \leq |h_M|^2$$

where h_m represents the channel coefficient between the mth scheduled source and the destination. Similarly, the K qualified relays are scheduled as

$$|g_1|^2 \leq |g_2|^2 \leq \cdots \leq |g_K|^2$$

where g_k denotes the channel coefficient between the destination and the relay scheduled at the $(K-k+1)$th time slot. This scheduling significantly improves the performance of the user with the worst channel conditions by exploiting the relays. The relay selection can be performed through either a central strategy or distributed scheme, which significantly reduces the feedback overhead. Moreover, the relays and the sources can be scheduled in a distributed way based on the backoff time and the quality of the channels. In this cooperative multiple access scheme, nonorthogonal transmissions are adopted as in [1, 23].

In the first time slot of each data frame, the first scheduled source transmits its message and all qualified relays decode the message to save in their memories. The received signal at the destination can be written as

$$y(1) = h_1 s_1(1) + n(1)$$

where $n(i)$ denotes the additive white Gaussian noise at the common destination. During the second time slot, the relay with the largest channel coefficient between the relay and the destination and the second scheduled source transmit the messages $s_1(1)$ and $s_2(1)$, respectively. The received signal at the destination is

$$y(2) = h_2 s_2(1) + g_K s_1(1) + n(2)$$

At the same time, the rest of the $K-1$ qualified relays receive the mixture of both $s_2(1)$ and $s_1(1)$ and successfully decode $s_2(1)$ using a successive decoding scheme. At the nth time slot, the source with $h_{n'}$ and relay with g_{K-n+1} ($n \leq K$) transmit $s_{n'}(\lceil \frac{n}{M} \rceil)$ (where $n' = (n \bmod M)$) and its previous observation, respectively. In the case of $n > K$, a noncooperative direct transmission scheme is adopted [10, 15]. The signal model for one data frame can be defined as follows:

$$\mathbf{y} = \mathbf{H}\mathbf{s} + \mathbf{n} \tag{4.32}$$

where $\mathbf{y} = \begin{bmatrix} y(1) & \cdots & y(N) \end{bmatrix}^T$, $\mathbf{s} = \begin{bmatrix} s_1(1) & \cdots & s_M(1) & \cdots & s_1(Q) & \cdots & s_M(Q) \end{bmatrix}^T$,

$\mathbf{n} = \begin{bmatrix} n(1) & \cdots & n(N) \end{bmatrix}^T$, and the channel matrix is

$$\mathbf{H} = \begin{bmatrix} h_1 & 0 & \cdots & \cdots & \cdots & \cdots & 0 \\ g_K & h_2 & 0 & \vdots & \vdots & \vdots & 0 \\ \vdots & \ddots & \ddots & \vdots & \vdots & \vdots & \vdots \\ 0 & 0 & g_1 & h_{(K+1)'} & 0 & \cdots & 0 \\ 0 & 0 & 0 & 0 & h_{(K+2)'} & 0 & 0 \\ \vdots & \vdots & \vdots & \vdots & \ddots & \ddots & \vdots \\ 0 & 0 & 0 & 0 & 0 & 0 & h_M \end{bmatrix}_{N \times N}$$

For example, the signal model with four qualified relays and three sources (i.e., $K = 4$ and $M = 3$) is represented as follows:

$$\begin{bmatrix} y(1) \\ y(2) \\ y(3) \\ y(4) \\ y(5) \\ y(6) \end{bmatrix} = \begin{bmatrix} h_1 & 0 & 0 & 0 & 0 & 0 \\ g_4 & h_2 & 0 & 0 & 0 & 0 \\ 0 & g_3 & h_3 & 0 & 0 & 0 \\ 0 & 0 & g_2 & h_1 & 0 & 0 \\ 0 & 0 & 0 & g_1 & h_2 & 0 \\ 0 & 0 & 0 & 0 & 0 & h_3 \end{bmatrix} \begin{bmatrix} s_1(1) \\ s_2(1) \\ s_3(1) \\ s_1(2) \\ s_2(2) \\ s_3(2) \end{bmatrix}$$

Next, the performance of the proposed cooperative multiple access protocol is analytically evaluated by deriving the diversity-multiplexing trade-off. However, it is difficult to derive the explicit expression of the outage probability and trade-off due to the nonregular channel matrix $\mathbf{H}$ in (4.32). To circumvent this problem, it is assumed that the qualified relays plus one $K + 1$ are an integer multiple of a number of sources M. This assumption results in the following channel matrix:

$$\mathbf{H} = \begin{bmatrix} h_1 & 0 & 0 & 0 \\ g_K & h_2 & 0 & 0 \\ \vdots & \ddots & \ddots & \vdots \\ 0 & 0 & g_1 & h_M \end{bmatrix}_{N \times N} \tag{4.33}$$

Note that the simulation results are provided in the last part of this section without this assumption. An achievable diversity-multiplexing trade-off of the proposed cooperative multiple access protocol is provided in the following theorem with the relationship assumption between K and M:

Theorem 4.16
Assume that all addressed channels are independent and identically distributed quasi-static Rayleigh fading, and the number of qualified relays is $K = QM - 1$. The following diversity-multiplexing trade-off is achievable:

$$d_K(r) = (1 - r) + [K - (K + M)r]^+ \tag{4.34}$$

where $(x)^+$ denote $\max\{x, 0\}$.

Proof: *Please refer to [10].*   ■

Since the number of qualified relays is dynamically varying according to the channel conditions, it would be more appropriate to define the diversity-multiplexing trade-off in terms of number of relays (i.e., L). The relationship between the number of qualified relays and the total number of relays is defined through the following lemma:

Lemma 4.1

Assume that all addressed channels are independent and identically distributed and quasi-static Rayleigh fading. Then,

$$P(K = k) \doteq \rho^{-(L-k)(1-r)}$$

as $\rho \to \infty$.

Proof: *Please refer to [10].*   ■

From Theorem 4.16 and Lemma 4.1, the overall outage probability, can be derived as

$$P(\mathcal{O}) = \sum_{k=0}^{L} P(\mathcal{O}|K = k)P(K = k) \qquad (4.35)$$

$$\doteq \sum_{k=0}^{L} \rho^{-(L-k+1)(1-r)-[k-(k+M)r]^{+}} \doteq \rho^{-d(r)}$$

where $d(r) = (1-r) + [L - (L+M)r]^{+}$ and the last relationship is obtained from the fact that the error probability with $K = L$ is the dominating factor. From this overall outage probability, it is revealed that the maximum possible diversity gain of the proposed cooperative multiple access scheme can be $L+1$ with a fixed data rate. This maximum diversity gain is achievable with the help of the opportunistic scheduling. The diversity-multiplexing trade-off can be written as

$$d(r) = \begin{cases} (L+1)\left(1 - r - \frac{M}{L+1}r\right), & \text{if } 0 \le r \le \frac{L}{L+M} \\ 1 - r, & \text{if } \frac{L}{L+M} < r \le 1 \end{cases}$$

Moreover, it can be shown that the achievable trade-off of the proposed scheme is approximately equal to the optimal MISO trade-off $d(r) \to d_{MISO}(r)$ with the large number of relays L and fixed number of sources M, where $d_{MISO}(r)$ is defined as

$$d_{MISO}(r) = (L+1)(1-r), \quad \text{if } 0 \le r \le 1$$

In contrast to the limited diversity gain of the schemes in [1, 12], the proposed relay-assisted cooperative multiple access scheme introduces another degree of freedom through exploiting the available relays and significantly improves the reception capability. However, the schemes proposed in [1, 12] based on source nodes

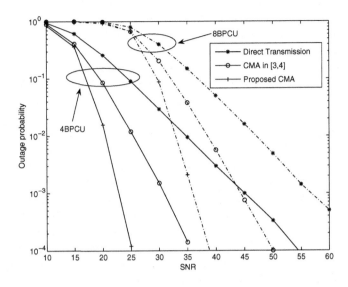

Figure 4.12: The outage probability for the proposed scheme, the superposition co-operative scheme [1, 12], and a noncooperative scheme.

cooperation outperform this relay-assisted scheme in terms of diversity-multiplexing trade-off.

In order to validate the performance of the proposed relay-assisted scheme, numerical results are provided without the assumptions $K = QM - 1$ and $K = L$, which

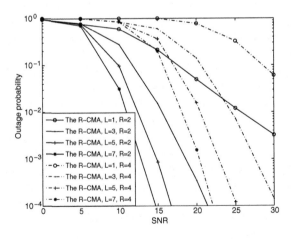

Figure 4.13: The outage probability for the proposed CMA scheme with different choices of the relay numbers L.

are used to derive the analytical results. In the first set of experiments, the outage probability of the proposed scheme is compared with the superimposed modulation-based schemes in [1, 12] and the noncooperative scheme, as shown in Figure 4.12. For this comparison, a network is considered with two sources and five relays, whereas the target data rates have been set to $R = 4$ and $R = 8$ bits per channel use (BPCU). As can be seen from Figure 4.12, the proposed scheme outperforms the compared schemes in terms of outage probabilities for all SNR. This performance gain is achieved by the diversity gain introduced through the exploitation of the relays, whereas the diversity gains of the compared schemes are limited by the source nodes.

In the second set of simulations, the outage probabilities with two sources are evaluated for different numbers of relays as depicted in Figure 4.13. As evidenced from Figure 4.13, the robustness of the proposed scheme improves with the number of relays. As the achievable trade-off of the proposed scheme is approximately equal to the optimal upper bound, the proposed relay-assisted scheme improves the diversity gain and the spectral efficiency by increasing the number of relays, which is provided in Theorem 4.16.

4.4   Cooperative Broadcast Networks

In this section, cooperative transmission protocols are developed for broadcast channels in the presence and absence of direct source-destination (S-D) links. The performance of the proposed protocols is evaluated through information theoretic metrics, namely, outage probability and diversity-multiplexing trade-off. In contrast to the conventional two-hop scheme, which achieves a diversity gain of 1/2, the proposed protocol in the absence of direct S-D links can achieve a multiplexing gain close to 1. On the other hand, the protocol developed in the presence of the S-D links outperforms the comparable scheme.

4.4.1   Cooperative Transmission Protocol without Direct S-D Links

In this subsection, a novel cooperative broadcast protocol is developed without S-D links. This protocol can achieve a multiplexing gain of $\frac{M}{M+1}$, where M denotes the number of destinations. In order to enhance the performance, two approaches have been considered by modifying the protocol. The first one is based on cognitive radio, which introduces the achievable diversity gain close to the maximum possible diversity gain. The other one is with the assumption of having enough reliable relays, which improves the multiplexing gain close to one.

To develop this protocol, a broadcast relay network is considered with a source node, M destination, and L relays, where the source transmits different messages to each destination through the relays, as shown in Figure 4.14. It is assumed that all users have the same target data rate R and time division duplexing. In addition,

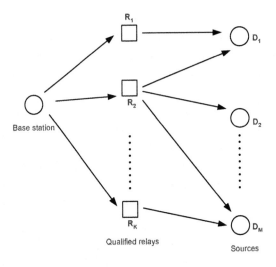

Figure 4.14: A relay-assisted broadcast network without the direct source-destination links.

half-duplex constraints are imposed on the nodes as in [1, 20]. Here, the direct link between source and destination is not considered as in [23], whereas the following subsection proposes a protocol by taking these direct links into account. In this protocol, the source first transmits information to the relays and the received signals are then forwarded to the destinations. All the channels considered in this relay network are assumed to be quasi-static Rayleigh fading and constant during a data frame.

In this proposed scheme, relays are initially chosen for involvement in the data transmission based on the satisfaction criteria $\left\{ |h_{SR_i}|^2 \geq \frac{2^{\frac{(M+1)}{M}R} - 1}{\rho} \right\}$, where ρ denotes the SNR, R denotes the target data rate to each destination, and h_{SR_i} denotes the channel coefficient between the source and the ith relay. It is assumed that K relays, $K \leq L$ are able to successfully decode the information from the source and are indicated as $\{R_1, \dots, R_K\}$. In addition, it is also assumed that $K \geq M$. In a scenario, that does not satisfy these assumptions, a two-hop relaying scheme can be employed as in [23]. Relay selection strategy and acquiring CSI are provided with more detail in [11].

The data transmission from the source to each destination takes place as follows in $M + 1$ time slots. In the first time slot, the message s_1, intended for the first user, is transmitted at the rate $\frac{(M+1)}{M}R$. This message is decoded by all qualified K relays and stored to decode the following messages using successive decoding. In the second time slot, the source transmits the message s_2, intended for the second user, at the same rate and the relay R_1^*, chosen based on the criteria $|h_{R_1^* D_1}|^2 = \max\{|h_{R_1 D_1}|^2, \dots, |h_{R_K D_1}|^2\}$, forwards the message s_1 to the first destination. The first destination only receives the message transmitted by R_1^* due to the reason of not having the direct link from the source. The rest of the qualified relays receive the mixture of s_1 and s_2 and successfully decode using a successive decoding technique. This transmission procedure continues at the source as well as relays, and at the last time slot $M+1$, R_M^* forwards the message s_M to the Mth destination, whereas all other nodes, including source, keep silent.

The proposed scheme is evaluated through outage probability and diversity-multiplexing trade-off. The notations $\mathcal{O}$ and $\mathcal{O}_K$ represent the overall outage event and the outage event with K qualified relays. Hence, the overall outage probability is defined as

$$P(\mathcal{O}) = \sum_{k=0}^{L} P(\mathcal{O}_k)P(K=k) \tag{4.36}$$

By considering only the user with the worst performance, the following theorem provides the achievable diversity-multiplexing trade-off of the proposed transmission protocol with assumption on K:

Theorem 4.17
Assume all wireless channels are independent identically Rayleigh faded and there are no direct links between the source and destinations. Provided that there are K qualified relays, $K \geq M$, the achievable outage probability is

$$P(\mathcal{O}_K) \doteq \rho^{-d_{two\_hop,K}(r)} \tag{4.37}$$

where $d_{two\_hop,K}(r) = (K - M + 1)\left(1 - \frac{M+1}{M}r\right)$, for $0 < r < \frac{M}{M+1}$.

Proof: *Please refer to [11].* ■

In the scenario of $K \leq M$, it would be more appropriate to develop a relationship between achievable diversity-multiplexing trade-off and the total number of relays, L, since the outage probability $P(\mathcal{O}_K)$ is 1. The probability of the event with K qualified relays is defined as [6]

$$P(K=k) = \frac{L!}{(L-k)!k!}\left(1 - e^{-\frac{2^{\frac{M+1}{M}R}-1}{\rho}}\right)^{L-k} e^{-k\frac{2^{\frac{M+1}{M}R}-1}{\rho}} \doteq \rho^{-(L-k)\left(1-\frac{M}{M+1}r\right)} \tag{4.38}$$

for $0 \leq k \leq L$. The following corollary provides the achievable diversity-multiplexing trade-off for the proposed scheme:

Corollary 4.1
By using the proposed cooperative transmission protocol, the achievable diversity-multiplexing trade-off for the user with the worst performance is

$$d_{two\_hop} = (L - M + 1)\left(1 - \frac{M+1}{M}r\right), \quad 0 < r < \frac{M}{M+1} \tag{4.39}$$

The proposed protocol for one user (i.e., $M = 1$) achieves the same performance as in the scheme in [23]. However, the proposed scheme outperforms the scheme in [23] with the multiplexing gain of $\frac{M}{M+1}$, which is larger than the achievable multiplexing gain, $\frac{1}{2}$, of the scheme in [23]. The multiplexing gain of the proposed protocol can achieve close to 1 with a large number of destinations.

As evidenced from the optimal diversity-multiplexing trade-off upper bound of single-input multiple-output systems (i.e., $d_{SIMO}(r) = L(1-r)$), the achievable diversity gain of the proposed protocol can be only a fraction of the maximum diversity gain L. In contrast to the optimal single-input multiple-output upper bound, the achievable multiplexing gain of the proposed scheme is less. In order to increase the diversity gain of this protocol, a novel scheme has been investigated based on a cognitive radio approach in [11]. In addition, a simple alternative scheme is proposed to increase the diversity gain in the following section.

4.4.1.1 An Alternative Scheme to Increase the Achievable Multiplexing Gain

As provided in Theorem 4.17, the proposed scheme can achieve the multiplexing gain of $\frac{M}{M+1}$. However, the multiplexing gain, $\frac{M}{M+1}$, approaches 1 as the number of destination, M, increases. Hence, it would be preferred to include a large number of destinations in the system, which increases the multiplexing gain of the proposed protocol. However, this approach reduces the achievable diversity gain, $L - M + 1$, while increasing the system complexity.

Compared to the schemes investigated in [1] and [20], the proposed scheme has a loss in the achievable multiplexing gain due to the requirement of an extra time slot for relay transmission. To circumvent this multiplexing gain loss, the overall length of the data frame can be increased. In the proposed scheme, only M relays are chosen to participate in the data transmission out of K qualified relays. In order to increase the multiplexing gain, all K relays can be employed so that the data frame length is increased to $K + 1$. At the nth time slot $(1 < n < (K+1))$, a relay transmits a message intended to the $[(n \mod M) - 1]$th destination while the source broadcasts a new message. This approach results in all destinations being divided into two groups, namely, $\mathscr{D}_1$ and $\mathscr{D}_2$ which include p destinations served $q + 1$ times and the $M - p$ destinations served q times, respectively. The integers p and q are defined as $K = qM + p$, $q = \lfloor \frac{K}{M} \rfloor$ and p is the residual.

1. *Without the cognitive precoding approach:* In this scheme, the previously used qualified relays cannot be employed again for data transmission. This makes each destination choose the best relay from the set of unused qualified relays. The outage probabilities of this scheme are defined as follows:

$$P(\mathscr{O}_{K,m}) = \begin{cases} P\left(\frac{K}{K+1}\left(\log \prod_{i=1}^{q+1}(1+\rho|h_{R_i^*D_m}|^2)\right) < (q+1)R\right) & m \in \mathscr{D}_1 \\ P(\mathscr{O}_{K,m}) = P\left(\frac{K}{K+1}\left(\log \prod_{i=1}^{q}(1+\rho|h_{R_i^*D_m}|^2)\right) < qR\right) & m \in \mathscr{D}_2 \end{cases}$$

$$(4.40)$$

These two outage probabilities are approximated at high SNR in the following lemma:

Lemma 4.2

Provided that there are K qualified relays and $K \geq M$, the achievable diversity-multiplexing trade-off obtained from the outage probabilities in (4.40) can be upper bounded as

$$d_{K,m} = \begin{cases} (K-p+1) - \frac{K+1}{K}(K+M-p)r, & m \in \mathcal{D}_1 \\ (K-M+1) - \frac{K+1}{K}(K-p)r, & m \in \mathcal{D}_2 \end{cases} \tag{4.41}$$

where $K = qM + p$.

Proof: *Please refer to [11].* ■

This scheme can be exploited in a cellular network with many mobile users or in a dense sensor network as the achievable multiplexing gain approaches 1 with large number of relays and constant M at high SNR. Moreover, the overall diversity-multiplexing trade-off of this scheme is 1, as in Theorem 4.17.

2. *With the cognitive precoding approach:* In this scheme, at least the second best relay of each destination can transmit the corresponding information as the qualified relays can be participate again in the data transmission. By considering the event of serving p destinations with $(q + 1)$ times, the outage probability for these users are defined as follows:

$$P(\mathscr{O}_{K,m}) = \begin{cases} P\left(\frac{K}{K+1}\left(\log(1+\rho|h_{R_m^* D_m}|^2)^{q+1}\right) < (q+1)R\right) & m \in \mathcal{D}_1 \\ P\left(\frac{K}{K+1}\left(\log(1+\rho|h_{R_m^* D_m}|^2)^q\right) < qR\right) & m \in \mathcal{D}_2 \end{cases} \tag{4.42}$$

In addition, the two outage probabilities can be approximated at high SNR in the following lemma:

Lemma 4.3

Provided that there are K qualified relays and $K \geq M$, the diversity-multiplexing trade-off achievable for the proposed cognitive approach can be upper bounded as

$$d_K = (K-1)\left(1 - \left(\frac{K+1}{K}\right)^{\frac{1}{q}} r\right) \tag{4.43}$$

where $K = qM + p$.

The proof of this lemma can be derived by considering the worst case relay selection.

In order to evaluate the performance, the proposed cooperative transmission protocol is compared with the conventional two-hop transmission scheme combining relay selection strategies in [2] and [23]. These comparable schemes require, in particular, $2M$ time slots, and each destination is served through the best relay in two

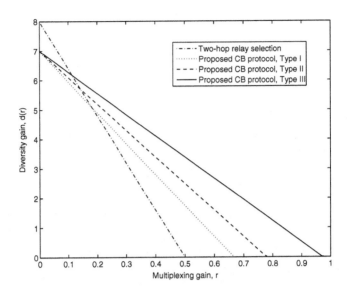

Figure 4.15: The diversity-multiplexing trade-off achieved by the proposed broadcast transmission protocols and the comparable schemes. It is assumed that there is no direct link between the source and destination nodes. The numbers of qualified relays and destination nodes are $K = 8$ and $M = 2$.

consecutive time slots. To simply depict the proposed schemes in the simulation results, the original proposed protocol is denoted as type I, whereas the alternative schemes investigated to increase the multiplexing gain without and with the cognitive precoding approaches are represented as type II and type III, respectively.

To assess the performance of the proposed scheme, use the outage probabilities depicted in Figure 4.25(a) for different SNRs with six relays ($L = 6$), three destinations ($M = 3$), and 4 bits per channel use (BPCU) target data rate ($R = 4$). Similarly, Figure 4.25(b) represents the outage probabilities with eight relays ($L = 8$), four destinations ($M = 4$), and 5 BPCU target data rate ($R = 5$). As evidenced from these simulation results, the proposed schemes outperform the conventional two-hop scheme with relay selection in terms of outage probabilities. This performance gain is due to the difference in the achievable multiplexing gain of $\frac{M}{M+1}$ in the proposed scheme and $1/2$ in the conventional scheme. In addition, these results show that the proposed alternative schemes increase the achievable multiplexing gain with $K > M$. The achievable diversity-multiplexing trade-off of the proposed schemes is represented in Figure 4.15 with eight relays ($L = 8$) and two destinations ($M = 2$). Compared to the conventional schemes, the proposed protocol achieves a better performance in terms of diversity-multiplexing trade-off at high multiplexing gain, as shown in Figure 4.15.

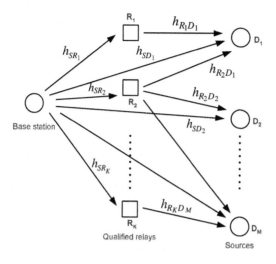

Figure 4.16: A relay-assisted broadcast network with direct source-destination links.

4.4.2 Cooperative Transmission Protocol with Direct S-D Links

In this section, a cooperative transmission protocol has been proposed for a relay-assisted broadcasting network where direct links between source and destinations are also available, as shown in Figure 4.16. In this scheme, destinations are opportunistically scheduled based on the channel conditions between the source and destinations. In particular, the destination with the worst channel condition between the source and destination is served first, whereas the destination with the best channel is served last. This scheduling helps the worst destination to have more relays than the best destination, which improves the system performance by enhancing the quality of the transmission for the destinations with poorer channel conditions. Moreover, it is assumed that the global CSI is available at the source. All the relays are half duplex and employ the amplify-and-forward strategy, in contrast to the ones with the decode-and-forward strategy in the previous section. Without loss of generality, the destinations are denoted $D_1, \cdots, D_M$ with the channel conditions $|h_{SD_1}|^2 \leq \cdots \leq |h_{SD_M}|^2$, where h_{SD_1} represents the channel coefficient between the source and D_1. Moreover, the relay chosen to forward the message at the $(i+1)$th time slot is denoted by R_i. In the following, the proposed protocol and signal model are described in detail.

In this scheme, the message $s(1) = \alpha_{11}s_1 + \alpha_{12}\tilde{s}_2$ is broadcasted by the source in the first time slot, where s_1 is the message intended for the first destination. The structure of $\tilde{s}_2$ and design of the scalars α_{11} and α_{12} are provided later in this section. The received signal at each relay can be written as

$$y_{R_i}(1) = h_{SR_i}s(1) + n_{R_i}(1)$$

where $n_{R_i(1)}$ is the additive white complex Gaussian noise at the ith relay with power

P_n. In the second time slot, the source and the chosen relay transmit, respectively, the messages $s(2) = \alpha_{13}s_1 + \alpha_{14}\tilde{s}_2$ and $\frac{y_{R_i}(1)}{\beta_{R_1}}$, where $\beta_{R_1} = \sqrt{|h_{SR_1}|^2 + \frac{1}{\rho}}$ ensures that the transmission power is normalized. The received signal at the first destination, D_1, can be represented as

$$
y_{d_1} = \begin{bmatrix} h_{SD_1} & h_{R_1D_1} \end{bmatrix} \begin{bmatrix} \alpha_{13} & \alpha_{14} \\ \dfrac{h_{SR_1}\alpha_{11}}{\beta_{R_1}} & \dfrac{h_{SR_1}\alpha_{12}}{\beta_{R_1}} \end{bmatrix} \begin{bmatrix} s_1 \\ \tilde{s}_2 \end{bmatrix} + n_{D_1} + \dfrac{h_{R_1D_1}}{\beta_{R_1}} n_{R_1} \tag{4.44}
$$

The precoding matrix, $\mathbf{P}_1$, should be designed such that the co-channel interference is removed and each destination receives its intended message:

$$
\mathbf{P}_1 = \begin{bmatrix} \alpha_{13} & \alpha_{14} \\ \alpha_{11} & \alpha_{12} \end{bmatrix} = \begin{bmatrix} \dfrac{h_{SD_1}^*}{\gamma_{11}} & \dfrac{h_{R_1D_1}}{\gamma_{12}} \\ \dfrac{h_{R_1D_1}^*\beta_{R_1}}{h_{SR_1}\gamma_{12}} & -\dfrac{h_{SD_1}\beta_{R_1}}{h_{SR_1}\gamma_{12}} \end{bmatrix} \tag{4.45}
$$

where $(\cdot)^*$ denotes the conjugate operation,

$$
\gamma_{11} = \sqrt{|h_{SD_1}|^2 + \frac{|h_{R_1D_1}|^2 \beta_{R_1}^2}{|h_{SR_1}|^2}}
$$

$$
\gamma_{12} = \sqrt{|h_{R_1D_1}|^2 + \frac{|h_{SD_1}|^2 \beta_{R_1}^2}{|h_{SR_1}|^2}} \tag{4.46}
$$

Since $\beta_{R_1}^2 \approx |h_{SR_1}|^2$ at high SNR, the coefficients γ_{11} and γ_{12} will be the same and they can be defined as $\gamma_1 = \gamma_{1i} = \sqrt{|h_{SD_1}|^2 + |h_{R_1D_1}|^2}$. Hence, the received signal at the first destination can be written in a simple form as follows:

$$
y_{d_1} = \left(\frac{|h_{SD_1}|^2}{\gamma_1} + \frac{|h_{R_1D_1}|^2}{\gamma_1} \right) s_{D_1} + n_{D_1} + \frac{h_{R_1D_1}}{\beta_{R_1}} n_{R_1} \tag{4.47}
$$

which ensures that the co-channel interference is completely suppressed in the received signal and the messages transmitted from the source and the relay are coherently combined due to the design of this precoding matrix. The following mutual information is achieved at the first destination:

$$
\mathscr{I}_1 = \frac{M}{M+1} \log \left(1 + \rho \frac{(|h_{SD_1}|^2 + |h_{R_1D_1}|^2)|h_{SR_1}|^2}{|h_{R_1D_1}|^2 + |h_{SR_1}|^2} \right)
$$

At the same time, the received signal at each relay R_i can be written as follows:

$$
y_{R_i} = h_{SR_i}\tilde{s}_2 + \tilde{n}_{R_i} \tag{4.48}
$$

which is obtained by coherently combining the received signal over the two time slots; details are provided in [11]. In addition, $\tilde{n}_{R_i}$ can be approximated as a white complex Gaussian noise with power P_n, the same as n_{R_1}, provided that the number of relays is sufficiently large.

Similarly, the achievable data rate at the mth destination served at the $(m+1)$th time slot can be represented as

$$\mathcal{I}_m = \frac{M}{M+1} \log \left(1 + \rho \frac{\left(|h_{SD_m}|^2 + |h_{R_m D_m}|^2 \right) |h_{SR_m}|^2}{|h_{R_m D_m}|^2 + |h_{SR_m}|^2} \right), \quad 1 \le m \le M-1 \quad (4.49)$$

Different from the previous time slots, the source and the relay R_M transmit the messages $\frac{h^*_{SD_M}}{\theta} s_M$ and $\frac{h^*_{R_M D_M}}{\theta} \left(s_M + \frac{\tilde{n}_{R_M}}{h_{SR_M}} \right)$, respectively, where $\theta^2 = |h_{SD_M}|^2 + |h_{R_M D_M}|^2 + \frac{1}{|h_{SR_M}|^2 \rho}$. Based on this, the data rate achieved at the Mth user can be written as

$$\mathcal{I}_M \approx \frac{M}{M+1} \log \left(1 + \rho \frac{\left(|h_{SD_M}|^2 + |h_{R_M D_M}|^2 \right)}{1 + \frac{|h_{R_M D_M}|^4}{\left(|h_{SD_M}|^2 + |h_{R_M D_M}|^2 \right) |h_{SR_M}|^2}} \right) \quad (4.50)$$

$$\ge \frac{M}{M+1} \log \left(1 + \rho \frac{\left(|h_{SD_M}|^2 + |h_{R_M D_M}|^2 \right) |h_{SR_M}|^2}{|h_{SR_M}|^2 + |h_{R_M D_M}|^2} \right)$$

where the lower bound is obtained to have a similar expression of the mutual information. To implement this protocol, appropriate relays need to be chosen from the available relays. This relay selection can be implemented based on distributive strategy; the details of this strategy can be found in [11].

To assess the performance of the proposed protocol, the outage probability and achievable diversity-multiplexing trade-off are evaluated. The event of not satisfying the data rate of the mth destination is defined as $\mathcal{O}_m$, and the outage probability of the mth user can be written as

$$P(\mathcal{O}_m) = P \left(\frac{M}{M+1} \log \left(1 + \rho \frac{\left(|h_{SD_m}|^2 + |h_{R_m D_m}|^2 \right) |h_{SR_m}|^2}{|h_{R_m D_m}|^2 + |h_{SR_m}|^2} \right) < R \right)$$

By considering the performance of the worst user, the outage performance of the proposed scheme can be provided through the following theorem:

Theorem 4.18
Assume all addressed wireless channels are independent and identically distributed Rayleigh faded and the destinations can hear the source directly. The outage probability achieved by the proposed cooperative broadcast protocol can be approximated at high SNR as

$$P(\mathcal{O}) \doteq \rho^{-d(r)} \quad (4.51)$$

where $d(r) = (L - M + 1) \left(1 - \frac{M+1}{M} r \right)$, for $0 < r < \frac{M}{M+1}$.

Proof: *Please refer to [11].* ▪

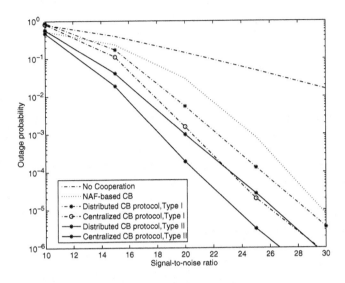

Figure 4.17: The outage probability achieved by the proposed broadcast transmission protocols and the comparable schemes vs. SNR. It is assumed that the destination nodes can hear the source directly. The numbers of relays and destination nodes are $L = 4$ and $M = 2$. The targeted data rate is $R = 3$ BPCU.

With large number of destination nodes, $\frac{M}{M+1}$ approaches 1. This yields the proposed protocol outperforming the scheme in [1] in terms of diversity gain for the multiplexing gain $\frac{1}{2} \leq r \leq 1$. To increase the multiplexing gain, the approaches proposed for the protocol without S-D links can be exploited for this scheme. This leads to an approximate diversity-multiplexing trade-off $d(r) \approx (L - M + 1)(1 - r)$ with a large enough number of relays.

To access the performance of the proposed scheme, a network consisting of four relays ($L = 4$) and two sources ($M = 2$) is considered with target data rate 3 BPCU ($R = 3$). To simplify the notation, the proposed protocol with direct S-D links and enhanced version serving destinations with multiple times are denoted as CB protocol type I and type II. In addition, the proposed protocol with distributed and centralized relay selection strategies is also considered in the simulations. The performances of the proposed schemes are compared with those of the noncooperative transmission and nonorthogonal amplify-and-forward protocols discussed in [1]. As shown in Figure 4.17, these proposed schemes outperform the compared schemes in terms of outage probabilities. This performance gain is obtained through the achievable diversity gain $(L - M + 1)$ up to the the maximum multiplexing gain $\frac{M}{M+1}$, whereas the noncooperative scheme achieves diversity gain larger than 1 for the multiplexing gain $\frac{1}{2} \leq r \leq 1$. Figure 4.18 depicts the achievable diversity-multiplexing trade-off for the proposed schemes. This figure confirms that the proposed scheme specifically outperforms the noncooperative scheme with the multiplexing gain $\frac{1}{2} \leq r \leq \frac{M}{M+1}$.

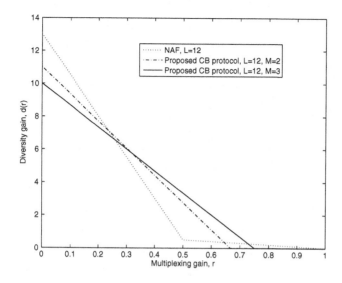

Figure 4.18: The diversity-multiplexing trade-off achieved by the proposed broadcast transmission protocols and the comparable schemes. It is assumed that the destination nodes can hear the source directly. The number of relays is $L = 12$.

4.5 Network Coding for Cooperative Networks

In this section, two network-coded cooperative schemes are proposed for multiple access networks and a two-way relay network based on the amplify-and-forward strategy and interference alignment scheme, respectively. These proposed schemes are evaluated by deriving the outage probability, ergodic capacity, and diversity-multiplexing trade-off.

4.5.1 Network Coding for Multiple Access Channels

In this subsection, a network coded amplify-and-forward cooperative scheme is investigated for multiple access channels. This spectrally efficient scheme is developed based on the distributed relay selection strategy. The proposed scheme is evaluated by deriving the explicit expression of information metrics, namely, outage probability and ergodic capacity. The analytical results and numerical results confirm that the proposed scheme outperforms in terms of ergodic capacity while achieving full diversity gain.

To propose a network-coded protocol, a multiple access network is considered with M sources communicating a common destination through the help of L relays, as shown in Figure 4.19. In this scheme, the channel coefficients between all sources and relays as well as the destination are assumed to be independent identical Rayleigh fading. During the first time slot, all sources transmit their own messages. The re-

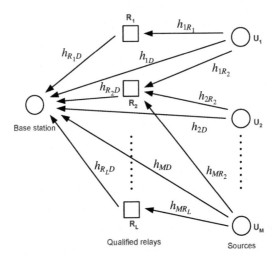

Figure 4.19: A relay-assisted multiple access network with direct source-destination links.

ceived signal at the destination can be expressed as

$$y_{D1} = \sum_{m=1}^{M} h_{mD} s_m + n_1 \tag{4.52}$$

where s_m is the message transmitted from the mth source, n_1 is the additive Gaussian noise at the destination, and h_{mD} is the coefficient for the channel between the mth source and the destination. Similarly, the received signal at the relays during the first time slot can be written as

$$y_{R_n} = \sum_{m=1}^{M} h_{mR_n} s_m + n_{R_n}, \quad n \in \{1, \ldots, L\} \tag{4.53}$$

At the end of the first time slot, each relay has a mixture of messages from all sources. In order to enhance the system performance, the concept of network coding is introduced in the proposed scheme. It is assumed that only $M-1$ relays have been chosen to forward their messages in turn during the next $M-1$ time slots based on the amplify-and-forward strategy. The received signal from the relay is

$$y_{D(l+1)} = h_{R_l D} \hat{y}_{R_l} + n_{l+1}, \quad l = 1, \ldots, M-1 \tag{4.54}$$

where $\hat{y}_{R_l} = y_{R_l}/\beta_l$, $\beta_l = \sqrt{\sum_{m=1}^{M} |h_{mR_l}|^2 + 1/\rho}$ and ρ is denoted as SNR. Note that β_0 is defined as $\beta_0 = \sqrt{\sum_{m=1}^{M} |h_{mD}|^2}$. By stacking all the received signal at the destination during the M time slots, the signal model can be represented as follows:

$$\mathbf{y} = \mathbf{DHs} + \mathbf{n}$$

where $\mathbf{y} = \begin{bmatrix} y_{D1} & \cdots & y_{DM} \end{bmatrix}^T$, $\mathbf{s} = \begin{bmatrix} s_1 & \cdots & s_M \end{bmatrix}^T$, $\mathbf{D} = \mathrm{diag}\{\beta_0, h_{R_1 D}, \ldots, h_{R_{M-1} D}\}$,

$$\mathbf{H} = \begin{bmatrix} h_{1D}/\beta_0 & \cdots & h_{MD}/\beta_0 \\ h_{1R_1}/\beta_1 & \cdots & h_{MR_1}/\beta_1 \\ \vdots & \vdots & \vdots \\ h_{1R_{M-1}}/\beta_{M-1} & \cdots & h_{MR_{M-1}}/\beta_{M-1} \end{bmatrix}, \quad \text{and} \quad \mathbf{n} = \begin{bmatrix} n_1 \\ n_2 + h_{R_1 D} n_{R_1}/\beta_1 \\ \vdots \\ n_M + h_{R_{M-1} D} n_{R_{M-1}}/\beta_{M-1} \end{bmatrix}.$$

The achievable sum rate of the proposed scheme can be expressed as

$$\mathscr{I} = \frac{1}{M} \log \det\{\mathbf{I}_M + \rho \mathbf{D} \mathbf{H} \mathbf{H}^H \mathbf{D}^H \mathbf{C}^{-1}\} \tag{4.55}$$

$$\approx \log \rho + \frac{1}{M} \log \det\{\mathbf{D}^H \mathbf{D} \mathbf{C}^{-1}\} + \frac{1}{M} \log \det\{\mathbf{H} \mathbf{H}^H\} \tag{4.56}$$

where the high SNR assumption is applied, $\mathbf{C} = \mathrm{diag}\{1, 1 + |h_{R_1 D}|^2/\beta_1^2, \cdots, 1 + |h_{R_{M-1} D}|^2/\beta_{M-1}^2\}$, and

$$\log \det\{\mathbf{D}^H \mathbf{D} \mathbf{C}^{-1}\} = \frac{1}{M} \log \beta_0^2 \prod_{l=1}^{M-1} \frac{|h_{R_l D}|^2 \beta_l^2}{|h_{R_l D}|^2 + \beta_l^2} \tag{4.57}$$

In this signal model, the matrix $\mathbf{H}$ is not a regular Gaussian random matrix and each row is normalized. It is important to note that relay selection strategies influence the distribution of $\mathbf{H}$ and determine the system performance. For this protocol, a distribution relay selection strategy has been provided with more detail in [13].

A question might arise in choosing the optimal number of sources and relays for this network-coded cooperative transmission scheme to achieve the best performance. It is difficult to derive a relationship between the optimal number of sources and the system performance. However, the following conjecture is provided to maximize the sum rate without analytical proof:

Conjecture 4.1
The sum rate achieved by the proposed transmission protocol can be maximized where there are only two sources participating in cooperation.

However, the simulation results have been provided to confirm this conjecture in [13]. The proposed network-coded protocol is evaluated through outage probability and diversity gain with an optimal number of sources (i.e., $M = 2$) and L relays. Similar to the definition provided in [12, 27], the outage event can be expressed as

$$\mathscr{O} \triangleq \bigcup_{\mathscr{A}} \mathscr{O}_{\mathscr{A}} \tag{4.58}$$

by considering the union over all possible subsets $\mathscr{A} \subseteq \{1, 2\}$, and $\mathscr{O}_{\mathscr{A}}$ is defined as

$$\mathscr{O}_{\mathscr{A}} \triangleq \left\{ \mathscr{I}(\mathbf{s}_{\mathscr{A}}; \mathbf{y} | \mathbf{s}_{\mathscr{A}^c}, \mathbf{H} = H, \mathbf{D} = D) \leq \sum_{i \in \mathscr{A}} R_i \right\} \tag{4.59}$$

The mutual information for the two-user case considered in this scheme can be represented as

$$\mathscr{I}_{\mathscr{A}_i} = \log\left[1 + \rho\left(|h_{iD}|^2 + \frac{|h_{RD}|^2|h_{iR}|^2}{|h_{iR}|^2 + |h_{RD}|^2}\right)\right], \quad i \in \{1,2\} \qquad (4.60)$$

$$\mathscr{I}_{\mathscr{A}_3} = \log\det\{\mathbf{I} + \rho\mathbf{D}\mathbf{H}\mathbf{H}^H\mathbf{D}^H\mathbf{C}^{-1}\} \qquad (4.61)$$

where $\mathscr{I}_{\mathscr{A}_n} = \mathscr{I}(\mathbf{s}_{\mathscr{A}_n}; \mathbf{y}|\mathbf{s}_{\mathscr{A}_n^c}, \mathbf{H} = H, \mathbf{D} = D)$. The outage probability of the proposed transmission scheme is provided in the following theorem at high SNR:

Theorem 4.19
Assume that all CSI are independent and identically distributed Rayleigh fading. For the scenario with two sources and L relays, the outage probability of the proposed network-coded transmission protocol can be approximated at high SNR as

$$P(\mathscr{O}) \doteq \frac{1}{\rho^{L+1}} \qquad (4.62)$$

Proof: *Please refer to [13].* ■

The above theorem reveals that the proposed protocol can achieve a full diversity gain $L + 1$. However, ergodic capacity is derived to explain the advantage of the proposed scheme over existing schemes.

Definition 4.1
Ergodic capacity is the long-term data rate that a system can support, i.e.,

$$\mathscr{C}_e = \int_0^\infty \mathscr{I}f_{\mathscr{I}}(\mathscr{I})d\mathscr{I}$$

where $f_{\mathscr{I}}(\cdot)$ is the probability density function (PDF) of the mutual information $\mathscr{I}$.

The following theorem provides the ergodic capacity of the proposed protocol based on sum rate:

Theorem 4.20
Assume all channels are independent and identically distributed Rayleigh fading. The ergodic capacity achieved by the proposed network-coded cooperative transmission protocol can be bounded as

$$\mathscr{E}\{\mathscr{I}_D\} + \frac{1}{2}\left(\sum_{k=1}^{L} C_L^k(-1)^k\log 4k\right) \leq \mathscr{E}\{\mathscr{I}\} \leq \mathscr{E}\{\mathscr{I}_D\} + \frac{1}{2}\left(\sum_{k=1}^{L} C_L^k(-1)^k\log k\right)(4.63)$$

where $\mathscr{E}\{\mathscr{I}_D\} \approx \log\rho - \mathbf{C}\log e$ is the ergodic capacity achieved by direct transmission.

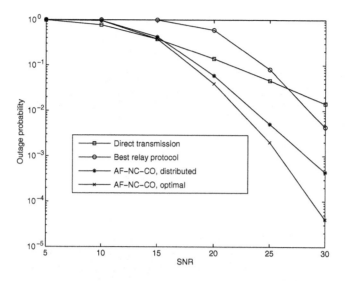

Figure 4.20: Outage probability vs. SNR. The data rate is set as $R = 4$ bits/Hz/s. The number of relays is $L = 2$.

Proof: *Please refer to [13].* ■

The capacity difference between the proposed scheme and the direct transmission scheme is quantified as follows:

$$\frac{1}{2}\left(\sum_{k=1}^{L} C_L^k(-1)^k \log 4k\right) \leq \mathscr{E} - \mathscr{E}_D \leq \frac{1}{2}\left(\sum_{k=1}^{L} C_L^k(-1)^k \log k\right)$$

The lower bound of the difference can be positive with enough relays. This yields that the proposed protocol can achieve a larger ergodic capacity than that of the direct transmission. To evaluate the performance of the proposed protocol, first a multiple access network is considered with two sources ($M = 2$) and two relays ($L = 2$). The target per user data rate has been set to 4 bits/s/Hz. Figure 4.20 compares the outage probabilities of the proposed scheme with the direct transmission and best-relay schemes in [2]. The proposed optimal and distributed relay selection-based schemes outperform the compared schemes. In addition, the performance between both proposed schemes is not much different, which yields that the distributed relay selection strategy does not cause a significant amount of performance loss of the proposed scheme.

4.5.2 Network Coding for Two-Way Relay Channels

In this section, a joint uplink and downlink transmission scheme is proposed for a two-way relay channel, where a base station and M single antenna mobile users

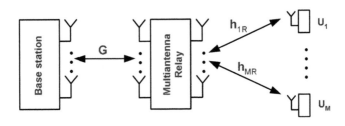

Figure 4.21: A two-way relay network where base station and relay consist of multiple antennas and each user is equipped with a single antenna.

exchange their information with the help of a multiantenna relay. This protocol is developed based on physical layer network coding and interference alignment techniques that require only two time slots to complete $2M$ uplink and downlink transmissions. In order to enhance the performance of the mobile users, precoding at the relay and base station is designed to completely remove the co-channel interference. To validate the performance of the proposed scheme, the expression of achievable multiplexing gain has been derived analytically, which confirms that the proposed scheme outperforms the existing time-sharing schemes.

To propose this protocol, a two-way relay network is considered with M single antenna mobile users and a multiantenna base station interchanging their information through a multiantenna relay, as shown in Figure 4.21. It is assumed that all channels are independent and identically distributed, and there is no direct link between the mobile users and the base station. In addition, the half-duplex constraint is imposed on all nodes, while employing time division duplexing. In order to design the precoding at the relay and base station, it is required to have the global CSI at both the relay and base station. The protocols to obtain this CSI are provided in more detail in [8, 16, 18].

In this network, the base station and M mobile users need to interchange their messages between them with the help of a multiantenna relay. In particular, M messages are required to send from the base station to the M mobile users, as well as from the M mobile users to the base station. The messages from the mth user to the base station and intended for the mth user from the base station are represented as u_m and s_m, respectively. Moreover, the same target data rate R is assumed between the base station and each user. By directly applying the physical layer network coding technique explored in [9, 29], $2M$ time slots are required to accomplish the information exchange between the mobile users and the base station. However, the proposed scheme in this section only requires two time slots to complete data transmission from both sides provided that the base station and the relay consist of M antennas.

During the first time slot, both the base station and all mobile users transmit their messages to the relay. Hence, the relay receives the following signal at the end of the first time slot:

$$\mathbf{r} = \mathbf{GPs} + \sum_{m=1}^{M} \mathbf{h}_{mR} u_m + \mathbf{n}_R \qquad (4.64)$$

where $\mathbf{s} = \begin{bmatrix} s_1 & \cdots & s_M \end{bmatrix}^T$ consist of all symbols intended for all mobile users and $\mathbf{P}$ is a $M \times M$ precoding matrix at the base station. In addition, $\mathbf{G}$ represents the channel matrix between the base station and the relay, whereas $\mathbf{h}_{mR}$ denotes the channel coefficients between the mth mobile user and the relay. The vector $\mathbf{n}_R$ denotes additive white Gaussian noise at the relay. In this transmission scheme, it is assumed that the maximum transmission power available at each antenna at the base station and each mobile user is to 1.

During the second time slot, the relay forwards the received messages based on the amplify-and-forward technique. Hence, the received signal at the mth mobile user during the second time slot can be written as

$$y_m = \mathbf{h}_{mR}^H \mathbf{W} \left(\mathbf{GPs} + \sum_{m=1}^{M} \mathbf{h}_{mR} u_m + \mathbf{n}_R \right) + n_m \qquad (4.65)$$

whereas the base station observes the following signal:

$$\mathbf{y}_{BS} = \mathbf{G}^H \mathbf{W} \left(\mathbf{GPs} + \sum_{m=1}^{M} \mathbf{h}_{mR} u_m + \mathbf{n}_R \right) + \mathbf{n}_{BS} \qquad (4.66)$$

The vector $\mathbf{n}_{BS}$ and n_m denote the additive white Gaussian noise at the base station and the mth mobile user, respectively.

It is important to note that each single antenna mobile user experiences the co-channel interference, which significantly degrades the performance of the symbol detection at the mobile user. To overcome this performance loss, the design of precoder at the relay and the base station should take into account this co-channel interference as well as the power constraints. Based on the novel interference alignment technique proposed in [3], the messages corresponding to the same user (i.e., s_i and u_i) are aligned at the relay in the same direction. This signal alignment can be performed by designing the relay precoding matrix as follows:

$$\mathbf{P} = \mathbf{G}^{-1} \mathbf{H} \mathbf{D}_s \qquad (4.67)$$

where $\mathbf{H} = \begin{bmatrix} \mathbf{h}_{1R} & \cdots & \mathbf{h}_{MR} \end{bmatrix}$ and the diagonal matrix $\mathbf{D}_s$ satisfy the transmission power constraint at the base station. Based on this precoding matrix $\mathbf{P}$, the signals corresponding to the same users are aligned at the relay as

$$\mathbf{r} = \mathbf{H}(\mathbf{D}_s \mathbf{s} + \mathbf{u}) + \mathbf{n}_R \qquad (4.68)$$

where $\mathbf{u} = \begin{bmatrix} u_1 & \cdots & u_M \end{bmatrix}^T$. The relay broadcasts these grouped signals without separating them. The total transmission power at the base station can be expressed in terms of $\mathbf{D}_s$ as follows:

$$\text{Tr}\left\{ \mathbf{PP}^H \right\} = \text{Tr}\left\{ \mathbf{G}^{-1} \mathbf{H} \mathbf{D}_s^2 \left(\mathbf{G}^{-1} \mathbf{H} \right)^H \right\} = \text{Tr}\left\{ \left(\mathbf{G}^{-1} \mathbf{H} \right)^H \mathbf{G}^{-1} \mathbf{H} \mathbf{D}_s^2 \right\} \qquad (4.69)$$

Based on this expression, the matrix $\mathbf{D}_s$ is chosen to satisfy the per antenna power constraint 1 as follows:

$$
\mathbf{D}_s = \begin{bmatrix} \frac{1}{\sqrt{\mathbf{h}_{1R}^H (\mathbf{G}^{-1})^H \mathbf{G}^{-1} \mathbf{h}_{1R}}} & \cdots & 0 \\ 0 & \ddots & 0 \\ 0 & \cdots & \frac{1}{\sqrt{\mathbf{h}_{MR}^H (\mathbf{G}^{-1})^H \mathbf{G}^{-1} \mathbf{h}_{MR}}} \end{bmatrix} \tag{4.70}
$$

Next, the relay precoding matrix should be also designed to completely remove the co-channel interference between the mobile users. Based on the proposed base station precoding matrix $\mathbf{P}$, the signal transmitted from the relay using the amplify-and-forward protocol can be represented as

$$
(\mathbf{Wr})^* = (\mathbf{W}[\mathbf{H}(\mathbf{D}_s \mathbf{s} + \mathbf{u}) + \mathbf{n}_R])^* \tag{4.71}
$$

At the end of the second time slot, each mobile user receives the following signal:

$$
y_m = \mathbf{h}_{mR}^H \mathbf{W}(\mathbf{H}(\mathbf{D}_s \mathbf{s} + \mathbf{u}) + \mathbf{n}_R) + n_m \tag{4.72}
$$

In order to completely remove the co-channel interference between mobile users, the relay amplification matrix $\mathbf{W}$ should fulfill the following condition:

$$
\mathbf{H}^H \mathbf{W} \mathbf{H} = \text{diag}\{\xi_1, \cdots, \xi_M\} \tag{4.73}
$$

where the value of ξ_m depends on the choice of the precoding matrix. A choice of the relay precoding matrix with stable transmission power is provided as follows:

$$
\mathbf{W} = (\mathbf{H}^H)^{-1} \mathbf{D}_r \mathbf{H}^{-1} \tag{4.74}
$$

where $\mathbf{D}_r$ is a diagonal matrix to satisfy the transmission power constraint. Based on this relay precoding structure, the total transmission power at the relay with high SNR assumption is obtained as follows:

$$
\begin{aligned}
P_{ow} &= \text{Tr}\left\{ \mathbf{WH}(\mathbf{D}_s \mathbf{D}_s^H + \mathbf{I}_M)\mathbf{H}^H \mathbf{W}^H + \frac{1}{\rho} \mathbf{W}(\mathbf{DD}^H + \mathbf{I}_M)\mathbf{W}^H \right\} \\
&\approx \text{Tr}\left\{ (\mathbf{H}^H)^{-1} \mathbf{D}_r (\mathbf{D}_s^2 + \mathbf{I}_M)\mathbf{D}_r^H \mathbf{H}^{-1} \right\}
\end{aligned} \tag{4.75}
$$

In addition, the following power normalization matrix $\mathbf{D}_r$ is considered to satisfy the unity per antenna power constraint:

$$
\mathbf{D}_r = \begin{bmatrix} \sqrt{\left([\mathbf{H}^H \mathbf{H}]_{1,1}^{-1}\right)^{-1} \Big/ 2} & \cdots & 0 \\ 0 & \ddots & 0 \\ 0 & \cdots & \sqrt{\left([\mathbf{H}^H \mathbf{H}]_{M,M}^{-1}\right)^{-1} \Big/ 2} \end{bmatrix}
$$

Based on this relay amplification matrix, the average total transmission power at the relay can be written as follows:

$$
\begin{aligned}
\mathcal{E}\{P_{ow}\} &\approx \mathcal{E}\left\{\mathrm{Tr}\left\{(\mathbf{H}^H\mathbf{H})^{-1}\mathbf{D}_r^2(\mathbf{D}_s^2+\mathbf{I}_M)\right\}\right\} \\
&= \frac{1}{2}\mathcal{E}\left\{\mathrm{Tr}\left\{(\mathbf{I}_M+\mathbf{D}_s^2)\right\}\right\} = \frac{M}{2}\left(1+\mathcal{E}\left\{\frac{1}{\mathbf{h}_{1R}^H(\mathbf{G}^{-1})^H\mathbf{G}^{-1}\mathbf{h}_{1R}}\right\}\right)
\end{aligned}
$$
(4.76)

where $[\mathbf{A}]_{m,m}$ denotes the mth element on the diagonal of the matrix $\mathbf{A}$. In addition, the proposed relay precoding matrix satisfies the transmission power constraint at the relay [8]. By using this relay precoding matrix, the received signal at the mth mobile user can be expressed as

$$
\begin{aligned}
y_m &= \mathbf{h}_{mR}^H\left[\mathbf{WH}(\mathbf{D}_s\mathbf{s}+\mathbf{u})+\mathbf{Wn}_R\right]+n_m \\
&= \mathbf{h}_{mR}^H(\mathbf{H}^H)^{-1}\mathbf{D}_r\left[((\mathbf{D}_s\mathbf{s}+\mathbf{u})+(\mathbf{H})^{-1}\mathbf{n}_R\right]+n_m \\
&= (d_{sm}s_m+u_m)+\tilde{\mathbf{h}}_m\mathbf{n}_R+d_{rm}^{-1}n_m
\end{aligned}
$$
(4.77)

where d_{sm} and d_{rm} are the mth elements at the diagonal of the matrices $\mathbf{D}_s$ and $\mathbf{D}_r$ and $\tilde{\mathbf{h}}_m$ is the mth row vector of $\mathbf{H}^{-1}$. This signal model completely removes the co-channel interference due to the choice of the precoding matrices at the relay and the base station. Similarly, the received signal at the base station can be expressed as

$$
\begin{aligned}
\mathbf{y}_{BS} &= \mathbf{G}^H\mathbf{W}(\mathbf{H}(\mathbf{D}_s\mathbf{s}+\mathbf{u})+\mathbf{n}_R)+\mathbf{n}_{BS} \\
&= \mathbf{G}^H(\mathbf{H}^H)^{-1}\mathbf{D}_r\mathbf{H}^{-1}(\mathbf{H}(\mathbf{D}_s\mathbf{s}+\mathbf{u})+\mathbf{n}_R)+\mathbf{n}_{BS} \\
&= \mathbf{G}^H\left((\mathbf{H}^H)^{-1}\mathbf{D}_r(\mathbf{D}_s\mathbf{s}+\mathbf{u})+(\mathbf{H}^H)^{-1}\mathbf{D}_r\mathbf{H}^{-1}\mathbf{n}_R\right)+\mathbf{n}_{BS}
\end{aligned}
$$
(4.78)

These precoding matrices introduce a complicated structure in the received signal at the base station while removing the co-channel interference at the mobile users. However, these proposed precoding matrices achieve the diversity order 1.

The performance of the proposed network coding protocol is evaluated based on the zero forcing detection approach due to its simplicity and the same performance as the MMSE-based scheme [17]. In addition, this receiver performance analysis is separately developed at the base station and the mobile users because of the different received signal models.

4.5.2.1 Performance Analysis for the Receiver Reliability at the Mobile Users

By exploiting its own information u_m, the received SNR at the mth mobile user can be written as follows:

$$
\begin{aligned}
SNR_{U_m} &= \frac{\rho}{\mathbf{h}_{mR}^H(\mathbf{G}^{-1})^H\mathbf{G}^{-1}\mathbf{h}_{mR}\left[\tilde{\mathbf{h}}_m\tilde{\mathbf{h}}_m^H+\dfrac{[\mathbf{H}^H\mathbf{H}]_{M,M}^{-1}}{2}\right]} \\
&= \frac{\rho}{\frac{3}{2}\mathbf{h}_{mR}^H(\mathbf{G}^{-1})^H\mathbf{G}^{-1}\mathbf{h}_{mR}\left[(\mathbf{H}^H\mathbf{H})^{-1}\right]_{m,m}}
\end{aligned}
$$
(4.79)

By using the fact $\tilde{\mathbf{h}}_m \tilde{\mathbf{h}}_m^H = [\mathbf{H}^H \mathbf{H}]_{M,M}^{-1}$, the expression of $[\mathbf{H}^H \mathbf{H}]_{M,M}^{-1}$ can be obtained as follows [25]:

$$\left[(\mathbf{H}^H \mathbf{H})^{-1}\right]_{m,m} = \frac{\det(\tilde{\mathbf{H}}_m^H \tilde{\mathbf{H}}_m)}{\det(\mathbf{H}^H \mathbf{H})} = \frac{1}{\mathbf{h}_{mR}^H \left(\mathbf{I}_M - \tilde{\mathbf{P}}_m\right) \mathbf{h}_{mR}}$$

where $\tilde{\mathbf{H}}_m = \begin{bmatrix} \mathbf{h}_{1R} & \cdots & \mathbf{h}_{(m-1)R} & \mathbf{h}_{(m+1)R} & \cdots & \mathbf{h}_{MR} \end{bmatrix}$ and $\tilde{\mathbf{P}}_m = \tilde{\mathbf{H}}_m (\tilde{\mathbf{H}}_m^H \tilde{\mathbf{H}}_m)^{-1} \tilde{\mathbf{H}}_m^H$. Moreover, the expression for $\left[(\mathbf{H}^H \mathbf{H})^{-1}\right]_{m,m}$ can be written by exploiting the idempotent matrix property of $\tilde{\mathbf{P}}_m$ and its one nonzero eigenvalue as follows:

$$\left[(\mathbf{H}^H \mathbf{H})^{-1}\right]_{m,m} = \frac{1}{\mathbf{h}_{mR}^H \mathbf{u}_m \mathbf{u}_m^H \mathbf{h}_{mR}} \tag{4.80}$$

where $\mathbf{u}_m$ is the eigenvector of $\left(\mathbf{I}_M - \tilde{\mathbf{P}}_m\right)$ corresponding to the eigenvalue 1. Based on this expression, the outage probability of the mth mobile user can be written by considering the fact that two time slots have been used for this scheme:

$$P(\mathscr{I}_{U_m} < 2R) = P\left(\frac{\mathbf{h}_{mR}^H \mathbf{u}_m \mathbf{u}_m^H \mathbf{h}_{mR}}{\mathbf{h}_{mR}^H (\mathbf{G}^{-1})^H \mathbf{G}^{-1} \mathbf{h}_{mR}} < \frac{3(2^{2R} - 1)}{2\rho}\right) \tag{4.81}$$

In addition, the achieved outage probability at the mth mobile user in the proposed protocol is provided in the following theorem:

Theorem 4.21
Through the downlink channels, at the mth mobile user, the achievable outage probability for the proposed network coding transmission protocol can be approximated as

$$P(\mathscr{I}_{U_m} < 2R) \leq -\frac{3M(M-1)}{4}\left(\frac{2^{2R}-1}{\rho}\right)\ln\left(\frac{2^{2R}-1}{\rho}\right) \tag{4.82}$$

when $\rho \to \infty$. The achievable diversity-multiplexing trade-off for the mth downlink transmission can be expressed as

$$d_{U_m}(r) = 1 - 2r$$

for the multiplexing gains $0 \leq r \leq \frac{1}{2}$.

   Proof: *Please refer to [8].* ■

The above theorem yields that the same outage probability is experienced by all users while achieving the diversity gain 1 in the downlink channels.

4.5.2.2 Performance Analysis for the Receiver Reliability at the Base Station

The received signal at the base station can be expressed as follows:

$$\mathbf{y}_{BS} = \mathbf{G}^H \left((\mathbf{H}^H)^{-1} \mathbf{D}_r(\mathbf{s} + \mathbf{u}) + (\mathbf{H}^H)^{-1} \mathbf{D}_r \mathbf{H}^{-1} \mathbf{n}_R\right) + \mathbf{n}_{BS} \tag{4.83}$$

Based on zero forcing approaches and extracting the transmitted information from the base station, the following can be obtained:

$$\mathbf{D}_r^{-1}\mathbf{H}^H(\mathbf{G}^H)^{-1}\mathbf{y}_{BS} = \mathbf{u} + \mathbf{H}^{-1}\mathbf{n}_R + \mathbf{D}_r^{-1}\mathbf{H}^H(\mathbf{G}^H)^{-1}\mathbf{n}_{BS} \qquad (4.84)$$

By using this detection scheme, the received SNR of the mth mobile user can be derived as follows:

$$
\begin{aligned}
SNR_{BS_m} &= \frac{\rho}{\left[\mathbf{H}^{-1}(\mathbf{H}^H)^{-1} + \mathbf{D}_r^{-1}\mathbf{H}^H(\mathbf{G}^H)^{-1}\mathbf{G}^{-1}\mathbf{H}\mathbf{D}_r^{-1}\right]_{m,m}} \\
&= \frac{\rho}{\left[(\mathbf{H}^H\mathbf{H})^{-1}\right]_{m,m} + \left[\mathbf{D}_r^{-1}\mathbf{H}^H\mathbf{U}\Lambda^{-1}\mathbf{U}^H\mathbf{H}\mathbf{D}_r^{-1}\right]_{m,m}} \\
&= \frac{\rho\mathbf{h}_{mR}^H\mathbf{v}\mathbf{v}^H\mathbf{h}_{mR}}{1 + \frac{1}{2}\mathbf{h}_{mR}^H\mathbf{U}\Lambda^{-1}\mathbf{U}^H\mathbf{h}_{mR}}.
\end{aligned}
\qquad (4.85)
$$

By exploiting the above SNR expression, the outage probability for the mth user information is given by the following theorem:

Theorem 4.22

Through uplink channels, at the base station, the achievable outage probability for the ith user's information by using the proposed network coding transmission protocol can be approximated as

$$P(\mathscr{I}_{BS_m} < 2R) \leq -\frac{M(M-1)}{2}\left(\frac{2^{2R}-1}{\rho}\right)\ln\left(\frac{2^{2R}-1}{\rho}\right) \qquad (4.86)$$

when $\rho \to \infty$. And the achievable diversity-multiplexing trade-off for the mth uplink transmission can be expressed as

$$d_{BS_m}(r) = 1 - 2r$$

for the multiplexing gains $0 \leq r \leq \frac{1}{2}$.

Proof: *Please refer to [8].* ■

In order to analyze the overall system performance, the sum rate and the worst transmission performance from $2M$ transmissions are evaluated for the proposed transmission protocol. Based on Theorems 4.21 and 4.22, the overall diversity-multiplexing trade off is provided by the following corollary:

Corollary 4.2

The overall diversity-multiplexing trade-off for the sum rate achieved by the proposed network coding protocol can be shown as follows:

$$d(r) = 1 - \frac{1}{M}r \qquad (4.87)$$

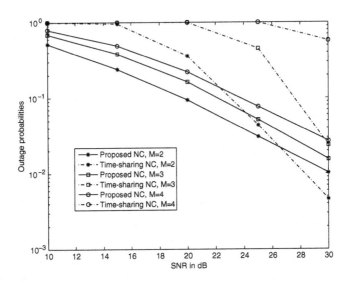

Figure 4.22: Outage probability vs. the signal-to-noise ratio. The target data rate for all users is $R = 1$ bit per channel use (BPCU). The base station and the relay have M antennas, and each of the M users has a single antenna.

for $0 \leq r \leq M$. The worst outage performance among the M uplink and M downlink transmissions is

$$P(\mathscr{I}_{min} < 2R) \leq -2M^2(M-1)\left(\frac{2^{2R}-1}{\rho}\right)\ln\left(\frac{2^{2R}-1}{\rho}\right)$$

where $\mathscr{I}_{min} = \min\{\mathscr{I}_{BS_1}, \cdots, \mathscr{I}_{BS_M}, \mathscr{I}_{U_1}, \cdots, \mathscr{I}_{U_M}\}$ and the high SNR assumption has been used.

Proof: *Please refer to [8].* ■

This corollary demonstrates that the achievable multiplexing gain of the proposed protocol is M, which is larger than that for existing network coding schemes. However, it is only possible to achieve diversity gain 1 by the proposed scheme. To improve this diversity gain, different techniques have been proposed with analytical results in [8].

To evaluate the performance of the proposed network coding protocol, simulation results are generated based on the Monte Carlo method and compared with the time-sharing physical layer network coding schemes in [9,29]. A single antenna is selected for both the base station and the relay by using the optimal antenna selection strategy. The channel coefficients between the relay and the base station as well as the mobile users are generated using zero mean circularly symmetric Gaussian random variables with the variances according to the SNR. In addition, the same target data rate R is assumed for all pairs of sources and destinations.

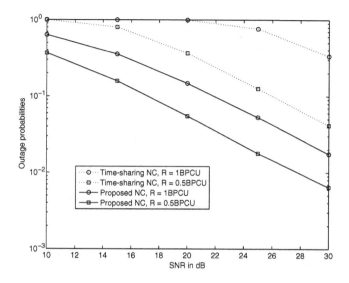

Figure 4.23: Outage probability vs. the signal-to-noise ratio. The number of users is M. The base station and the relay have $M = 3$ antennas. Each of the M users has a single antenna.

The outage performance of the proposed protocol with target rate $R = 1$ is compared with time-sharing network coding schemes in Figure 4.22 for different M. This outage performance is evaluated for the worst user performance. Specifically with a large number of users, the proposed scheme outperforms the time-sharing scheme as shown in Figure 4.22. This performance gain is achieved due to the requirement of two time slots, which does not depend on the number of users. However, the time-sharing scheme achieves larger diversity gain than the proposed scheme.

Figure 4.23 depicts the outage probabilities with three mobile users for different target data rates. As seen from Figure 4.23, the proposed scheme outperforms the time-sharing scheme due to the maximum achievable multiplexing gain M, whereas the maximum achievable multiplexing gain of the time-sharing protocol is 1. Figure 4.24 compares ergodic capacities of the proposed scheme with those of the time-sharing scheme for different numbers of mobile users. The proposed scheme achieves larger ergodic capacity than the time-sharing scheme, as observed in Figure 4.24.

4.6 Summary

In this chapter, different cooperative schemes have been explored for a relay network consisting of a source and a destination with multiple relays. These strategies were developed based on the available a priori channel information, and the performance has been evaluated by deriving explicit expressions of the outage probability and the diversity-multiplexing trade-off. Next, two cooperative multiple access protocols

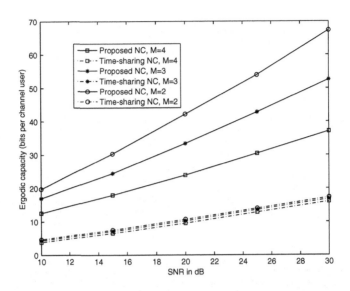

Figure 4.24: Ergodic capacity vs. the signal-to-noise ratio. The number of users is M. The base station and the relay have $M = 3$ antennas.

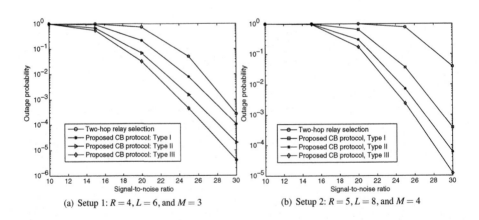

Figure 4.25: The outage probability achieved by the proposed broadcast transmission protocols and the comparable schemes vs. SNR. It is assumed that there is no direct link between the source and destination nodes.

have been proposed based on source cooperation and relay assistance. It has been shown that these schemes outperform the existing schemes in terms of outage probabilities. Then, two transmission schemes have been investigated for broadcast networks with and without direct source-destination links. The outage probability and achievable diversity-multiplexing trade-off of these broadcast protocols have been derived analytically by considering the user with the worst performance. Simulation results have been provided to validate the performance. Finally, two network-coded cooperative schemes have been proposed for a relay-assisted multiple access network and two-way relay network based on the amplify-and-forward strategy and interference alignment scheme, respectively. Simulation results have been provided to support the analytical results and to demonstrate the merits of the proposed network-coded protocols.

References

1. K. Azarian, H. E. Gamal, and P. Schniter. On the achievable diversity-multiplexing tradeoff in half-duplex cooperative channels. *IEEE Trans. Inform. Theory*, 51:4152–4172, 2005.

2. Aggelos Bletsas, Ashish Khisti, David P. Reed, and Andrew Lippman. A simple cooperative diversity method based on network path selection. *IEEE J. Select. Areas Commun.*, 24:659–672, 2006.

3. V. R. Cadambe and S. A. Jafar. Interference alignment and the degrees of freedom for the k user interference channel. *IEEE Trans. Inform. Theory*, 54:3425C–3441C, 2008.

4. T. M. Cover and A. A. E. Gamal. Capacity theorems for the relay channel. *IEEE Trans. Inform. Theory*, 24:572–584, 1979.

5. H. A. David and H. N. Nagaraja. *Order statistics*. 3rd ed. John Wiley, 2003.

6. Z. Ding, Y. Gong, T. Ratnarajah, and C. Cowan. On the performance of opportunistic cooperative wireless networks. *IEEE Trans. Commun.*, 56:1236–1240, 2008.

7. Z. Ding, Y. Gong, T. Ratnarajah, and C. Cowan. On the performance of opportunistic cooperative wireless networks. *IEEE Trans. Commun.*, 56(8):1236–1240, 2008.

8. Zhiguo Ding, I. Krikidis, J. Thompson, and K.K. Leung. Physical layer network coding and precoding for the two-way relay channel in cellular systems. *IEEE Trans. Signal Process.*, 59(2):696–712, 2011.

9. Zhiguo Ding, Kin K. Leung, Dennis L. Goeckel, and Don Towsley. On the study of network coding with diversity. *IEEE Trans. Wireless Commun.*, 8:1247–1259, 2009.

10. Zhiguo Ding, K.K. Leung, D.L. Goeckel, and D. Towsley. A relay assisted cooperative transmission protocol for wireless multiple access systems. *IEEE Trans. Commun.*, 58(8):2425–2435, 2010.

11. Zhiguo Ding, K.K. Leung, D.L. Goeckel, and D. Towsley. Cooperative transmission protocol for wireless broadcast channels. *IEEE Trans. Wireless Commun.*, 9(12):3701–3713, 2010.

12. Zhiguo Ding, T. Ratnarajah, and Colin Cowan. On the diversity-multiplexing tradeoff for wireless cooperative multiple access systems. *IEEE Trans. Signal Processing*, 55:4627–4638, 2007.

13. Zhiguo Ding, T. Ratnarajah, and Kin K. Leung. On the study of network coded AF transmission protocol for wireless multiple access channels. *IEEE Trans. Wireless Commun.*, 8:118–123, 2009.

14. A. Goldsmith. *Wireless communications*. Cambridge University Press, Cambridge, 2005.

15. Y. Gong, Z. Ding, T. Ratnarajah, and C. Cowan. Novel channel estimator for a superposition-based cooperative system. In *Proceedings of European Signal Processing Conference (EUSIPCO-06)*, September 2006, pp. 169–172.

16. Y. Gong, Z. Ding, T. Ratnarajah, and C. Cowan. Turbo channel estimation and equalization for a superposition-based cooperative system. *IET Proc. Commun.*, 3:1790–1799, 2009.

17. M. Joham, W. Utschick, and J. A. Nossek. Linear transmit processing in MIMO communications systems. *IEEE Trans. Signal Processing*, 53:2700–2712, 2005.

18. S. Lalos, A. A. Rontogiannis, and K. Berberidis. Frequency domain channel estimation for cooperative communication networks. *IEEE Trans. Signal Processing*, 58:3400–3405, 2003.

19. J. N. Laneman and G. W. Wornell. Distributed space-time-coded protocols for exploiting cooperative diversity in wireless networks. *IEEE Trans. Inform. Theory*, 49:2415–2425, 2003.

20. J. Nicholas Laneman, David N. C. Tse, and Gregory W. Wornell. Cooperative diversity in wireless networks: Efficient protocols and outage behavior. *IEEE Trans. Inform. Theory*, 50:3062–3080, 2004.

21. Erik G. Larsson and Branimir R. Vojcic. Cooperative transmit diversity based on superposition modulation. *IEEE Commun. Lett.*, 9:778–780, 2005.

22. K. Loa, C. Wu, S. Sheu, Y. Yuan, M. Chion, D. Huo, and L. Xu. IMT-Advanced relay standards. *IEEE Commun. Mag.*, 40–48, 2010.

23. R. U. Nabar, H. Bolcskei, and F. W. Kneubuhler. Fading relay channels: Performance limits and space-time signal design. *IEEE J. Select. Areas Commun.*, 22:1099–1109, 2004.

24. S. Peters and R. W. Heath Jr. The future of WiMAX: Multihop relaying with IEEE 802.16j. *IEEE Commun. Mag.*, 104–111, 2009.

25. M. Rupp, C. Mecklenbrauker, and G. Gritsch. High diversity with simple space time block codes and linear receivers. *Proc. GLOBECOM*, 2:302–306, 2003.

26. J. Sydir and R. Taori. An evolved cellular system architecture incorporating relay stations. *IEEE Commun. Mag.*, 115–121, 2009.

27. David N. C. Tse, Pramod Viswanath, and Lizhong Zheng. Diversity-multiplexing tradeoff in multiple-access channels. *IEEE Trans. Inform. Theory*, 50:1859–1874, 2004.

28. Y. Yang, H. Hu, J. Xu, and G. Mao. Relay technologies for WiMAX and LTE-Advanced mobile systems. *IEEE Commun. Mag.*, 100–105, 2009.

29. S. Zhang, S. Liew, and P. Lam. Physical layer network coding. In *Proceedings of 12th Annual International Conference on Mobile Computing and Networking (ACM MobiCom 2006)*, September 2006, pp. 63–68.

30. B. Zhao and M. C. Valenti. Practical relay networks: A generalization of hybrid-ARQ. *IEEE J. on Select. Areas Commun.*, 23:7–18, 2005.

31. Y. Zhao, R. Adve, and T. J. Lim. Outage probability at arbitrary SNR with cooperative diversity. *IEEE Commun. Lett.*, 9:700–702, 2005.

32. L. Zheng and D. N. C. Tse. Diversity and multiplexing: A fundamental tradeoff in multiple antenna channels. *IEEE Trans. Inform. Theory*, 49:1073–1096, 2003.

Chapter 5

Base Station Joint Transmission with Limited Backhaul Data Transfer for Multicell Networks

Jian Zhao

Institute for Infocomm Research, Singapore

Tony Q. S. Quek

Singapore University of Technology and Design, and Institute for Infocomm Research, Singapore

Zhongding Lei

Institute for Infocomm Research, Singapore

CONTENTS

5.1 Introduction

Coordinated Multipoint (CoMP) transmission has been incorporated into the latest 3GPP LTE-Advanced releases to combat the intercell interference (ICI) and improve the signal quality of cell edge users in multicell wireless systems [1]. The idea is to aggregate neighboring base stations (BSs) and form the *CoMP cooperating set*. Transmission strategies for the BSs within this set are coordinated to effectively mitigate the interference at the mobile stations (MSs). Coordinated scheduling/beamforming (CS/CB) and joint transmission (JT) are the two major approaches in CoMP [2]. In the CS/CB approach, the data stream for each MS is only available at and transmitted from a single BS; however, the user data for each MS can be distributed to multiple cooperating BSs in the JT approach and be simultaneously transmitted from them.

In terms of system throughput improvement, the JT approach is more attractive since it forms a virtual large antenna array for each MS. However, this is at the expense of high signaling overhead through the backhaul. This high signaling overhead is incurred by distributing each user's data to multiple BSs in the cooperating set. In a low-mobility environment, the amount of user data to be transported in the backhaul significantly outweighs that of the channel state information (CSI). In order to alleviate the backhaul requirement, it is desirable to distribute the user data only to a subset of the cooperating BSs. This is termed *BS clustering*.

Different beamformer design methods for the CS/CB approach under various performance criteria have been proposed in [3–6]. Reference [3] discusses coordinated beamformer design to minimize the maximum BS antenna power without user data exchange between BSs. A decentralized method to minimize the sum BS transmit power subject to given quality of service (QoS) requirements has been proposed in [4] for both the CS/CB and JT approaches. In [5], each BS maximizes the rate for intracell users and cancels intercell interference in a decentralized manner. Fast iterative algorithms to maximize the minimum user rate have been proposed in [6]. The optimal assignment of one BS for each user has been addressed in [7]. The JT approach has been considered in [8–11]. Reference [8] considers fully synchronous transmission from cooperating BSs. Forming clusters in multicell networks and performing intercluster coordination have been considered in [9] to mitigate interference to cluster edge users. In [10], a zero-forcing scheme has been proposed, which requires all the users' data to be distributed on all the BSs. BS clustering has been considered in [11] using heuristic methods. All those works assign the BSs to each MS in a predefined manner.

Weighted sum-rate maximization (WSRM) in multicell CoMP networks has been discussed in [12–14]. Reference [12] has designed distributed WSRM algorithms that iteratively update the transmit power and the user assignment at each BS using local information and limited BS data exchange. Algorithms with co-channel user selection and power allocation across tones have been presented in [13] for WSRM in multicell multicarrier networks. Both [12] and [13] assume no user data exchange between BSs. Centralized and distributed WSRM algorithms can be found in [14] for multicell networks, where all the users' data are available at all the transmitting BSs.

A number of recent research efforts have considered CoMP networks with finite-capacity backhaul [15–20]. The information-theoretic analysis for uplink CoMP cellular networks using a linear Wyner-type cellular model with no fading has been presented in [15], where cooperative decoding at the BSs is enabled by local and finite-capacity backhaul links. This work has been extended in [16] to consider Rayleigh fading channels, where a joint decoding scheme based on distributed Wyner-Ziv compression [21] is compared with a partial local decoding scheme at the BSs. An analysis of various uplink CoMP schemes under a constrained backhaul infrastructure and imperfect CSI has been performed in [17]. The optimal training length impacted by the finite backhaul capacity in an uplink CoMP network has been studied in [18] using large system analysis. A two-cell downlink CoMP transmission scenario with a finite-capacity backhaul has been considered in [19], where a rate-splitting approach has been proposed; i.e., the data symbols transmitted at each BS are composed of private messages that are only available to the BS itself and shared messages that are available at both BSs. A corresponding achievable rate region with respect to the backhaul constraints has been characterized. Those works [15–19] focus on the derivation and analysis of achievable rate regions for cellular CoMP networks under finite-capacity backhaul constraints.

In this chapter, we consider a multicell network with multiantenna BSs and address the problem of jointly designing the BS clusters and the transmit beamformers subject to per BS power constraints at BSs and given QoS requirements at each MS. Our aim is to minimize the backhaul user data transfer from the data center to the BSs for the JT approach. We show that such a problem can be cast into an ℓ_0-norm minimization problem that minimizes the nonzero elements of a data routing matrix. Unfortunately, this problem is NP-hard. Inspired by recent developments in compressive sensing [22,23], we propose two algorithms to obtain solutions that approach the optimal ones. The first algorithm is based on the ℓ_1-relaxation and the reweighted ℓ_1-norm minimization, which requires solving a series of convex ℓ_1-norm minimization problems using, e.g., the interior-point method. The second algorithm is based on iterative link removal. The idea is that we first solve the ℓ_2-norm relaxation of the joint clustering and beamforming problem and then iteratively remove the links that correspond to the smallest link transmit power. This algorithm utilizes the uplink-downlink duality [3, 6, 24] in multicell systems and employs a projected subgradient algorithm with fast iterative function evaluation to find the results in each intermediate step. Furthermore, this algorithm can also be implemented in a semidistributed manner under certain assumptions. Simulation results show that the proposed algorithms can significantly reduce the user data transfer in the backhaul; meanwhile, the per BS power constraints at the BSs and the QoS requirements at the MSs are satisfied by the obtained results. Therefore, this chapter provides a generic formulation of the problem on how to minimize the backhaul data exchange between BSs under given QoS constraints, as well as analytical near-optimum algorithms to tackle it.

The remainder of this chapter is organized as follows: In Section 5.2, we introduce the system model and formulate our joint clustering and beamforming problem into an ℓ_0-norm minimization problem. The reweighted ℓ_1-norm minimization-based method is presented in Section 5.3. The iterative link removal-based method is presented in Section 5.4. Section 5.5 provides the convergence behavior and the simulation results in a multicell scenario. Finally, conclusions are drawn in Section 5.6.

Notation: We use bold uppercase letters to denote matrices and bold lowercase letters to denote vectors. $\mathscr{CN}(0, \sigma^2)$ denotes a circularly symmetric complex normal zero mean random variable with variance σ^2. $(\cdot)^T$ and $(\cdot)^H$ stand for the transpose and conjugate transpose, respectively. $\text{vec}(\mathbf{A})$ is a column vector composed of the entries of $\mathbf{A}$ taken column-wise. $\mathbb{C}$, $\mathbb{R}$ and $\mathbb{Z}$ stand for the complex, real, and integer numbers, respectively. $\{\mathbf{w}_{ij}\}$ denotes the set made of $\mathbf{w}_{ij}, \forall i, j$. $\lfloor x \rfloor$ is the largest integer not greater than x. $[\mathbf{X}]_{ij}$ denotes the (i, j)th element of $\mathbf{X}$. $\text{diag}(\{\mathbf{A}_i\})$ denotes a block diagonal matrix formed by $\mathbf{A}_i, \forall i$.

5.2   System Model and Problem Formulation

5.2.1   System Model

We consider the CoMP downlink transmission scenario as shown in Figure 5.1. There are B BSs cooperating to transmit to K MSs using the same time and frequency channel. Here $x_i \in \mathbb{C}$ denotes the complex data symbol for the ith MS, where $i \in \{1, \cdots, K\}$, and $\mathrm{E}\left\{|x_i|^2\right\} = 1$. The data center has access to the data of all the MSs. The backhaul is the channel connecting the data center to the BSs, which can be optical fibers or out-of-band microwave links. The control information and user data at each BS are obtained directly from the data center via the backhaul [25, 26].

We assume that each BS is equipped with M transmit antennas and each MS has a single receive antenna. The transmit beamformer for the ith MS on the jth BS is denoted as $\mathbf{w}_{ij} \in \mathbb{C}^M$, where $i \in \{1, \cdots, K\}$ and $j \in \{1, \cdots, B\}$. The transmit power constraint at the jth BS is P_j, such that $\sum_{i=1}^{K} \mathbf{w}_{ij}^H \mathbf{w}_{ij} \leq P_j$. The channel from the jth BS to the ith MS is denoted as $\mathbf{h}_{ij}^H$, where $\mathbf{h}_{ij} \in \mathbb{C}^M$. The received signal y_i at the ith MS can be expressed as

$$y_i = \sum_{j=1}^{B} \mathbf{h}_{ij}^H \mathbf{w}_{ij} x_i + \sum_{k \neq i}^{K} \sum_{j=1}^{B} \mathbf{h}_{ij}^H \mathbf{w}_{kj} x_k + n_i \tag{5.1}$$

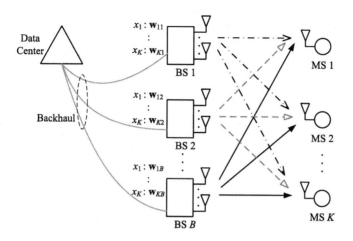

Figure 5.1: CoMP downlink transmission scenario.

The first term on the right-hand side of (5.1) represents the received useful signal, the second term represents the interference, and $n_i \sim \mathcal{CN}(0, \sigma^2)$ is the additive white Gaussian noise (AWGN) at the ith MS. The signal-to-interference-plus-noise ratio (SINR) at the ith MS can then be written as

$$\text{SINR}_i = \frac{\left| \sum_{j=1}^{B} \mathbf{h}_{ij}^H \mathbf{w}_{ij} \right|^2}{\sigma^2 + \sum_{k \neq i}^{K} \left| \sum_{j=1}^{B} \mathbf{h}_{ij}^H \mathbf{w}_{kj} \right|^2} \tag{5.2}$$

The SINR is an important QoS metric. SINR requirements have to be satisfied for the successful transmission of data. We denote the SINR requirement for the ith MS as γ_i. For ease of discussion, we only consider fixed and feasible target SINR requirements. In a commercial cellular network, the actual transmission rate will usually be adjusted with dynamic link adaptation.

To minimize the signaling overhead in the backhaul, we want to distribute the user data only to the minimum number of cooperating BSs, subject to the SINR constraint of each MS. We denote the set of BS indices that the user data x_i is distributed to as $\mathcal{B}_i$, where $\mathcal{B}_i \subseteq \{1, \cdots, B\}$. We assume the data rate to transfer the data x_i, $\forall i \in \{1, \cdots, K\}$, through the backhaul to each of the BS in $\mathcal{B}_i$ to be equal and denote it as R. Thus, the sum of user data rate R_{sum} transferred from the data center through the backhaul is given by

$$R_{\text{sum}} = R \cdot \sum_{i=1}^{K} |\mathcal{B}_i| \tag{5.3}$$

where $|\mathcal{B}_i|$ denotes the cardinality of the set $\mathcal{B}_i$. For the given SINR requirement, we obtain from (5.3) that minimizing R_{sum} is equivalent to minimizing $\sum_{i=1}^{K} |\mathcal{B}_i|$.

5.2.2 Problem Formulation

To capture the data transfer in the backhaul, we define a data routing matrix $\mathbf{P} \in \mathbb{R}_+^{K \times B}$, where the (i, j)th element of $\mathbf{P}$ is given by

$$[\mathbf{P}]_{ij} = \mathbf{w}_{ij}^H \mathbf{w}_{ij} \tag{5.4}$$

and it represents the power allocated at the jth BS for the data x_i. The matrix $\mathbf{P}$ also shows which MS's data are routed from the data center to each of the BSs. When $[\mathbf{P}]_{ij} = 0$, the data x_i is not routed to BS j. When $[\mathbf{P}]_{ij} > 0$, we say there exists a *link* between the jth BS and the ith MS, and the data center needs to distribute x_i to the BSs with $[\mathbf{P}]_{ij} > 0$ before the BSs transmit to the MSs. Therefore, $\mathscr{B}_i$ corresponds to the indices of the nonzero elements of the ith row in $\mathbf{P}$ and $\sum_{i=1}^{K} |\mathscr{B}_i|$ is equivalent to counting the number of nonzero elements in $\mathbf{P}$. Since the size of $\mathbf{P}$ is fixed and equals $K \times B$, minimizing $\sum_{i=1}^{K} |\mathscr{B}_i|$ is equivalent to maximizing the number of zero elements, i.e., the *sparsity*, in $\mathbf{P}$. We can mathematically formulate the R_{sum} minimization problem in (5.3) as

$$\begin{aligned}
\min_{\{\mathbf{w}_{ij}\}} \quad & \|\text{vec}(\mathbf{P})\|_0 \\
\text{subject to} \quad & \text{SINR}_i \geq \gamma_i, \ \forall i \in \{1, \ldots, K\} \\
& \sum_{i=1}^{K} \mathbf{w}_{ij}^H \mathbf{w}_{ij} \leq P_j, \ \forall j \in \{1, \ldots, B\}
\end{aligned} \tag{5.5}$$

where $\|\cdot\|_0$ represents the ℓ_0-norm, which denotes the number of nonzero elements of a vector. Note that the optimization problem (5.5) needs to jointly determine the BS subsets for the routing of each user's data, as well as to design the transmit beamformers at the BSs.

Remark 5.1
The problem (5.5) is a combinatorial optimization problem and is NP-hard.

The proof for a related problem where all the constraints are linear can be found in [27]. Using similar arguments as in [27,28], the problem (5.5) is NP-hard. A brute-force solution to a combinatorial optimization problem like (5.5) is by exhaustive search. That is, for each number $S \in \mathbb{Z}$, where $0 \leq S \leq KB$, we must check all the $\binom{KB}{S}$ possible assignments of zero elements in $\mathbf{P}$. For each assignment, we must search for the $\{\mathbf{w}_{ij}\}$ satisfying the constraints of (5.5). In the end, we pick out the maximum S with feasible $\{\mathbf{w}_{ij}\}$. However, the complexity of exhaustive search grows exponentially with the size of $\mathbf{P}$, which is not applicable in real-world applications.

The difficulty to solve (5.5) lies in the combinatorial nonconvex objective function involving the ℓ_0-norm. One approach to obtain approximate solutions for nonconvex optimization problems is convex relaxation [29], i.e., replacing the nonconvex objective function with its convex envelope. The convex envelope of the ℓ_0-norm is the ℓ_1-norm. Optimization based on the ℓ_1-norm has been shown to exactly recover sparse signals in ℓ_0-norm problems and closely approximate compressible signals with high probability using relatively few measurements if a certain property is satisfied [22,30]. This is known as *basis pursuit* [22,30] that conveniently reduces a sparse estimation problem to a linear program in signal detection, which is the essence of compressive sensing. However, simply substituting the objective function of (5.5) by $\|\text{vec}(\mathbf{P})\|_1$ does not produce sparse solutions in general, and such a relaxation is not useful at first glance. Therefore, we provide in the next section an equivalent formulation of (5.5), where sparse solutions can be obtained.

5.3   Reweighted ℓ_1-Norm Minimization-Based Method

We observe that $[\mathbf{P}]_{ij} > 0$ or $= 0$ is equivalent to $\|\mathbf{w}_{ij}\|_2 > 0$ or $= 0$, respectively. By introducing slack variables s_{ij}, the problem (5.5) can be equivalently reformulated as

$$\begin{aligned} \min_{\{\mathbf{w}_{ij}\},\{s_{ij}\}} \quad & \|\text{vec}(\mathbf{S})\|_0 \\ \text{subject to} \quad & \text{SINR}_i \geq \gamma_i, \ \forall i \in \{1,\ldots,K\} \\ & \|\mathbf{w}_{ij}\|_2 \leq s_{ij}, \ \forall i,j \\ & \sum_{i=1}^{K} s_{ij}^2 \leq P_j, \ \forall j \in \{1,\ldots,B\} \end{aligned} \tag{5.6}$$

where $[\mathbf{S}]_{ij} = s_{ij}$. The first and second constraints are second-order cone (SOC) constraints, the third constraint is a sum-of-squares constraint, and all those constraints are convex. The problem (5.6) is equivalent to (5.5). Applying the ℓ_1-norm relaxation to (5.6), we obtain

$$\begin{aligned} \min_{\{\mathbf{w}_{ij}\},\{s_{ij}\}} \quad & \sum_{i,j} s_{ij} \\ \text{subject to} \quad & \text{Constraints of (5.6)} \end{aligned} \tag{5.7}$$

Now the objective function becomes linear and (5.7) is a convex optimization problem. Furthermore, sparse solutions of $\{s_{ij}\}$ can usually be obtained from (5.7) [22, 23]. The following proposition states which s_{ij}'s solution should be zero.

Proposition 5.1
The solutions of $\{s_{ij}\}$ in (5.7) are obtained at $s_{ij}^* = 0$ if $\rho_{ij} < 1$, where ρ_{ij} is the dual variable associated with the constraint "$\|\mathbf{w}_{ij}\|_2 \leq s_{ij}$" in (5.7), $\forall i \in \{1,\cdots,K\}$ and $j \in \{1,\cdots,B\}$.

Proof: *We introduce additional slack variables $\mathbf{z}_i \in \mathbb{C}^K, \forall i \in \{1,\cdots,K\}$. The problem (5.7) can then be expressed in the following form:*

$$\begin{aligned} \min_{\{\mathbf{w}_{ij}\},\{s_{ij}\}} \quad & \sum_{i,j} s_{ij} \\ \textit{subject to} \quad & \|\mathbf{z}_i\|_2 \leq \tfrac{1}{\sqrt{\gamma_i}} \sum_{j=1}^{B} \mathbf{h}_{ij}^H \mathbf{w}_{ij}, \ \forall i \in \{1,\cdots,K\} \\ & \|\mathbf{w}_{ij}\|_2 \leq s_{ij}, \ \forall i,j \\ & \sum_{i=1}^{K} s_{ij}^2 \leq P_j, \ \forall j \in \{1,\ldots,B\} \\ & \left[\sum_{j=1}^{B} \mathbf{h}_{ij}^H \mathbf{w}_{1j}, \cdots, \sum_{j=1}^{B} \mathbf{h}_{ij}^H \mathbf{w}_{(i-1)j}, \sigma, \sum_{j=1}^{B} \mathbf{h}_{ij}^H \mathbf{w}_{(i+1)j}, \cdots, \sum_{j=1}^{B} \mathbf{h}_{ij}^H \mathbf{w}_{Kj}\right]^T \\ & \qquad\qquad\qquad\qquad = \mathbf{z}_i, \quad \forall i \in \{1,\cdots,K\} \end{aligned}$$

Note that $\mathbf{z}_i$ is complex, except that its ith element is real. By introducing $\lambda_i, \rho_{ij}, \mu_j \in \mathbb{R}_+$ for the first three constraints and $\mathbf{v}_i \in \mathbb{C}^K$ for the fourth constraint, respectively, the Lagrangian of the above problem can be expressed as

$$\mathscr{L}\left(\{\mathbf{w}_{ij}\}, \{s_{ij}\}\right)$$

$$= \sum_{i,j} s_{ij} + \sum_i \left(\lambda_i \|\mathbf{z}_i\|_2 - \frac{\lambda_i}{\sqrt{\gamma_i}} \sum_{j=1}^{B} \mathbf{h}_{ij}^H \mathbf{w}_{ij} \right) + \sum_{i,j} \left(\rho_{ij} \|\mathbf{w}_{ij}\|_2 - \rho_{ij} s_{ij} \right)$$

$$+ \sum_i \left(\sum_{k \neq i} \left(v_{ik}^* \sum_{j=1}^{B} \mathbf{h}_{ij}^H \mathbf{w}_{kj} \right) + v_{ii} \sigma - \mathbf{v}_i^H \mathbf{z}_i \right) + \sum_j \mu_j \left(\sum_i s_{ij}^2 - P_j \right) \quad (5.8)$$

$$= \sum_{i,j} \left(\mu_j s_{ij}^2 + (1 - \rho_{ij}) s_{ij} \right) + \sum_i \left(\lambda_i \|\mathbf{z}_i\|_2 - \mathbf{v}_i^H \cdot \mathbf{z}_i \right) + \sum_{i,j} \rho_{ij} \|\mathbf{w}_{ij}\|_2$$

$$+ \sum_i \left(-\frac{\lambda_i}{\sqrt{\gamma_i}} \sum_{j=1}^{B} \mathbf{h}_{ij}^H \mathbf{w}_{ij} + \sum_{k \neq i} v_{ki}^* \sum_{j=1}^{B} \mathbf{h}_{kj}^H \mathbf{w}_{ij} \right) + \sum_i v_{ii} \sigma - \sum_j \mu_j P_j. \quad (5.9)$$

Here v_{ki} denotes the ith element of $\mathbf{v}_k$. Furthermore, s_{ij} is chosen from the range $s_{ij} \geq 0$.

According to the complementary slackness conditions [29], the dual variable $\mu_j = 0$ when the power constraint at the jth BS is not satisfied with equality. In this case, we must have $\rho_{ij} \leq 1$; otherwise, the term $(1 - \rho_{ij})s_{ij}$ is unbounded below because $s_{ij} \geq 0$. Furthermore, if $\rho_{ij} < 1$, the Lagrangian $\mathscr{L}\left(\{\mathbf{w}_{ij}\}, \{s_{ij}\}\right)$ is minimized when $s_{ij} = 0$. When the power constraint at the jth BS is satisfied with equality, i.e., $\mu_j > 0$, we also have that $s_{ij} = 0$ minimizes the Lagrangian $\mathscr{L}\left(\{\mathbf{w}_{ij}\}, \{s_{ij}\}\right)$ if $\rho_{ij} < 1$. Therefore, the optimal point of s_{ij} is obtained at $s_{ij} = 0$ when $\rho_{ij} < 1$.

The solutions obtained from (5.7) are suboptimal solutions of (5.6) due to convex relaxation. Improved solutions can be obtained by applying a weighting factor on each s_{ij} in the objective function of (5.7). This is called *reweighted ℓ_1-norm minimization methods* [31], originally proposed to enhance the data acquisition in compressive sensing. The weighting factors should be chosen such that the solutions with large s_{ij} values are penalized. A good choice of the weighting factor β_{ij} in the $(n+1)$-th iteration is $\beta_{ij}^{(n+1)} = 1/\left(\delta + \left| s_{ij}^{(n)} \right| \right)$, where $s_{ij}^{(n)}$ is the solution of s_{ij} in the nth iteration, and δ is a small nonnegative parameter to ensure numerical stability. The reweighted ℓ_1-norm minimization-based method is summarized in Algorithm 2. Furthermore, we have the following proposition for its convergence.

Proposition 5.2

The sequence $\left\{ \mathbf{w}_{ij}^{(n)}, s_{ij}^{(n)} \right\}$ generated by Algorithm 5.2 is feasible for (5.6). Furthermore, the sequence $\sum_{i,j} \log(s_{ij}^{(n)} + \delta)$ converges to a local minimum of $f(\{s_{ij}\}) = \sum_{i,j} \log(s_{ij} + \delta)$, which is a surrogate function for the objective function $\|\text{vec}(\mathbf{S})\|_0$ in (5.6).

Algorithm 5.2 Reweighted ℓ_1-norm minimization-based method

1: Set the iteration counter $n = 0$ and $\beta_{ij}^{(0)} = 1, \forall i \in \{1, \cdots, K\}$ and $j \in \{1, \cdots, B\}$.
2: Solve the weighted ℓ_1-norm minimization problem

$$\left\{ \mathbf{w}_{ij}^{(n+1)}, s_{ij}^{(n+1)} \right\} = \arg\min \sum_{i,j} \beta_{ij}^{(n)} s_{ij}$$

subject to: Constraints of (5.6).

3: Update the weighting factor:

$$\beta_{ij}^{(n+1)} = \frac{1}{\delta + \left| s_{ij}^{(n)} \right|}, \quad \forall i, j$$

4: Calculate the relative error:

$$r = \frac{\left\| \text{vec} \left(\mathbf{S}^{(n)} - \mathbf{S}^{(n-1)} \right) \right\|_2}{\left\| \text{vec} \left(\mathbf{S}^{(n)} \right) \right\|_2}$$

Terminate if $r < \varepsilon$ or $n \geq N_{\max}$, where ε is a predefined threshold and $N_{\max}$ is the specified maximum number of iterations. Otherwise, set $n = n + 1$ and go to step 2.

Proof: *In Algorithm 5.2, $\left\{ \mathbf{w}_{ij}^{(n)}, s_{ij}^{(n)} \right\}$ are obtained by solving the weighted ℓ_1-norm minimization problems that are subject to the same constraints as (5.6). Therefore, $\left\{ \mathbf{w}_{ij}^{(n)}, s_{ij}^{(n)} \right\}$ are also feasible for (5.6). For a given nonnegative real vector $\mathbf{x} \in \mathbb{R}_+^M$, its ℓ_0-norm can be approximated as [32]*

$$\|\mathbf{x}\|_0 = \lim_{\delta \to 0} \sum_{i=1}^M \frac{\log(1 + |x_i| \delta^{-1})}{\log(1 + \delta^{-1})} \tag{5.10}$$

If we choose δ to be sufficiently small and replace the objective function of (5.6) with the RHS of (5.10), we obtain the following problem by neglecting some constant numbers:

$$\min_{\{\mathbf{w}_{ij}\}, \{s_{ij}\}} \quad \Sigma_{i,j} \log(s_{ij} + \delta)$$

$$\text{subject to} \quad \text{Constraints of (5.6)} \tag{5.11}$$

The cost function is a log sum function. It is concave and below its tangent. Specifically, let $\left\{ \mathbf{w}_{ij}^{(n)}, s_{ij}^{(n)} \right\}$ denote the solution in the nth iteration. Due to the concavity of the log function and $s_{ij} \geq 0$ for feasible solutions, the first-order approximation of $\log(s_{ij} + \delta)$ yields

$$\log(s_{ij} + \delta) \leq \log(s_{ij}^{(n)} + \delta) + \frac{1}{s_{ij}^{(n)} + \delta} \left(s_{ij} - s_{ij}^{(n)} \right) \tag{5.12}$$

The RHS of (5.12) majorizes the log *function, and can be used as a surrogate function for the original cost function in (5.11). Specifically, replacing the RHS of (5.12) into (5.11), the solution we get in the $(n+1)$th iteration is*

$$\left\{ \mathbf{w}_{ij}^{(n+1)}, s_{ij}^{(n+1)} \right\} = \arg\min \sum_{i,j} \frac{s_{ij}}{s_{ij}^{(n)} + \delta} \quad \text{subject to: Constraints of (5.6)} \quad (5.13)$$

We can see that the weighting factor of s_{ij} in the objective function of (5.13) is the same as to the $\beta_{ij}^{(n)}$ in Algorithm 5.2. By iteratively minimizing the surrogate function, we can obtain better approximate solutions of (5.11), which is along the idea of the majorization-minimization method [49].

For $s_{ij}^{(n)}$ and $s_{ij}^{(n+1)}$, we have

$$\sum_{i,j} \log(s_{ij}^{(n+1)} + \delta) \leq \log(s_{ij}^{(n)} + \delta) + \frac{s_{ij}^{(n+1)} - s_{ij}^{(n)}}{s_{ij}^{(n)} + \delta}$$

$$\leq \log(s_{ij}^{(n)} + \delta) + \frac{s_{ij}^{(n)} - s_{ij}^{(n)}}{s_{ij}^{(n)} + \delta} = \sum_{i,j} \log(s_{ij}^{(n)} + \delta) \quad (5.14)$$

The two inequalities follow from (5.12) and (5.13), respectively. So $\sum_{i,j} \log(s_{ij}^{(n)} + \delta)$ is a monotonically decreasing sequence, lower bounded by $KB\log(\delta)$. Therefore, the optimal objective value of (5.11) must converge.

The reweighted ℓ_1-norm minimization-based method achieves better sparsity results than (5.7) because although the ℓ_1-norm is the best *convex* relaxation of the ℓ_0-norm, some *concave* functions can offer tighter approximations [22, 29]. In the proof, the choice of the weighting factors is explained using the log-sum surrogate function of the ℓ_0-norm [31,32]. Because the log-sum penalty function is concave, the global minimum solution may not always be attained. It is important to choose a suitable starting point, and the solution of the unweighted ℓ_1-norm minimization (5.7) offers a good initialization for the reweighted ℓ_1 minimization algorithm [31, 33]. Furthermore, we observe in the simulation that the resulting sparsity of $\{s_{ij}\}$ using Algorithm 5.2 typically converges within 10 iterations. The biggest improvement in sparsity comes from the first several iterations, and the iterations afterwards offer marginal improvement in sparsity. Therefore, Algorithm 5.2 can be terminated after the first several iterations, depending on the requirement in the sparsity of the final output.

5.4 Iterative Link Removal-Based Method

The reweighted ℓ_1-norm minimization-based method needs to solve a series of convex optimization problems using, e.g., interior-point methods, which can be computationally intensive. In this section, we propose another approach. The basic idea is that we first solve the ℓ_2-norm relaxation of the joint clustering and beamforming problem (5.6), and then iteratively remove the links that correspond to the smallest link transmit power. Methods employing fixed-point iterations [6, 34] are utilized to find the solution in each intermediate step, which enjoy low complexity and are usually faster than the interior-point methods. We show that the iterative link removal-based method can also be implemented in a semidistributed manner with a lightweight central controller under certain assumptions. The computational complexity analysis will be presented at the end of this section.

5.4.1 ℓ_2-Norm Relaxation of Joint Clustering and Beamforming

We first consider the ℓ_2-norm relaxation of the joint clustering and beamforming problem (5.6) subject to SINR and per-BS power constraints, which can be expressed as

$$
\begin{aligned}
&\min_{\{\mathbf{w}_{ij}\}} && \textstyle\sum_{i,j} \mathbf{w}_{ij}^H \mathbf{w}_{ij} \\
&\text{subject to} && SINR_i \geq \gamma_i,\ \forall i \in \{1,\ldots,K\} \\
& && \textstyle\sum_i \mathbf{w}_{ij}^H \mathbf{w}_{ij} \leq P_j,\ \forall j \in \{1,\ldots,B\}
\end{aligned}
\tag{5.15}
$$

Note that the power constraints P_j, $\forall j \in \{1,\ldots,B\}$, in (5.15) are constants. The problem (5.15) is different from the min-max BS antenna power problem considered in [3, 24], where the per-BS transmit power constraints are variables to be optimized. The problem (5.15) is a SOC programming problem that is convex, and it can be solved using the interior-point methods. However, that may be computationally intensive. Therefore, we exploit the special structure of (5.15) and propose a low-complexity approach that employs the fixed-point methods. Our proposed approach is based on the following theorem, which generalizes the uplink-downlink duality to multicell environments under *explicitly given* per BS power constraints.

Theorem 5.1

The dual problem of the downlink transmit optimization problem (5.15) is

$$
\begin{aligned}
&\max_{\{v_j\}} \min_{\{q_i\},\{\hat{\mathbf{w}}_{ij}\}} && \textstyle\sum_{i=1}^K q_i \sigma^2 - \sum_{j=1}^B v_j P_j \\
&\quad\text{subject to} && SINR_i^U \geq \gamma_i,\ q_i \geq 0,\quad \forall i \in \{1,\ldots,K\} \\
& && v_j \geq 0,\quad \forall j \in \{1,\ldots,B\}
\end{aligned}
\tag{5.16}
$$

where

$$
SINR_i^U = \frac{q_i \left| \sum_{j=1}^B \hat{\mathbf{w}}_{ij}^H \mathbf{h}_{ij} \right|^2}{\sum_{k\neq i}^K q_k \left| \sum_{j=1}^B \hat{\mathbf{w}}_{ij}^H \mathbf{h}_{kj} \right|^2 + \sum_{j=1}^B (1+v_j) \left\| \hat{\mathbf{w}}_{ij} \right\|_2^2}
\tag{5.17}
$$

This is an uplink multiuser receive beamforming problem in the dual-uplink channel with the same SINR constraints and uncertain noise variances, where v_j is the dual variable associated with the jth BS power constraint in (5.15), and $1+v_j$ can be interpreted as the dual-uplink noise variance at the jth BS in (5.16). Here q_i is the dual *variable associated with the ith SINR constraint in (5.15), and it can be interpreted as the dual-uplink transmit power of the ith MS. Here $\hat{\mathbf{w}}_{ij}$ represents the uplink receive beamformer at the jth BS for the data of MS i. Note that the optimal $\hat{\mathbf{w}}_{ij}$ is a scaled version of the optimal $\mathbf{w}_{ij}$ for the downlink.*

Proof: *Concatenate the channel vectors and beamformers to the ith MS as vectors $\mathbf{h}_i = \left[\mathbf{h}_{i1}^H, \mathbf{h}_{i2}^H, \cdots, \mathbf{h}_{iB}^H\right]^H$ and $\mathbf{w}_i = \left[\mathbf{w}_{i1}^H, \mathbf{w}_{i2}^H, \cdots, \mathbf{w}_{iB}^H\right]^H$, where $i \in \{1, \ldots, K\}$. We introduce dual variables $\{q_i\}$ and $\{v_j\}$ for the SINR constraints and per-BS power constraints of (5.15), respectively. The Lagrangian for (5.15) is obtained as follows:*

$$\mathscr{L}(\{\mathbf{w}_i\}, \{q_i\}, \{v_j\})$$

$$= \sum_i \mathbf{w}_i^H \mathbf{w}_i + \sum_i q_i \left(\sigma^2 + \sum_{k \neq i}^K \left|\mathbf{h}_i^H \mathbf{w}_k\right|^2 - \frac{\left|\mathbf{h}_i^H \mathbf{w}_i\right|^2}{\gamma_i}\right) + \sum_j v_j \left(\sum_i \mathbf{w}_i^H \mathbf{Q}_j \mathbf{w}_i - P_j\right)$$

$$= \sum_i \mathbf{w}_i^H \left(\mathbf{I} + \sum_{j=1}^B v_j \mathbf{Q}_j\right) \mathbf{w}_i + \sum_i \left(\sum_{k \neq i}^K q_k \left|\mathbf{w}_i^H \mathbf{h}_k\right|^2 - \frac{q_i \left|\mathbf{w}_i^H \mathbf{h}_i\right|^2}{\gamma_i}\right) + \sum_i q_i \sigma^2 - \sum_j v_j P_j$$

where $\mathbf{Q}_j$ is defined as an all-zero $MB \times MB$ matrix, except that the $(jM - M + 1)$-th to the jM-th diagonal elements are all 1's. The Lagrange dual function is obtained as

$$g(\{q_i\}, \{v_j\})$$

$$= \inf_{\{\mathbf{w}_i\}} \mathscr{L}(\{\mathbf{w}_i\}, \{q_i\}, \{v_j\})$$

$$= \begin{cases} \sum_{i=1}^K q_i \sigma^2 - \sum_{j=1}^B v_j P_j, & \text{if } \left(\mathbf{I} + \sum_{j=1}^B v_j \mathbf{Q}_j\right) + \sum_{k \neq i}^K q_k \mathbf{h}_k \mathbf{h}_k^H \geq \frac{q_i \mathbf{h}_i \mathbf{h}_i^H}{\gamma_i}, \forall i \\ -\infty, & \text{otherwise} \end{cases}$$

Therefore, the Lagrange dual problem can then be expressed as

$$\begin{aligned} \max_{\{v_j\}, \{q_i\}} \quad & \sum_{i=1}^K q_i \sigma^2 - \sum_{j=1}^B v_j P_j \\ \text{subject to} \quad & \left(\mathbf{I} + \sum_{j=1}^B v_j \mathbf{Q}_j\right) + \sum_{k \neq i}^K q_k \mathbf{h}_k \mathbf{h}_k^H \geq \frac{q_i \mathbf{h}_i \mathbf{h}_i^H}{\gamma_i}, \quad \forall i \\ & v_j \geq 0, \ q_i \geq 0, \quad \forall i \in \{1, \ldots, K\}, j \in \{1, \ldots, B\} \end{aligned} \qquad (5.18)$$

Introducing the dual-uplink receive beamformers $\{\hat{\mathbf{w}}_i\}$, the first constraint can be rewritten as [24]

$$SINR_i^U = \frac{q_i \left|\hat{\mathbf{w}}_i \mathbf{h}_i\right|^2}{\sum_{k \neq i}^K q_k \left|\hat{\mathbf{w}}_i \mathbf{h}_k\right|^2 + \hat{\mathbf{w}}_i^H \left(\mathbf{I} + \sum_{j=1}^B v_j \mathbf{Q}_j\right) \hat{\mathbf{w}}_i} \leq \gamma_i \qquad (5.19)$$

Furthermore, $\{\hat{\mathbf{w}}_i\}$ is proportional to $\{\mathbf{w}_i\}$, $\forall i \in \{1, \cdots, K\}$, at the optimum solution following discussions similar to those in [24]. Since the SINR constraint must be attained with equality for the optimal solution, the inequality in (5.19) can be reversed, and the maximization with respect to $\{q_i, \hat{\mathbf{w}}_i\}$ can be replaced by the minimization, which gives the Lagrange dual problem of (5.16).

The problem (5.16) consists of an outer maximization with regard to the dual-uplink noise variances $\{v_j\}$ and an inner minimization with regard to the uplink power $\{q_i\}$ and beamformers $\{\hat{\mathbf{w}}_{ij}\}$. The optimal objective value of the inner minimization is a function of $\{v_j\}$, i.e.,

$$\varphi(\{v_j\}) = \min_{\{q_i\},\{\hat{\mathbf{w}}_{ij}\}} \quad \Sigma_i q_i \sigma^2 - \Sigma_j v_j P_j \tag{5.20}$$
$$\text{subject to} \quad \text{SINR}_i^U \geq \gamma_i, \\ q_i \geq 0, \quad \forall i \in \{1,\ldots,K\}$$

For given $\{v_j\}$, the term $\Sigma_j v_j P_j$ is a fixed constant. Evaluating $\varphi(\{v_j\})$ for given $\{v_j\}$ is a uplink joint sum power minimization and beamforming problem with fixed BS noise variances [35]. In the following, we apply a fixed-point algorithm to solve (5.20).

Denoting all the BSs as a BS group, we define $\mathbf{h}_i = \left[\mathbf{h}_{i1}^H, \mathbf{h}_{i2}^H, \cdots, \mathbf{h}_{iB}^H\right]^H$, where $i \in \{1,\ldots,K\}$. That is the dual-uplink channel vector from the ith MS to the BS group. Concatenating the BS downlink transmit beamformers together, we obtain $\mathbf{w}_i = \left[\mathbf{w}_{i1}^H, \mathbf{w}_{i2}^H, \cdots, \mathbf{w}_{iB}^H\right]^H$, which is the BS group downlink beamformer to the ith MS. Likewise, we concatenate the BS receive beamformers together as $\hat{\mathbf{w}}_i = \left[\hat{\mathbf{w}}_{i1}^H, \hat{\mathbf{w}}_{i2}^H, \cdots, \hat{\mathbf{w}}_{iB}^H\right]^H$. Define $\mathbf{Q}_j$ as an all-zero $MB \times MB$ matrix, except that the $(jM - M + 1)$th to the jMth diagonal elements are all 1s, and we can express SINR_i^U as

$$\text{SINR}_i^U = \frac{q_i \hat{\mathbf{w}}_i^H \mathbf{h}_i \mathbf{h}_i^H \hat{\mathbf{w}}_i}{\hat{\mathbf{w}}_i^H \Sigma_i \hat{\mathbf{w}}_i} \tag{5.21}$$

where

$$\Sigma_i = \sum_{k \neq i}^{K} q_k \mathbf{h}_k \mathbf{h}_k^H + \left(\mathbf{I} + \sum_{j=1}^{B} v_j \mathbf{Q}_j\right) \tag{5.22}$$

Since the optimal receive beamformer $\hat{\mathbf{w}}_i$ that maximizes SINR_i^U is the minimum mean square error (MMSE) receiver [34, 36], we have

$$\hat{\mathbf{w}}_i = \Sigma_i^{-1} \mathbf{h}_i \tag{5.23}$$

and the corresponding SINR is then given by

$$\text{SINR}_i^{U,\,\max} = q_i \mathbf{h}_i^H \Sigma_i^{-1} \mathbf{h}_i \tag{5.24}$$

At the optimal solution, it is necessary that the SINR constraints in (5.20) are satisfied with equality [37], i.e., $\text{SINR}_i^{U,\,\max} = \gamma_i$. Therefore, the optimal uplink transmit power satisfies the following set of equations:

$$q_i^* = \frac{\gamma_i}{\mathbf{h}_i^H \Sigma_i^{-1} \mathbf{h}_i} \triangleq f_i(\mathbf{q}^*), \quad \forall i \in \{1, \cdots, K\} \tag{5.25}$$

where $\mathbf{q}^* = [q_1^*, q_2^*, \cdots, q_K^*]^T$. Using fixed-point methods, we can determine the optimal uplink transmit power as follows:

$$q_i^{(l+1)} = f_i(\mathbf{q}^{(l)}), \quad \forall i \in \{1, \cdots, K\} \tag{5.26}$$

Starting from some initial value of $\mathbf{q}^{(0)}$, the fixed-point equation (5.26) is guaranteed to converge to a unique fixed point because $f_i(\mathbf{q})$ is a standard function [34, 38]. Note the $f_i(\mathbf{q})$ in (5.25) is different from that of [3, Eq. (29)] in the expressions of the covariance matrix and the coefficient, and a faster convergence speed by our formulation is observed in simulations.

After the optimal $\mathbf{q}^*$ is obtained, the optimal receive beamformer $\hat{\mathbf{w}}_i$ can be calculated according to (5.23). The optimal downlink beamformer $\mathbf{w}_i$ is a scaled version of $\hat{\mathbf{w}}_i$, i.e., $\hat{\mathbf{w}}_i = \sqrt{\rho_i}\mathbf{w}_i, \forall i \in \{1, \cdots, K\}$. The scaling factor ρ_i can be calculated using the fact that the downlink SINR constraints in (5.15) must be satisfied with equality for the optimal solution, i.e.,

$$\frac{1}{\gamma_i}\rho_i\left|\mathbf{h}_i^H\hat{\mathbf{w}}_i\right|^2 = \sigma^2 + \sum_{k \neq i}^{K}\rho_k\left|\mathbf{h}_i^H\hat{\mathbf{w}}_k\right|^2, \quad \forall i \in \{1, \ldots, K\} \tag{5.27}$$

Denoting $\boldsymbol{\rho} = [\rho_1, \cdots, \rho_K]^T$, we can rewrite (5.27) as $\mathbf{F}\boldsymbol{\rho} = \mathbf{1}\sigma^2$, where $\mathbf{1}$ is a $K \times 1$ all-one vector and the elements of $\mathbf{F}$ are defined as

$$[\mathbf{F}]_{ik} = \begin{cases} \frac{1}{\gamma_i}\left|\mathbf{h}_i^H\hat{\mathbf{w}}_i\right|^2, & i = k; \\ -\left|\mathbf{h}_i^H\hat{\mathbf{w}}_k\right|^2, & i \neq k \end{cases} \tag{5.28}$$

So the scaling vector $\boldsymbol{\rho}$ can be obtained as

$$\boldsymbol{\rho} = \mathbf{F}^{-1}\mathbf{1}\sigma^2 \tag{5.29}$$

The downlink transmit power at the jth BS for the ith MS data stream is

$$p_{ij} = \left\|\mathbf{w}_{ij}\right\|_2^2 = \rho_i\left\|\hat{\mathbf{w}}_{ij}\right\|_2^2 \tag{5.30}$$

where $\hat{\mathbf{w}}_{ij}$ is the $M \times 1$ vector formed by the $(jM - M + 1)$th to the jMth elements of the $\hat{\mathbf{w}}_i$ obtained from (5.23) and the downlink transmit power at the jth BS is

$$p_j = \sum_{i=1}^{K} p_{ij} \tag{5.31}$$

Hence, this solves the inner minimization for given $\{v_i\}$.

The outer maximization can be solved using the projected subgradient method [39–41]. Projected subgradient methods following, e.g., the square summable but not summable step size rule have been proven to converge to the optimal value [39–41]. Similar to [24], it can be shown that $\varphi(\{v_j\})$ is concave in $\{v_j\}$, and a subgradient of $\varphi(\{v_j\})$ with respect to v_j is $p_j - P_j$. With a step size t_n, the dual-uplink noise variance $\{v_j\}$ can be updated as

$$v_j^{(n+1)} = \max\left\{v_j^{(n)} + t_n(p_j - P_j), 0\right\}, \quad \forall j = \{1, \cdots, B\} \tag{5.32}$$

Since (5.15) is convex and Slater's condition holds, the respective primal and dual objective values in (5.15) and (5.16) must be equal at the global optimum point. A stopping criterion for updating (5.32) can be that $\left|\varphi\left(\{v_j^{(n)}\}\right) - \sum_j p_j^{(n)}\right| \leq \varepsilon$. If the stopping criterion cannot be satisfied after a predefined maximum number N_m iterations, the iterations can be aborted and this indicates that (5.15) is infeasible.

5.4.2   Iterative Link Removal

The solutions obtained from (5.15) are not sparse in general. Therefore, we propose to trim the links, which correspond to the entries of $\mathbf{P}$, to obtain sparse solutions. We call this process *iterative link removal*. After the problem (5.15) is solved, the data streams can be sorted according to the p_{ij} in (5.30) from the lowest to the highest. We use $(i,j)^*$ to denote the position of the (i,j)th data stream in the sorted list, and we propose to gradually "remove" the data streams according to their p_{ij} from the lowest to the highest. That is, for each $0 \leq S \leq KB, S \in \mathbb{Z}$, we set the corresponding transmission power for data streams $(i,j)^* \leq S$ to zero, i.e., $\mathbf{w}_{ij} = \mathbf{0}, \forall (i,j)^* \leq S$. Here $S = 0$ corresponds to full BS cooperation, i.e., no removal. Meanwhile, we solve the following problem to see if the SINR and power constraints are still feasible.

$$
\begin{aligned}
\min_{\{\mathbf{w}_{ij}\}} \quad & \textstyle\sum_{i,j} \mathbf{w}_{ij}^H \mathbf{w}_{ij} \\
\text{subject to} \quad & \mathsf{SINR}_i \geq \gamma_i, \ \forall i \in \{1,\dots,K\} \\
& \textstyle\sum_i \mathbf{w}_{ij}^H \mathbf{w}_{ij} \leq P_j, \ \forall j \in \{1,\dots,B\} \\
& \mathbf{w}_{ij} = \mathbf{0}, \ \forall (i,j)^* \leq S, \text{ except that } (i,j) \text{ is the last link for MS } i
\end{aligned}
$$

$$(5.33)$$

This problem is the same as to (5.15) except for the last constraint. Here, S corresponds to the sparsity of the routing matrix $\mathbf{P}$ in (5.4). By increasing S, we can find the largest S that satisfies the constraints of (5.33). For each user index i, we must keep at least one $\mathbf{w}_{ij} \neq \mathbf{0}$ within the removal process; i.e., we always skip the last transmission link to MS i, which corresponds to the link from its serving cell. The problem (5.33) is also a SOC programming problem and its feasibility can be determined using, e.g., the interior-point methods. However, by exploiting the similarity between (5.33) and (5.15), we can also determine the feasibility of (5.33) with slight modification of the algorithm in Section 5.4.1.

As mentioned previously, we denote the set of BS indices that have the data stream for MS i as $\mathscr{B}_i$ for the given value of S. Following the same discussion as in Section 5.4.1, the dual-uplink SINR for the data stream of MS i can be expressed as

$$
\mathsf{SINR}_i^{\mathsf{U}} = \frac{q_i \left| \sum_{j \in \mathscr{B}_i} \hat{\mathbf{w}}_{ij}^H \mathbf{h}_{ij} \right|^2}{\sum_{k \neq i}^K q_k \left| \sum_{j \in \mathscr{B}_i} \hat{\mathbf{w}}_{ij}^H \mathbf{h}_{kj} \right|^2 + \sum_{j \in \mathscr{B}_i} (1+v_j) \left\| \hat{\mathbf{w}}_{ij} \right\|_2^2}
\tag{5.34}
$$

Note the dual-uplink SINR for MS i only depends on the dual-uplink channel to those BS $j \in \mathscr{B}_i$. For $m = 1, \cdots, K$, let $\tilde{\mathbf{h}}_{mj} = \mathbf{h}_{mj}, \forall j \in \mathscr{B}_i$, and $\tilde{\mathbf{h}}_{mj} = \mathbf{0}, \forall j \notin \mathscr{B}_i$. We introduce the equivalent channel $\tilde{\mathbf{h}}_m = [\tilde{\mathbf{h}}_{m1}^H, \cdots, \tilde{\mathbf{h}}_{mB}^H]^H$, and use $\tilde{\mathbf{h}}_m$ instead of $\mathbf{h}_m, \forall m \in \{1, \cdots, K\}$, when calculating $\hat{\mathbf{w}}_i$ and q_i according to (5.22), (5.23), and (5.25). It only affects the inner minimization, and the outer maximization process remains unchanged. Whether (5.33) is feasible is indicated by whether the algorithm converges, e.g., whether the stopping criterion can be satisfied within N_m iterations.

The iterative link removal-based method is summarized in Algorithm 5.3. The bisection method [41, 42] is applied to speed up the link removal process.

Algorithm 5.3 Iterative link removal-based method

Start: Solve the problem (5.15) using the fixed-point iteration (5.26) and subgradient update (5.32);

if not convergent **then**

    The problem (5.15) is infeasible. Exit to link adaptation;

else

    Sort the data streams according to their power p_{ij};

    Initialize: $S_{min} = 0, S_{max} = KB$, and $S = \lfloor (S_{min} + S_{max})/2 \rfloor$.

    while $S_{max} - S_{min} \geq 2$ **do**

        Solve problem (5.33) for S using the fixed-point iteration (5.26) and subgradient update (5.32) with equivalent channel $\tilde{\mathbf{h}}_m, \forall m = 1, \cdots, K$, for calculating $q_i, \hat{\mathbf{w}}_i$ and $\mathbf{w}_i$;

        if infeasible **then**

            $S_{max} = S, S = \lfloor (S_{min} + S_{max})/2 \rfloor$;

        else

            $S_{min} = S, S = \lfloor (S_{min} + S_{max})/2 \rfloor$;

        end if

    end while

    return S and its corresponding $\{\mathbf{w}_{ij}\}$.

end if

5.4.3 Implementation Issues

Similar to the reweighted ℓ_1-norm minimization-based method, the iterative link removal-based method can be implemented at a centralized controller that performs all the computations. However, if the cross-covariance terms between the channel vectors of different BSs are small for each user compared to the autocovariance terms, and thus can be neglected, i.e., $\mathbf{h}_{kj}\mathbf{h}_{kj'}^H \approx \mathbf{0}, \forall j \neq j'$, which is usually valid for multi-cell networks due to the distance between different BSs, the iterative link removal-based method can also be implemented in a semidistributed manner in time-division duplexing (TDD) mode, where the downlink channels can be estimated from the uplink transmission.

Consider the calculation of $\mathbf{\Sigma}_i$ in (5.22). Since the cross-covariance terms between the channels of different BSs are small for each user k, we have $\mathbf{h}_k\mathbf{h}_k^H \approx$ diag$\left(\left\{\mathbf{h}_{kj}\mathbf{h}_{kj}^H\right\}\right)$ and $\mathbf{\Sigma}_i \approx$ diag$\left(\left\{\sum_{k \neq i}^K q_k\mathbf{h}_{kj}\mathbf{h}_{kj}^H + (1+v_j)\mathbf{I}\right\}\right) \triangleq$ diag$\left(\left\{\mathbf{\Sigma}_{ij}\right\}\right)$, where the set of matrices inside diag$(\cdot)$ are indexed by $j = 1, \cdots, B$. That means the matrix $\mathbf{\Sigma}_i$ becomes *block-diagonal*. Moreover, the jth component block, i.e., the covariance matrix $\mathbf{\Sigma}_{ij}$, only involves the channel vector $\mathbf{h}_{kj}$, and thus it can be estimated locally at the jth BS. Therefore, the calculation of $\hat{\mathbf{w}}_i$ in (5.23) can be divided into the calculation of $\hat{\mathbf{w}}_{ij}$ at the jth BS, where

$$\hat{\mathbf{w}}_{ij} \approx \mathbf{\Sigma}_{ij}^{-1}\mathbf{h}_{ij}, \quad \forall j = 1, \cdots, B \tag{5.35}$$

Furthermore, in the fixed-point iteration (5.25), we have

$$\frac{\gamma_i}{\mathbf{h}_i^H \boldsymbol{\Sigma}_i^{-1} \mathbf{h}_i} \approx \frac{\gamma_i}{\sum_{j=1}^B \mathbf{h}_{ij}^H \boldsymbol{\Sigma}_{ij}^{-1} \mathbf{h}_{ij}} \tag{5.36}$$

where the term $\mathbf{h}_{ij}^H \boldsymbol{\Sigma}_{ij}^{-1} \mathbf{h}_{ij}$ can be computed locally at the jth BS.

As a result, the semidistributed implementation of the iterative link removal-based method in TDD mode can be described as follows: (1) Each MS-i chooses an initial uplink transmit power $q_i^{(0)}$. (2) Each BS-j estimates the channels $\mathbf{h}_{ij}$ from all the MSs and calculates $\boldsymbol{\Sigma}_{ij}$ locally. (3) Each BS-j calculates and sends the value of $\mathbf{h}_{ij}^H \boldsymbol{\Sigma}_{ij}^{-1} \mathbf{h}_{ij}$ to the central controller. (4) The controller sums up the values of $\mathbf{h}_{ij}^H \boldsymbol{\Sigma}_{ij}^{-1} \mathbf{h}_{ij}$ from all the BSs and updates q_i according to (5.26) and (5.36) using fixed-point iterations. (5) With the $\hat{\mathbf{w}}_{ij}$ computed locally at the BS-j using (5.35) corresponding to the final q_i, the scaling factor ρ_i can be obtained using the well-known distributed downlink power control algorithm [43] with the fixed effective channel for achieving the given SINR requirements γ_i. Therefore, the downlink transmit power p_{ij} in (5.30) can also be obtained at the BS-j. (6) The update of v_j in (5.32) can be done on a per BS basis. (7) The BSs send the calculated p_{ij} to the controller and the controller sorts them to decide the removal sequence.

In the semidistributed implementation process, only scalars, e.g., the value of $\mathbf{h}_{ij}^H \boldsymbol{\Sigma}_{ij}^{-1} \mathbf{h}_{ij}$, need to be communicated from the BSs to the central controller. Full knowledge of the channel vectors is not required therein. Moreover, the central controller only performs simple calculations like summation, scalar division, and sorting. Most of the computation loads, e.g., the matrix inversion, are distributed among the BSs. Therefore, the central controller can be "lightweight." The drawback of the semidistributed implementation may be the delay induced in the iterations, which also depends on the convergence speed of the subgradient updates.

5.4.4 Complexity Analysis

Algorithm 5.2 needs to solve a series of SOC programs. For each of the SOC programs, the computational complexity using the interior-point algorithm is $O((KB)^{3.5} M^3 \log \varepsilon_1^{-1})$, where ε_1 is the required accuracy of the duality gap termination [44]. In the worst case, $N_{\max}$ iterations must be performed. Therefore, the worst case complexity for Algorithm 5.2 is $O(N_{\max}(KB)^{3.5} M^3 \log \varepsilon_1^{-1})$. In Algorithm 5.3, the most computation-intensive step in the inner-minimization is the matrix inversion in (5.23) for all the K users. Each matrix inversion has the complexity of $O((BM)^3)$ [45]. Using the convergence results of the fixed-point algorithm in [6, 34, 46] and considering N_m subgradient updates required in the worst case, the complexity of solving (5.16) is obtained as $O(N_m K(BM)^3 \log c^{-1})$, where c^{-1} is a constant that determines the convergence rate of the fixed-point iterations. By applying the bisection method [29] to remove the links, the complexity of Algorithm 5.3 is therefore obtained as $O(N_m K(BM)^3 \log(KB) \log \varepsilon^{-1})$. Compared to the complexity of $O(2^{KB}(KBM)^3 \sqrt{K+B} \log \varepsilon_1^{-1})$ required by the exhaustive search method [47], both of the two algorithms save a lot of computational complexity.

5.5 Simulation Results

5.5.1 Convergence Behavior

We consider the case that the number of BSs and the number of MSs are $B = K = 5$. Each BS has $M = 3$ transmit antennas, and each MS has a single receive antenna. The SINR requirement for each MS receiver is set to $\gamma_i = 6.02\text{dB}$, $\forall i \in \{1, \cdots, K\}$. The noise variances at the MSs are normalized to 1, i.e., $\sigma^2 = 1$. Furthermore, the per BS transmit power constraint P_j is set according to $P_j/\sigma^2 = 10\text{dB}$, $\forall j \in \{1, \cdots, B\}$. A randomly generated channel is used to simulate the convergence results.

The convergence behavior for the reweighted ℓ_1-norm minimization-based method, i.e., Algorithm 5.2, is shown in Figure 5.2, where the parameter δ is set to be $10^{-2}, 10^{-3}$, and 10^{-4}, respectively. The number of active links corresponds to $\sum_{i=1}^{K} |\mathcal{B}_i|$ in (5.3). Figure 5.2 shows that it decreases with the number of iterations. The first several iterations lead to the biggest improvement. As iterations go on, there is no further improvement after the tenth iteration. In the first iteration, i.e., in the unweighted case (5.7), 20 links are required to be active. That means $KB - 20 = 5$ entries in the data routing matrix **P** are zeros, which shows a sparse solution is obtained. In the final iteration, only six links are required to be active. Compared to full cooperation, which requires all the links to be active, we save $1 - 6/25 = 76\%$ of the user data transfer in the backhaul. In general, a smaller δ leads to better conver-

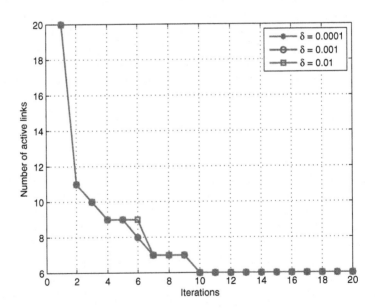

Figure 5.2: Convergence behavior for the reweighted ℓ_1-norm minimization-based method, $B = K = 5$. $M = 3$ antennas at each BS, the SINR requirement $\gamma_i = 6.02$ dB, $\forall i \in \{1, \cdots, K\}$.

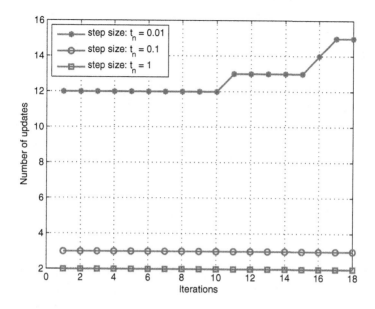

Figure 5.3: Number of subgradient updates for dual noise $\{v_j\}$ in each iteration for the iterative link removal-based method, $B = K = 5$. $M = 3$ antennas at each BS, the SINR requirement $\gamma_i = 6.02$ dB, $\forall i \in \{1, \cdots, K\}$. The stopping term $\varepsilon = 10^{-4}$.

gence behavior. This is because a small δ approximates the ℓ_0-norm better. However, the convergence for the different δ values does not differ much in Figure 5.2. When $\delta = 10^{-2}$, only one more active link is required at the sixth iteration compared to the cases of $\delta = 10^{-3}$ and 10^{-4}.

When the iterative link removal-based method, i.e., Algorithm 5.3, is applied to the considered system, the stopping term is chosen to be $\varepsilon = 10^{-4}$ and the maximum number of subgradient updates is set to be $N_m = 80$. The bisection method is not applied here to better show the convergence results. Only one active link is removed here in each iteration. We choose the step size t_n in (5.32) to be $t_n = 0.01, 0.1$, and 1, respectively. Starting from $KB = 25$ links, 17 links are removed and only 8 links remain active in the end for the considered choices of the step size t_n. This result is slightly worse than the final result obtained by Algorithm 5.2. Figure 5.3 shows that the required number of subgradient updates is sensitive to the choice of the step size t_n. A smaller step size will lead to a larger number of updates. If the step size is chosen to be 0.01, more than 12 updates are required for each iteration. However, when we choose the step size to be 0.1 or 1, only two to three iterations are required per iteration. On the other hand, the step size t_n cannot be chosen to be too big because a large t_n will affect the accuracy of the results. We find it advisable to set $0.1 < t_n < 1$ in our simulations to achieve a good balance in the accuracy and program speed.

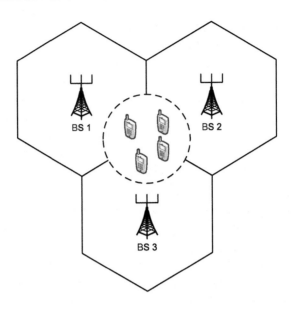

Figure 5.4: Cellular network simulation scenario.

5.5.2 Cellular Network Simulations

We consider a three-cell network as shown in Figure 5.4, where each BS is equipped with multiple antennas. The intercell distance between neighboring BSs is 0.5 km. The transmission power constraint at each BS is 30 dBm. The transmit antenna gain at each BS is 5 dB. The pathloss model from the BS to the MS is

$$L(\text{dB}) = 128 + 37.6 \cdot \log_{10} D \tag{5.37}$$

where D is in the unit of km. The lognormal shadowing parameter is 10 dB. The available transmission spectrum is 10 MHz, and the noise figure at each MS is 10 dB. The power of the noise plus out-of-cooperating cell interference is set to be −83.98 dBm. The users are randomly and uniformly distributed within a disk of 100 m radius at the center of the triangle formed by the three BSs. We perform 50 channel realizations for each user location, and 20 different user locations are chosen in each simulation. Only those channel realizations that can support the required SINR requirement are admitted. Table 5.1 shows a summary of the above simulation parameters. Five methods are compared: the reweighted ℓ_1-norm minimization-based method (Algo1), the iterative link removal-based method (Algo2), the channel strength-based clustering (CSBC) method [48], full cooperation, and exhaustive search (ExSearch). CSBC is similar to Algo2, but it removes the links according to the channel strength. Full cooperation distributes all the user data to all the BSs. ExSearch has been discussed in Section 5.2.2, which produces the optimum solution in minimizing the backhaul data transfer but requires enormous computations. We set the maximum number of iterations N_{max} for Algorithm 5.2 to be 20 and the relative error threshold ε to be

Table 5.1   Summary of Simulation Parameters

Number of cells	3
Intercell distance	0.5 km
BS transmission power constraint	30 dBm
BS transmit antenna gain	5 dB
Lognormal shadowing	10 dB
Transmission spectrum	10 MHz
Noise variance plus out-of-cooperating cell interference	-83.98 dBm
Disk radius for MS location	100 m
Channel realizations for each MS location	50
MS location	20
MS location distribution	Uniform
Number of antennas per MS	1

10^{-4}. The step size t_n for the dual-uplink noise variance update in Algorithm 5.3 is set to be 0.5 and the stopping term $\varepsilon = 10^{-4}$. The maximum number of subgradient updates is set to be $N_m = 80$. The SINR requirements for different MSs are the same, i.e., $\gamma_i = \gamma, \forall i \in \{1, \cdots, K\}$.

Figures 5.5 and 5.6 show the average number of active links and their corresponding average sum transmission power with respect to the number of antennas M at each BS. The number of MSs that are simultaneously served by those BSs is $K = 6$. The SINR requirement for each MS is set to be $\gamma = 10.3$ dB, which is the reference SINR for 16-QAM transmission with spectral efficiency of 2.41 bps/Hz [26]. The average number of active links is shown in Figure 5.5. Full cooperation requires all the $KB = 18$ links to be active irrespective of M. This induces a large backhaul signaling overhead. For the other four methods, the number of required active links decreases dramatically with the increase of M. When $M = 2$, Algo1 and Algo2 achieve almost the same results for the average active links, which is about 16.5. CSBC and ExSearch require one more and one less link on average, respectively. When $M = 3$, the required links for the four methods decrease to only 11, 12, 13, and 10.5, respectively. When $M \geq 6$, the number of required links for the three methods reduces to about 6. That is the minimum required number of active links, which equals K. Compared to full cooperation, the proposed methods save about $1 - 6/18 = 66.7\%$ of the backhaul user data transfer. This is due to the fact that more antennas at the BSs provides more spatial degrees of freedom, and proper choice of beamformers reduces the interuser interference. The same trend can be observed for the average sum transmit power in Figure 5.6. The sum transmit power decreases with the increase of M. Full cooperation serves as a lower bound for the sum transmit power because the other three methods perform partial cooperation. To our surprise, Algo1 requires both fewer active links in Figure 5.5 and lower sum transmit power in Figure 5.6 than Algo2 and CSBC when $3 \leq M \leq 8$. This shows that the sparsity pattern in the data routing matrix also affects the required sum transmit power.

Figures 5.7 and 5.8 show the results when the number of antennas per BS is fixed to $M = 4$ and the SINR requirement is set to $\gamma = 10.3$ dB. This system can serve roughly up to $BM = 12$ single-antenna MSs. Due to the huge complexity of

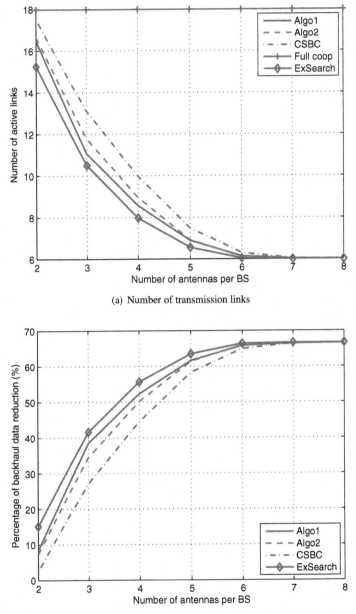

(a) Number of transmission links

(b) Percentage of backhaul data transfer reduction compared to full cooperation

Figure 5.5: Simulation results for $B = 3$ BSs, $K = 6$ MSs, SINR requirement $\gamma = 10.3$ dB, where $\gamma_i = \gamma$, $\forall i \in \{1, \cdots, K\}$. Number of antennas M at each BS changes.

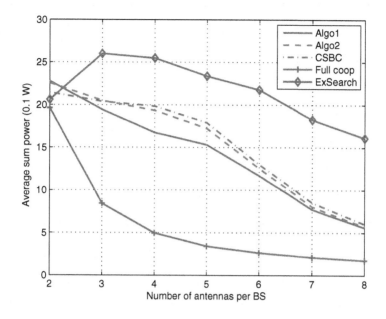

Figure 5.6: Sum transmission power (in 0.1 W unit) for $B = 3$ BSs, $K = 6$ MSs, SINR requirement $\gamma = 10.3$ dB, where $\gamma_i = \gamma$, $\forall i \in \{1, \cdots, K\}$. Number of antennas M at each BS changes.

ExSearch, the simulation for ExSearch is only performed up to $K = 6$ users. Full cooperation shows a linear curve according to BK in Figure 5.7. When $K \leq 4$, it is far below the system's serving capacity and the average number of active links approximately equals K for Algo1, Algo2, CSBC, and ExSearch. The number of required active links increases with K. When $K = 8$, the first three methods require about 15, 16, and 17 active links, respectively. In this case, Algo1 saves about $1 - 15/24 = 37.5\%$ of the backhaul user data transfer compared to full cooperation. When $K = 10$, the system capacity is almost reached. Algo1 and Algo2 require about 26 active links, and CSBC requires 28 active links. About 13.3% of the backhaul user data transfer can be saved by Algo1. In Figure 5.8, the curves of average sum power for Algo2 and CSBC are quite similar. Algo1 shows lower sum power, especially when $4 \leq K \leq 8$.

The simulation results for a system of $K = 4$ users and each BS with $M = 4$ antennas are shown in Figure 5.9 and 5.10. The SINR requirement is shown as the x-axis. When the SINR requirement $\gamma \leq 8.1$ dB, the number of active links for Algo1, Algo2, CSBC, and ExSearch in Figure 5.9 is about 4, which equals K. As γ increases, the performance gaps between different methods become obvious. The gap between Algo2 and CSBC remains about 0.6 when 14.1dB $\leq \gamma \leq 18.7$ dB. Algo1 remains very close to ExSearch in the whole SINR range. When $\gamma = 18.7$ dB, Algo1 requires about 8 active links, which is about 1.4 below that of Algo2. In this case, Algo1 saves about 33.3% of the backhaul user data transfer compared to full cooperation. The

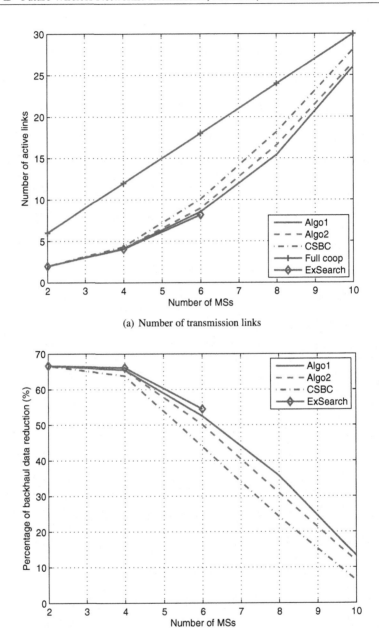

(a) Number of transmission links

(b) Percentage of backhaul data transfer reduction compared to full cooperation

Figure 5.7: Simulation results for $B = 3$ **BSs,** $M = 4$ **antennas per BS, SINR requirement** $\gamma = 10.3$ **dB. Number of users** K **changes.**

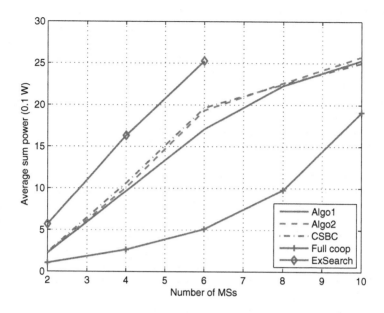

Figure 5.8: Sum transmission power (in 0.1 W unit) for $B = 3$ BSs, $M = 4$ antennas per BS, SINR requirement $\gamma = 10.3$ dB. Number of users K changes.

required sum transmit power for Algo1 is slightly below those of Algo2 and CSBC when $\gamma \leq 11.7$ dB in Figure 5.10. However, when $\gamma \geq 14.1$ dB, the sum transmit power for Algo1 surpasses those of the other two methods. The difference is about 0.3 W when $\gamma = 18.7$ dB.

5.6 Conclusion

We considered the problem of minimizing the user data transfer in the backhaul subject to SINR and per-BS power constraints in the CoMP JT downlink transmission scenario. We showed that it can be formulated into an ℓ_0-norm minimization problem. However, such a problem is combinatorial and NP-hard. Furthermore, solving such a problem needs to jointly determine the BS choices for user data and the BS beamformers. Inspired by recent results in compressive sensing, we proposed two algorithms to obtain near-optimum solutions using convex relaxation. This first algorithm applies ℓ_1-relaxation to an equivalent form of the original problem. We showed the conditions under which the obtained solution is sparse. The obtained solution is further improved using the reweighted ℓ_1-norm minimization method, which requires solving series of convex ℓ_1-norm minimization problems. Then, we proposed an iterative link removal-based algorithm employing the projected subgradient updates and iterative fixed-point methods. We also showed that this algorithm can be imple-

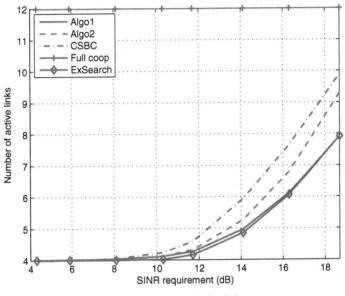

(a) Number of transmission links

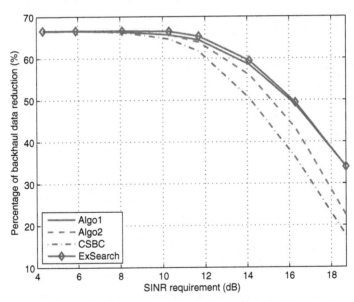

(b) Percentage of backhaul data transfer reduction compared to full cooperation

Figure 5.9: Simulation results for $B = 3$ BSs, $K = 4$ MSs, $M = 4$ antennas per BS. SINR requirement γ at each MS changes.

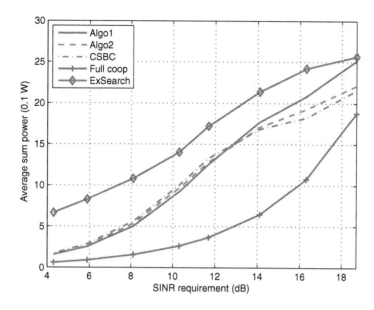

Figure 5.10: Sum transmission power (in 0.1 W unit) for $B = 3$ BSs, $K = 4$ MSs, $M = 4$ antennas per BS. SINR requirement γ at each MS changes.

mented in a semidistributed manner under certain assumptions. The computational complexity of the proposed algorithms is analyzed. Simulations showed that both of the proposed methods can significantly reduce the user data distribution to BSs in the backhaul. In terms of backhaul user data transfer, the results of the reweighted ℓ_1-norm minimization-based method are very near the optimum solutions in our simulations, which save the backhaul user data transfer by 13% to 67%. The iterative link removal-based method is also an attractive choice due to its low implementation complexity and good performance, which is only slightly worse than the reweighted ℓ_1-norm minimization-based method.

References

1. 3GPP TR 36.819, Coordinated multi-point operation for LTE physical layer aspects, v11.2.0, September 2013.

2. R. Irmer, H. Droste, P. Marsch, M. Grieger, G. Fettweis, S. Brueck, H. Mayer, L. Thiele, and V. Jungnickel, Coordinated multipoint: Concepts, performance, and field trial results, *IEEE Commun. Mag.*, vol. 49, no. 2, pp. 102–111, 2011.

3. H. Dahrouj and W. Yu, Coordinated beamforming for the multicell multi-antenna wireless system, *IEEE Trans. Wireless Commun.*, vol. 9, no. 5, pp. 1748–1759, 2010.

4. A. Tölli, H. Pennanen, and P. Komulainen, Decentralized minimum power multi-cell beamforming with limited backhaul signaling, *IEEE Trans. Wireless Commun.*, vol. 10, no. 2, pp. 570–580, 2011.

5. W. Ho, T. Q. S. Quek, S. Sun, and R. Heath Jr., Decentralized precoding for multicell MIMO downlink, *IEEE Trans. Wireless Commun.*, vol. 10, no. 6, pp. 1798–1809, 2011.

6. D. W. H. Cai, T. Q. S. Quek, C. W. Tan, and S. H. Low, Max-min SINR coordinated multipoint downlink transmission—Duality and algorithms, *IEEE Trans. Signal Process.*, vol. 60, no. 10, pp. 5384–5395, 2012.

7. M. Bengtsson, Jointly optimal downlink beamforming and base station assignment, in *Proceedings of the IEEE International Conference Acoustics, Speech, Signal Processing*, Salt Lake City, UT, May 2001, pp. 2961–2964.

8. S. Shamai and B. Zaidel, Enhancing the cellular downlink capacity via co-processing at the transmitting end, in *Proceedings of 53th IEEE Vehicle Technology Conference*, Rhodes, Greece, May 2001, pp. 1745–1749.

9. J. Zhang, R. Chen, J. Andrews, A. Ghosh, and R. Heath, Networked MIMO with clustered linear precoding, *IEEE Trans. Wireless Commun.*, vol. 8, no. 4, pp. 1910–1921, 2009.

10. R. Zhang, Cooperative multi-cell block diagonalization with per-base-station power constraints, *IEEE J. Select. Areas Commun.*, vol. 28, no. 9, pp. 1435–1445, 2010.

11. C. T. K. Ng and H. Huang, Linear precoding in cooperative MIMO cellular networks with limited coordination clusters, *IEEE J. Select Areas Commun.*, vol. 28, no. 9, pp. 1446–1454, 2010.

12. H. Zhang, L. Venturino, N. Prasad, P. Li, S. Rangarajan, and X. Wang, Weighted sum-rate maximization in multi-cell networks via coordinated scheduling and discrete power control, *IEEE J. Select Areas Commun.*, vol. 29, no. 6, pp. 1214–1224, 2011.

13. L. Venturino, N. Prasad, and X. Wang, Coordinated scheduling and power allocation in downlink multicell OFDMA networks, *IEEE Trans. Veh. Technol.*, vol. 58, no. 6, pp. 2835–2848, 2009.

14. T. Bogale and L. Vandendorpe, Weighted sum rate optimization for downlink multiuser MIMO coordinated base station systems: Centralized and distributed algorithms, *IEEE Trans. Signal Process.*, vol. 60, no. 4, pp. 1876–1889, 2012.

15. O. Simeone, O. Somekh, H. Poor, and S. Shamai, Local base station cooperation via finite-capacity links for the uplink of linear cellular networks, *IEEE Trans. Inf. Theory*, vol. 55, no. 1, pp. 190–204, 2009.

16. A. Sanderovich, O. Somekh, H. Poor, and S. Shamai, Uplink macro diversity of limited backhaul cellular network, *IEEE Trans. Inf. Theory*, vol. 55, no. 8, pp. 3457–3478, 2009.

17. P. Marsch and G. Fettweis, Uplink CoMP under a constrained backhaul and imperfect channel knowledge, *IEEE Trans. Wireless Commun.*, vol. 10, no. 6, pp. 1730–1742, 2011.

18. J. Hoydis, M. Kobayashi, and M. Debbah, Optimal channel training in uplink network MIMO systems, *IEEE Trans. Signal Process.*, vol. 59, no. 6, pp. 2824–2833, 2011.

19. R. Zakhour and D. Gesbert, Optimized data sharing in multicell MIMO with finite backhaul capacity, *IEEE Trans. Signal Process.*, vol. 59, no. 12, pp. 6102–6111, 2011.

20. J. Zhao, T. Q. S. Quek, and Z. Lei, Coordinated multipoint transmission with limited backhaul data transfer, *IEEE Trans. Wireless Commun.*, vol. 12, pp. 2762–2775, 2013.

21. M. Gastpar, The Wyner-Ziv problem with multiple sources, *IEEE Trans. Inf. Theory*, vol. 50, no. 11, pp. 2762–2768, 2004.

22. D. Donoho, Compressed sensing, *IEEE Trans. Inf. Theory*, vol. 52, no. 4, pp. 1289–1306, 2006.

23. M. Yuan and Y. Lin, Model selection and estimation in regression with grouped variables, *J. R. Stat. Soc. B*, vol. 68, no. 1, pp. 49–67, 2006.

24. W. Yu and T. Lan, Transmitter optimization for the multi-antenna downlink with per-antenna power constraints, *IEEE Trans. Signal Process.*, vol. 55, no. 6, pp. 2646–2660, 2007.

25. E. Dahlman, S. Parkvall, J. Sköld, and P. Beming, *3G evolution: HSPA and LTE for mobile broadband*, Academic Press, Burlington, MA, 2008.

26. A. Ghosh and R. Ratasuk, *Essentials of LTE and LTE-A*, Cambridge University Press, Cambridge, 2011.

27. B. Natarajan, Sparse approximate solutions to linear systems, *SIAM J. Comput.*, vol. 24, no. 2, pp. 227–234, 1995.

28. E. Matskani, N. Sidiropoulos, Z. Luo, and L. Tassiulas, Convex approximation techniques for joint multiuser downlink beamforming and admission control, *IEEE Trans. Wireless Commun.*, vol. 7, no. 7, pp. 2682–2693, 2008.

29. S. Boyd and L. Vandenberghe, *Convex optimization*, Cambridge University Press, Cambridge, 2004.

30. E. Candès, J. Romberg, and T. Tao, Robust uncertainty principles: Exact signal reconstruction from highly incomplete frequency information, *IEEE Trans. Inf. Theory*, vol. 52, no. 2, pp. 489–509, 2006.

31. E. Candès, M. Wakin, and S. Boyd, Enhancing sparsity by reweighted ℓ_1 minimization, *J. Fourier Anal. Appl.*, vol. 14, no. 5, pp. 877–905, 2008.

32. B. Sriperumbudur, D. Torres, and G. Lanckriet, A majorization-minimization approach to the sparse generalized eigenvalue problem, *Machine Learning*, vol. 85, no. 1, pp. 3–39, 2011.

33. M. Fazel, H. Hindi, and S. Boyd, Log-det heuristic for matrix rank minimization with applications to Hankel and Euclidean distance matrices, in *Proceedings of American Control Conference*, Denver, CO, June 2003, pp. 2156–2162.

34. D. W. H. Cai, T. Q. S. Quek, and C. W. Tan, A unified analysis of max-min weighted SINR for MIMO downlink system, *IEEE Trans. Signal Process.*, vol. 59, no. 8, pp. 3850–3862, 2011.

35. F. Rashid-Farrokhi, L. Tassiulas, and K. Liu, Joint optimal power control and beamforming in wireless networks using antenna arrays, *IEEE Trans. Commun.*, vol. 46, no. 10, pp. 1313–1324, 1998.

36. H. Van Trees, *Optimum array processing*, John Wiley & Sons, New York, 2002.

37. A. Wiesel, Y. Eldar, and S. Shamai, Linear precoding via conic optimization for fixed MIMO receivers, *IEEE Trans. Signal Process.*, vol. 54, no. 1, pp. 161–176, 2006.

38. R. Yates, A framework for uplink power control in cellular radio systems, *IEEE J. Select Areas Commun.*, vol. 13, no. 7, pp. 1341–1347, 1995.

39. N. Shor, K. Kiwiel, and A. Ruszczynski, *Minimization methods for non-differentiable functions*, Springer-Verlag, Berlin, 1985.

40. J. Goffin, On convergence rates of subgradient optimization methods, *Math. Program.*, vol. 13, no. 1, pp. 329–347, 1977.

41. D. P. Bertsekas, A. Nedic, and A. E. Ozdaglar, *Convex analysis and optimization*, Athena Scientific, Belmont, MA, 2003.

42. D. P. Bertsekas, *Nonlinear programming*, 2nd ed., Athena Scientific, Belmont, MA, 1999.

43. G. Foschini and Z. Miljanic, A simple distributed autonomous power control algorithm and its convergence, *IEEE Trans. Veh. Technol.*, vol. 42, no. 4, pp. 641–646, 1993.

44. Y. Ye, *Interior point algorithms: Theory and analysis*, John Wiley & Sons, New York, 1997.

45. R. A. Horn and C. R. Johnson, *Matrix analysis*, Cambridge University Press, Cambridge, 1986.

46. H. R. Feyzmahdavian, M. Johansson, and T. Charalambous, Contractive interference functions and rates of convergence of distributed power control laws, *IEEE Trans. Wireless Commun.*, vol. 11, no. 12, pp. 4494–4502, 2012.

47. C. H. Papadimitriou and K. Steiglitz, *Combinatorial optimization: Algorithms and complexity*, Dover Publications, Mineola, New York, 1998.

48. J. Zhao and Z. Lei, Clustering methods for base station cooperation, in *Proceedings of IEEE Wireless Communications Networking Conference*, Paris, 2012, pp. 1–6.

49. D. Hunter and K. Lange, A tutorial on MM algorithms, *American Statistician*, vol. 58, no. 1, pp. 30–37, 2004.

Chapter 6

Media Access Control Protocol in Wireless Networks

Yun Li and Bin Cao

Chongqing University of Posts and Telecommunications of China

CONTENTS

The media access control (MAC) lies in the data link (DL) layer. It is an important function that controls different nodes to access the same shared communication media. In this chapter, we first introduce the MAC protocols for different wireless network standards including IEEE 802.11, IEEE 802.15.4, IEEE 802.15.6, IEEE 802.16, and IEEE 802.22. Then we highlight some new MAC mechanisms, such as energy-efficient MAC, cognitive MAC, and relay MAC.

6.1   IEEE 802.11 MAC Protocol

IEEE 802.11 is a typical standard for WLAN whose indoor coverage is tens of meters and outdoor coverage is hundreds of meters [1]. In 1997, IEEE published the first IEEE 802.11 standard, which aims to define the physical layer (PHY) and medium access control (MAC) layer. In the first IEEE 802.11 standard, the PHY layer can provide three solutions: a frequency-hopping spread spectrum (FHSS), a direct-sequence spread spectrum (DSSS) PHY in the unlicensed 2.4 GHz band, and an infrared PHY at 316–353 THz [2]. The first version can only provide a basic data rate of 1 Mb/s with an optional 2 Mb/s mode, and no corresponding commercial infrared implementations to support.

After that, IEEE puts forward a series of high-speed WLAN standards, such as IEEE 802.11a/b/f/n, and the recent 802.11ac and 802.11ad. The main difference among these WLAN standard lies in PHY layer, but they have a similar MAC protocol. In the following, we demonstrate the basic function of the IEEE 802.11 MAC protocol.

6.1.1   MAC Architecture

In order to schedule the data transmission, two well-known functions are adopted in 802.11, called distributed coordination function (DCF) and point coordination function (PCF), respectively [3]. The MAC architecture is shown in Figure 6.1. Comparing with DCF, PCF can guarantee the quality of service (QoS), such as the constraints of transmission delay and jitter, which is an optional scheme with a central coordination entity in the original 802.11 standard. The PCF is an optional capability that is connection oriented and provides contention-free (CF) frame transfer. The PCF relies on the point coordinator (PC) to perform polling, enabling polled stations to transmit without contending for the channel. The function of the PC is performed by the AP within each basic service set (BSS). Stations within the BSS that are able to operate function in the CF period (CFP); they are known as CF-aware stations. And implementer decides the method of polling tables to be maintained and the polling sequence.

6.1.2   MAC Frame

IEEE 802.11 supports three different types of frames [3]: management, control, and data. The management frames are used for station association and disassociation

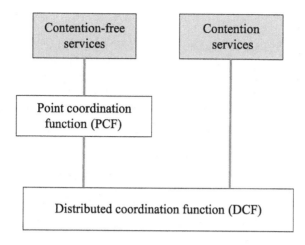

Figure 6.1: IEEE 802.11 MAC mechanism architecture.

with the AP, timing and synchronization, authentication and deauthentication. Control frames can be used for handshaking and positive acknowledgments during the CP, and to end the CFP during the CP and CFP. Data frames can be used for the transmission of data, during the CFP. Data frames can be combined with polling and acknowledgments. The standard of IEEE 802.11 frame format is illustrated in Figure 6.2. Note that the frame body (MSDU) is a variable-length field consisting of the data payload and seven octets for encryption/decryption if the optional Wired Equivalent Privacy (WEP) protocol is implemented. To identify a station, the IEEE standard 48-bit MAC addressing can be used. The two duration octets indicate the duration time (in microseconds) of allocated channel for a successful transmission. The type bits identify the frame as control, management, or data. The subtype bits further identify the type of frame (e.g., Clear to Send control frame). And error detection can be performed by a 32-bit cyclic redundancy check (CRC).

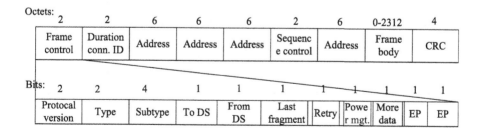

Figure 6.2: Standard IEEE 802.11 frame format. (From Crow, Brian P., et al. IEEE 802.11 wireless local area networks. *IEEE Communications Magazine*, **35.9 (1997): 116–126.)**

6.1.3 DCF

Different from PCF, DCF is mandatory in the IEEE 802.11 MAC protocol. The specification requires that all stations must support the DCF. The DCF operates solely in the ad hoc network, and either operates solely or coexists with the PCF in an infrastructure network. As shown in Figure 6.1, the DCF sits directly on top of the physical layer and supports contention services. For fairness, all stations with an MSDU queued for transmission must contend for access to the channel separately.

Comparing with carrier sense multiple access with collision detection (CSMA/CD) in IEEE 802.3, IEEE 802.11 DCF uses carrier sense multiple access with collision avoidance (CSMA/CA). The reason is that CSMA/CD requires the senders to detect the collision while transmitting, but the mobile equipment is half duplex, which cannot detect and transmit at the same time. In IEEE 802.11 DCF, to detect the activity in the channel, carrier sensing is performed at the PHY layer with physical carrier sensing and the MAC layer with virtual carrier sensing, respectively.

In DCF, CSMA/CA is used for data transmission, and carrier sense should be performed at the MAC layer to detect whether the channel is busy or idle. When a user sends data, other users should set a network allocation vector (NAV) accordingly. NAV is a timer, that indicates the channel is busy until it becomes zero. There is a defined duration field in the MAC data frame for NAV information; the sender should write its transmission time into the duration field and broadcast it. The other users receiving the data from the sender can obtain the NAV and thus keep silence to avoid collision.

When a user has a new packet to send, it can access the channel to transmit when the channel is idle and the duration time is equal to a distributed interframe space (DIFS). Otherwise, the user should keep silence and monitor the channel until it becomes idle for a DIFS. And then, to minimize the probability of collision with packets being transmitted by other users, the users should perform a random backoff before data transmission; the interval is randomly set from 0 to contention window (CW) 1, and this is the collision avoidance feature of the protocol.

DCF adopts a binary exponential backoff scheme to set the backoff time before transmission. At the first transmission attempt, CW is equal to the predefined minimum value, and it is doubled after each unsuccessful transmission until it reaches the maximum value. Figure 6.3 illustrates the DCF backoff scheme as follows.

When the channel is sensed idle, the backoff time is decreased sequentially, and the corresponding user can transmit data if the backoff time becomes zero. Before that, if the channel is sensed busy again, the backoff time should be paused until the channel becomes idle for more than a DIFS.

Figure 6.4 uses an example to illustrate the above operation. There are two stations A and B want to transmit data on the same wireless channel. When the channel is idle after a DIFS, station A performs a backoff scheme with the value of backoff time 8, and the channel is sensed busy when it decreases to 5. The reason is that station B transmits data at this time. In this case, station A should freeze the backoff time as 5 until the channel is sensed idle for a DIFS.

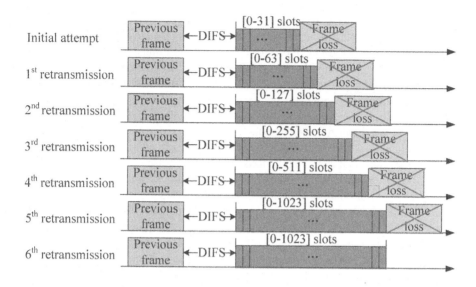

Figure 6.3: Binary exponential backoff scheme. (From Cao, Bin, Gang Feng, Yun Li, and Chonggang Wang. Cooperative media access control with optimal relay selection in error-prone Wireless networks, *IEEE Transaction on Vehicular Technology***, 63.1 (2014): 252–265.)**

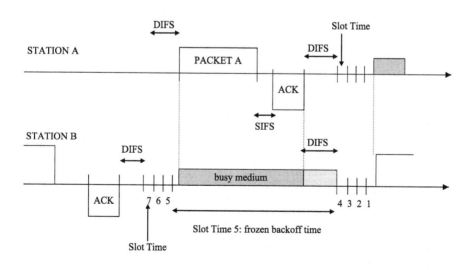

Figure 6.4: Example of basic access mechanism. (From Bianchi, Giuseppe. Performance analysis of the IEEE 802.11 distributed coordination function. *IEEE Journal on Selected Areas in Communications***, 18.3 (2000): 535–547.)**

Since the CSMA/CA does not rely on the capability of the stations to detect a collision by hearing their own transmission, an ACK is transmitted by the destination to indicate the successful packet reception after a period of time called short interframe space (SIFS). Considering the short duration time of SIFS compared with DIFS, no other station is able to detect the channel idle for a DIFS until the end of the ACK. The station should reschedule the packet transmission when the ACK is not received or the transmission of a different packet on the channel is detected.

The two-way handshaking technique for the packet transmission, called basic access mechanism, is shown in Figure 6.4. DCF defines an additional four-way handshaking technique that is optionally used for a packet transmission. This mechanism, called RTS (Request to Send)/CTS (Clear to Send), uses RTS and CTS as the control frame for data transmission. Based on the backoff rules explained above, a station waits until it senses the channel is idle for a DIFS, and then transmits a special short-frame RTS before packet transmission. After that, it can transmit the packet if the CTS frame sent from the destination is correctly received. Figure 6.5 uses an example to illustate the four-way handshaking mechanism as follows.

The information of length of the packet to be transmitted is carried by frames RTS and CTS, which can be obtain by any neighbor station to update a network NAV to avoid collision. According to overhearing, the hidden station from either the transmitting or the receiving station can receive the RTS or CTS frame, and it can suitably delay further transmission, and thus avoid collision.

Moreover, the RTS/CTS mechanism can effectively improve the system performance, especially in the large packet transmission, and this is because the length of

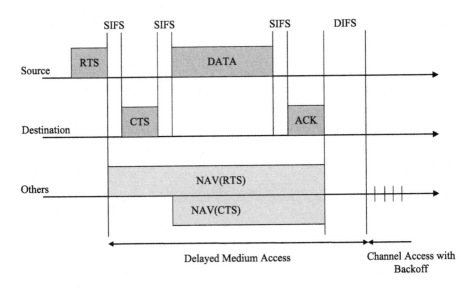

Figure 6.5: RTS/CTS access mechanism. (From Bianchi, Giuseppe. Performance analysis of the IEEE 802.11 distributed coordination function. *IEEE Journal on Selected Areas in Communications***, 18.3 (2000): 535–547.)**

the frames involved in the contention process would be reduced. In the RTS/CTS mechanism, packet collisions are replaced by RTS collisions. As the length of a RTS frame is shorter than the length of a packet frame, the collision can be early detected and the busy duration time should be shorter than that in the mechanism without RTS/CTS.

6.2 IEEE 802.15 MAC Protocol

Wireless personal area networks (PANs), such as wireless sensors networks (WSNs), has been widely used with IEEE 802.15 series protocol, which has been working on the PHY and MAC layers. In this section, we first introduce the IEEE 802.15.4 [6], and then focus on IEEE 802.15.6 [7].

6.2.1 IEEE 802.15.4

IEEE 802.15.4 is designed for various applications [8], including industrial control and monitoring, smart badges and tags, automotive sensing, sensing and location determination at disaster sites, and precision agriculture, such as the sensing of soil moisture, pesticide, herbicide, and pH levels [9], especially in home automation and networking.

As a new standard uniquely designed for low-rate wireless personal area networks and wireless sensor networks, the release of IEEE 802.15.4 is regarded as a milestone that specifies MAC and PHY [10]. The targets of 802.15.4 are providing low data rate, low power consumption, low-cost wireless networking, and offering device-level wireless connectivity. Due to the very short communication range (10 m or less), 802.15.4 networks can usually be a one-hop star, or a self-configuring, multihop network when lines of communication exceed 10 m. During the association procedure, a device in an 802.15.4 network can use either a 64-bit IEEE address or a 16-bit short address, and a single 802.15.4 network can accommodate 64k devices at most. The wireless links of 802.15.4 are the three license-free industrial, scientific, and medical (ISM) frequency bands. There are 16 channels in the 2.4 GHz band with data rates of 250 kb/s, 10 channels in the 915 MHz band with data rates of 40 kb/s, and 1 channel in the 868 MHz band with data rates of 20 kb/s, respectively [10].

6.2.2 IEEE 802.15.4 MAC Protocol

The MAC sublayer provides two services, namely, the MAC data service and the MAC management service, and it is an interface between the service-specific convergence sublayer (SSCS) and the PHY layer. The corresponding tasks of the MAC sublayer are listed as follows [10]:

■ *Generating network beacons:* If the device is a coordinator, it should determine whether to adopt beacon-enabled mode or not for data transmission. In beacon-enabled mode, a superframe structure is used, which is bounded by network beacons and divided into aNumSuperframeSlots (default value is

16) equally sized slots. In order to synchronize the attached devices and for other purposes, a coordinator sends out beacons periodically.

■ *Synchronizing to the beacons:* If the device is the one attached to a coordinator operating in a beacon-enabled mode, for data polling, energy saving, and detection, it can track the beacons to synchronize with the coordinator.

■ *Supporting personal area network (PAN) association and disassociation:* For self-configuration, association and disassociation functions are provided in the 802.15.4 MAC sublayer. Based on this function, not only a star network can be set up automatically, but also a self-configuring, peer-to-peer network can be created.

■ *Carrier sense multiple access with collision avoidance (CSMA/CA) mechanism:* For channel access, 802.15.4 uses the CSMA/CA mechanism, like most other protocols designed for wireless networks. Compared with 802.11, the Request to Send (RTS) and Clear to Send (CTS) mechanisms are not included in 802.15.4, because the data rate is very low in wireless personal area networks.

■ *Guaranteed time slot (GTS) mechanism:* When a beacon-enabled mode is used with the GTS mechanism, portions of the active superframe can be allocated by a coordinator to a device for transmission without competition. Compared with the contention access period (CAP), these portions of the active superframe are the contention-free period (CFP).

■ *Providing a reliable link between two peer MAC entities:* In order to improve the reliability of the link between two peers, the MAC sublayer employs various mechanisms to solve this problem, such as the frame acknowledgment and retransmission, data verification by using a 16-bit CRC, as well as CSMA/CA.

Gernally, star and peer-to-peer topologies are supported by 802.15.4 and the logical structures are shown in Figure 6.6. A one-hop star topology is typically employed in personal computer peripherals and wireless body area networks, and peer-to-peer topology is usually used in the cluster-tree and mesh networking topologies in which the scenarios are more complex. As mentioned, 802.15.4 networks can choose to work in beacon-enabled or non-beacon-enabled mode. A network coordinator schedules the communications and sends a beacon periodically for synchronization and association procedures in the beacon-enabled mode. There are no regular beacons in the non-beacon-enabled mode, but the beacons are sent by the coordinator for the soliciting device. Non-beacon-enabled mode uses unslotted CSMA for communication and decentralized access.

A superframe structure in the beacon-enabled mode is shown in Figure 6.7. In Figure 6.7, aBaseSuperframeDuration = 960 symbols, or 15.36 ms. The parameters BO and SO denote the beacon order and the superframe order, respectively. These values are determined by the coordinator and are restricted to the range

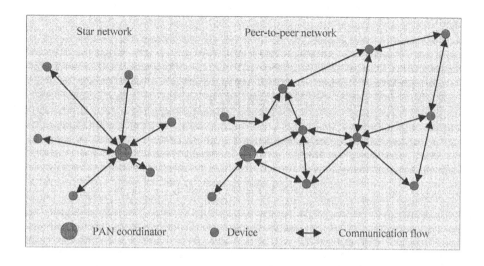

Figure 6.6: Star and peer-to-peer networks. (From Callaway, Ed, et al. Home networking with IEEE 802.15.4: a developing standard for low-rate wireless personal area networks. *IEEE Communications Magazine,* **40.8 (2002): 70–77.)**

$0<=SO<=BO<=14$ [12]. In each superframe, the first signal is the beacon, which is periodically sent by the coordinator with the information of beacon interval (BI), superframe duration (SD), and contention-free period (CFP). During the SD, nodes compete for medium access using slotted CSMA/CA in the contention access period (CAP). After that, all the sensor nodes would enter the sleep mode in the inactive period to power down and conserve energy.

For transmission, a node should perform backoff for a random number of backoff slots and uniformly choose a value between 0 and $2^{BE} - 1$, where the parameter BE is the backoff exponent, which is initially set to 3. Each backoff slot lasts 20 symbol durations (or 320 s) and is denoted by a slot. As well as 802.11, the purpose of random backoff is to reduce the collision probability among contending nodes. When the channel is clear of activity for a contention window (CW) duration, the node can attempt to transmit data, where CW duration is defined to be of two backoff slots (640 s) in the 802.15.4 standard. If the channel is sensed to be in busy, the backoff slots should be reset with the BE+1, and the node should wait for transmission until the channel can be sensed again. The BE is incremented if the channel is still busy, until it is equal to aMaxBE (which has a default value of 5).

6.2.3 IEEE 802.15.6 MAC Protocol

In recent years, a number of researchers from both academia and industry have paid attention to the wireless body area network (WBAN). The reason is that it has great potential to revolutionize the future of healthcare technology, and supports medical and consumer electronics (CE) applications. In order to successfully implement

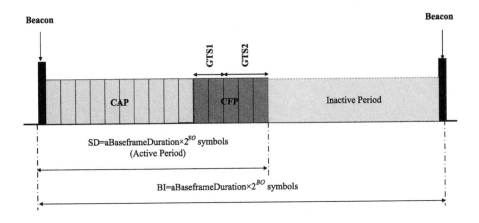

Figure 6.7: A superframe structure in 802.15.4. (From Cao, Bin, Yu Ge, Chee Wee Kim, Gang Feng, Hwee Pink Tan, and Yun Li. An experimental study for inter-user interference mitigation in wireless body sensor networks. *IEEE Sensors Journal*, 13.10 2013: 3585–3595.)

WBAN, the corresponding standard should be provided to address both medical and CE applications.

To this end, IEEE 802.15.6 [7] was established as the standardization of WBAN in November 2007. The IEEE 802.15.6 standard was developed for miniaturized low-power devices, which can be deployed on or implanted inside a human body to serve a variety of medical, consumer electronics, and entertainment applications [13]. In February 2012, the final version of IEEE 802.15.6 was published. The IEEE 802.15.6 defines the medium access control (MAC) layer, which supports the narrowband (NB), ultra-wideband (UWB), and human body communications (HBC) physical (PHY) layers.

In WBANs, all nodes can be organized into one- or two-hop star topology based on the definition of the IEEE 802.15.6 standard. The single central node that can control the entire operation of each WBAN is called the coordinator or hub. The WBAN includes one hub and a number of nodes with the transmission range from zero to mMaxBANSize. A node can be selected as the relay to forward the data frames between a node and the hub in a two-hop start WBAN. Similar to 802.15.4, the IEEE 802.15.6 standard divides the time axis or channel into beacon periods or superframes of equal length [14]. The duration of slots is equal, and the number of slots is from 0 to 255. In beacon mode, the beacons are sent from the hub to define the superframe boundaries and allocate the slots. In nonbeacon modes, the superframe boundaries are defined by polling frames without beacon. Usually, the hub want to all nodes is inactive when it transmits beacons in each superframe, and the hub can shift or rotate the offsets of the beacon periods to shift the schedule allocation slots.

The MAC frame format, communication modes, and access mechanisms defined in the IEEE 802.15.6 standard are shown as follows.

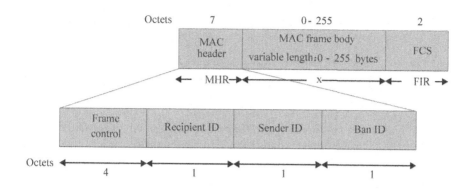

Figure 6.8: IEEE 802.15.6 MAC frame format. (From Ullah, Sana, Manar Mohaisen, and Mohammed A. Alnuem. A review of ieee 802.15.6 MAC, PHY, and security specifications. *International Journal of Distributed Sensor Networks***, 2013 (2013).)**

- *MAC frame format:* The general MAC frame format that concludes a 56-bit header, variable-length frame body, and 18-bit frame check sequence (FCS) is shown in Figure 6.8. The maximum length of the frame body is 255 octets. In the MAC header, there are 32-bit frame control, 8-bit recipient identification (ID), 8-bit sender ID, and 8-bit WBAN ID fields. The type of frame, that is, beacon, acknowledgment, or other control frame, is carried in the frame control field. The corresponding ID fields contain the address information of the recipient and the sender of the data frame, respectively. The WBAN ID contains information on the WBAN in which the transmission is active. In the MAC frame body, the message freshness information required for nonce construction and replay detection is carried by the first 8-bit field. Data frames and information about the authenticity and integrity of the frame are carried in the frame payload field and the last 32-bit message integrity code (MIC), respectively.

- *Network operation modes: Beacon mode with beacon period superframe boundaries:* In this mode, the hub transmits beacons in active superframes. The active superframes might be followed by several inactive superframes if there is no data frame need to send. The superframe structure is illustrated in Figure 1.9(a); it is divided into exclusive access phases (EAP1 and EAP2), random access phases (RAP1 and RAP2), a managed access phase (MAP), and a contention access phase (CAP), respectively. The aims of two EAPs are to transfer high-priority or emergency traffic. In contrast, nonrecurring traffic occupies the RAPs and CAP for transmission. Scheduled and unscheduled bilink allocations and scheduled uplink and downlink allocations use the period of MAP.

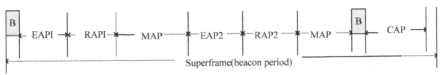

(a) Beacon mode with superframe boundaries

■ *Nonbeacon mode with superframe boundaries:* As illustrated in Figure 6.9(b), the hub operates during the MAP period in this mode.

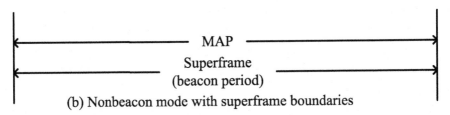

(b) Nonbeacon mode with superframe boundaries

■ *Nonbeacon mode without superframe boundaries:* As illustrated in Figure 6.9(c), the hub provides unscheduled type II polled or posted allocations or a combination of both in this mode.

■ *Access mechanisms:* There are three categories of access mechanisms in each period of the super frame: (1) random access mechanism, which adopts either CSMA/CA or a slotted Aloha procedure to allocate resources; (2) improvised and unscheduled access (connectionless contention-free access), which chooses unscheduled polling/posting to allo- cate resources; and (3) scheduled access and variants (connection-oriented contention- free access), which is also called 1-periodic or *m*-periodic allocations, which schedules

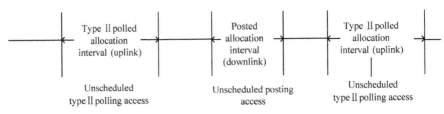

(c) Nonbeacon mode without superframe boundaries

Figure 6.9: IEEE 802.15.6 communication modes. (From Ullah, Sana, Manar Mohaisen, and Mohammed A. Alnuem. A review of ieee 802.15.6 MAC, PHY, and security specifications. *International Journal of Distributed Sensor Networks*, 2013 (2013).)

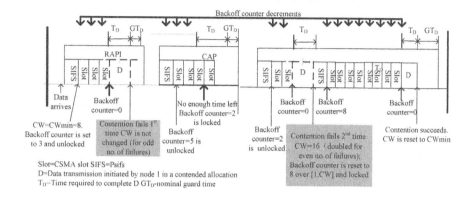

Figure 6.10: CSMA/CA procedure in IEEE 802.15.6 standard. (From Kwak, Kyung Sup, Sana Ullah, and Niamat Ullah. An overview of IEEE 802.15. 6 standard. Applied Sciences in Biomedical and Communication Technologies (ISABEL). 2010 3rd International Symposium on IEEE, 2010.)

the allocation of slots in one or multiple upcoming superframes. We simply illustrate the basic procedures of the CSMA/CA protocol as follows, and the details of these mechanisms can be referred to in the standard [14, 15].

In CSMA/CA, a node uniformly sets its backoff counter over the interval [1;CW], and CW is in the range (CWmin;CWmax). The values of CWmin and CWmax vary depending on the user priorities. The node starts decrementing the backoff counter by 1 for each idle CSMA slot of duration equal to pCSMASlotLength. When the backoff counter reaches zero, the node can transmit data. If any frame transmission is on the channel, the backoff counter should be frozen because the channel is busy, until it is idle again. When the node fails to receive an acknowledgment or group acknowledgment, this indicates that data transmission failed. In this case, the CW should be doubled if it does not reach CWmax.

In Figure 6.10, the CSMA/CA procedure defined in the IEEE 802.15.6 standard is shown. Waiting for SIFS = pSIFS duration, the node unlocks the backoff counter in RAP1. And then, it stars to send data when the backoff counter becomes zero. However, the CW is not doubled for an odd number of failures, and it remains unchanged when the node fails to receive an acknowledgment or the contention fails. In CAP, the node sets the backoff counter to 5 and locks it at 2 since the time between the end of the slot and the end of the CAP is not enough for completing the data transmission and the nominal guard time, represented by GTn. In the RAP2 period, the backoff counter is unlocked. The CW sets double when the node fails to receive an acknowledgment or the contention fails again, and the backoff counter is set to 8. The CW should be reset to CWmin since the data transmission is successful.

6.3 IEEE 802.16 MAC Protocol

Due to the fast growth of the Internet, demand for higher-speed Internet access is caused. Nowadays, wired technologies like DSL or cable can offer broadband connections for the increased requirements, but only the areas with the highest density of population are covered. Moreover, in rural areas having little wired infrastructure, how to satisfy the requirements of users has become an urgent need. In order to overcome the last-mile problem, wireless broadband access has been regarded as the best possible solution. Also known as WiMAX (Worldwide Interoperability for Microwave Access), IEEE 802.16 is introduced to enable the last-mile broadband wireless access [16]. In IEEE 802.16, the coverage is about 5 miles with a bandwidth of up to 70 Mbps, and the deployment does not require expensive base stations.

IEEE 802.16 defines the MAC and PHY layer sepecifications, and supports PMP (point-to-multipoint) and meshes these two transmission modes [17]. Mesh mode can be seen as an optional mode that is an extension of the PMP mode, and it can improve the performance as more subscribers are added to the system using multi-hop routes. Considering mesh mode slot allocation and reservation mechanisms, the 802.16 MAC protocol becomes a hot research area.

IEEE 802.16 can work in both licensed and unlicensed portions of the frequency spectrum. In December 2001, the first version of the 802.16 standard was approved, which can make high data rates available to users having line-of-sight (LOS) connectivity. With the support of the mesh mode, WiMAX provides broadband connections in wider areas, even to users with non-line-of-sight (NLOS) connections.

In the licensed spectrum between 10 and 66 GHz, IEEE 802.16 employs a single carrier (SC) scheme in the PHY layer [17]. For NLOS communication, IEEE 802.16a was provided in 2003 as the previous standard that offers a OFDM physical layer and support for orthogonal frequency division multiple access (OFDMA) in the MAC layer. At the end of 2004, IEEE 802.16d-2004 was provided, called Fixed WiMAX. In order to support mesh topologies, this revision enhances the MAC layer in addition to PMP topologies, and the supported frequencies are set to be within 2 and 11 GHz.

There are two major components in the IEEE 802.16 reference model: the data/control plane and the management plane [18–20]. The data/control plane includes two layers: the physical and MAC layers. Furthermore, the MAC layer is divided into the service specific converge sublayer (CS), MAC common part sublayer (MAC CPS), and security sublayer, which are illustated in Figure 6.11. The information is encapsulated or de-encapsulated at the MAC level, which is defined by the control and data planes. The functions such as classification, security, application QoS, and connection settings, among others, are provided by the management plane. In the following, the layers are described:

■ *Service-specific convergence sublayer (CS):* The service data units (SDUs) from the external networks are received through the MAC CS service access point. Therefore, the CS can provide the functions such as the classification of external SDUs and their associations with the appropriate service flow (SFID) and connection ID (CID).

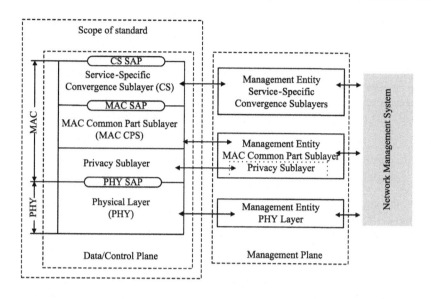

Figure 6.11: IEEE 802.16 reference model. (From Morales, Ana, and Marla Villapol. Reviewing the service specification of the IEEE 802.16 MAC layer connection management: a formal approach. *CLEI Electronic Journal,* **16.2 (2013): 2–2.)**

■ *Common part sublayer (MAC CPS):* MAC CPS controls the access to the medium, and it is also responsible for data encapsulation/de-encapsulation, and data packing and fragmentation. Moreover, the functions of the MAC CPS layer are data control error, including error detection and retransmission strategy, traffic control, and QoS provision. The MAC layer is connection-oriented, so all the services, including connectionless services, are mapped to a connection.

■ *Security sublayer:* The security sublayer can deliver privacy to subscribers in the wireless network, and it is also responsible for authentication and secure key exchange and encryption on the connections established between either a SS or a MS and the BS.

■ *Physical layer (PHY):* PHY is defined to work on the 10–66 GHz band, which is responsible for adaptive burst profiling, in which some transmission parameters, such as modulation and coding schemes, may be changed on either a per connection or per subscriber basis to adapt to channel changing conditions and to provide varying levels of service. Meanwhile, it supports time division duplexing (TDD) and frequency division duplexing (FDD) techniques, and the data encapsulated in MPDUs are carried in TDD or FDD PHY frames.

■ *MAC Layer:* In the MAC layer, there are some improvements in the 802.16e standard. Compared with the 802.16d, the standard enables mobility sup-

Table 6.1 MAC Protocol Sublayers

Sublayer	Function
CS	Guarantee QoS for flows
CPS	Access control bandwidth and power management scheduling
PS	Set up secure connections among SSs

Source: Kas, Miray, et al. A survey on scheduling in IEEE 802.16 mesh mode.

port by defining seamless handover and power conservation mechanisms for portable devices.

In the 802.16e standard, one mandatory and two optional handoff methods are supported to enable a seamless handover of ongoing connections from one base station to another. It also provides multicast and broadcast services. Besides, the battery life in end devices would be preserved by defining a series of sleep and idle mode power management functions.

The MAC protocol of WiMAX consists of three sublayers: convergence sublayer (CS), common part sublayer (CPS), and privacy sublayer (PS). Table 1.1 gives the functions of these sublayers.

WiMAX MAC is connection oriented and the unidirectional CID (connection identifier) identifies each link. Because higher-layer protocol addresses like IP addresses are mapped onto CIDs and SFIDs (service flow identifiers) in CS, every transmission is contained by a queue with its service type. Unlike CS for ATM and packet networks, CSs for IP and Ethernet are decided to be implemented by the WiMAX Forum [22]. PHS (payload header suppression) is another functionality standard in CS [23].

The core of the MAC layer, CPS, carries the functionalities of ranging, scheduling, bandwidth management, construction and transmission of MAC PDUs. This part is investigated in three subsections in the 802.16d standard: PMP, mesh, and data/control plane. Considering the CPS constitutes the core of MAC functionality, more details about PMP and mesh modes will be offered next.

The last sublayer of 802.16 MAC is PS. It provides private access to the subscribers across a fixed wireless network through encryption.

In the MAC layer, one important function is scheduling, which is defined as the allocation of limited resources to tasks over time [24]. In centralized scheduling, the decision maker controls the resource and is aware of all jobs and their requests. In distributed scheduling, nodes/users/agents compete for the resources with possibly conflicting goals. As the competing agents in IEEE 802.16, SSs maintain some local information regarding their needs and inform others via exchanging messages.

One of the most important components of an 802.16 mesh network is scheduling, which affects the overall performance of the system severely. A special sequence of time slots can be a representation of the scheduling problem for 802.16, in which each possible transmission is assigned a time slot so that the transmissions on the same slot are collision-free while the QoS requirements are fulfilled efficiently and the total time to calculate the schedule is minimized.

In IEEE 802.16, the frame scheduling mechanisms are usually classified into centralized and distributed scheduling. The coordinated and uncoordinated scheduling mechanisms are subclasses of distributed scheduling. The coordinated distributed scheduling claims to provide collision-free transmission of MSH-DSCH messages, which is the main difference between coordinated and uncoordinated distributed scheduling. In coordinated distributed scheduling, in order to enable their messages to not collide with messages from other nodes within their two-hop neighborhood, all nodes arrange their transmissions through a pseudorandom algorithm.

MSH-DSCH messages in uncoordinated scheduling may collide, and it is less reliable than coordinated scheduling. While in fast ad hoc setup of schedules and low-duty-cycle traffic scenarios [25], it is usually preferred. For intranet traffic (the traffic among SSs) and centralized scheduling for Internet traffic (the traffic between an SS and a mesh BS or a gateway), distributed scheduling is advised.

6.4   IEEE 802.22 MAC Protocol

6.4.1   Introduction to IEEE 802.22

Various wireless application services have been widely used such as mobile communications, Wi-Fi, and TV broadcast; this indicates that modern society is very dependented on the radio spectrum [26]. Since the deployment of applications in unlicensed bands (e.g., ISM and UNII) is unencumbered by regulation, these bands play a key role in wireless services. As a result, plenty of new applications are generated, including last-mile wireless PANs/LANs/MANs. The wireless service requirement is explosive, requiring the suitable radio spectrum for wireless communication more and more. Therefore, regulatory bodies (e.g., the FCC [27]) have to consider opening further expensive bands for the increasing new wireless applications. Although the unlicensed bands are very busy and competitive for data transmission, licensed bands such as the TV bands are significantly underutilized [28, 29].

In order to solve the current low usage of the radio spectrum, cognitive radios (CRs) [30–32] have been seen as the solution, which can enable flexible, efficient, and reliable spectrum use by adapting the radios' operating characteristics to real-time conditions of the environment. One of the amazing advantages of CRs is the potential to utilize the large amount of unused spectrum in an intelligent way without interfering with other incumbent devices in frequency bands already licensed for specific uses. Therefore, CRs are suitable for the rapid and significant advancements in radio technologies (e.g., software-defined radios, frequency agility, power control, etc.), and can be characterized by wideband spectrum sensing, real-time spectrum allocation and acquisition, and real-time measurement dissemination.

For unlicensed operation in the TV broadcast bands, the IEEE 802.22 was formed for wireless regional area networks (WRANs) in November 2004 [33], which is responsible for the specific task of developing an air interface (i.e., PHY and MAC) based on CRs.

Different from the BS coverage range of existing IEEE 802 standards, 802.22 WRAN can exceed 100 km when there is no transmission power constraint. Usually, the current specified coverage range of 802.22 WRAN is 33 km at 4 W cus- tomer premises equipment (CPE) effective isotropic radiated power (EIRP). WRANs have a much larger coverage range than the existing networks, which is primarily due to their higher power and the favorable propagation characteristics of TV frequency bands, which are shown in Figure 6.12. Meanwhile, unique technical challenges are generated by the enhanced coverage range, as well as opportunities.

The unique and most critical requirement for the 802.22 air interface is flexibility and adaptability, which originate from the fact that 802.22 operates in a spectrum where incumbents have to be protected by all means. Further, since 802.22 operation is unlicensed and a BS serves a large area, coexistence among collocated 802.22 cells (henceforth referred to as self-coexistence) is of paramount importance.

6.4.2 IEEE 802.22 MAC Protocol

In order to flexibly and efficiently transmit data, a novel MAC mechanism is provided by the IEEE 802.22, which supports cognitive capabilities for both reliable protection of incumbent services in the TV band and self-coexistence among 802.22 systems. The IEEE 802.22 MAC is applicable to any region in the world and does not require country-specific parameter sets. Compared to IEEE 802.16e, Table 1.2 shows the major features provided in IEEE 802.22.

■ *Superframe Structure:* According to the definition of the IEEE 802.22 stan- dard, an 802.22 system is a typical point-to-multipoint network, where the medium access of a number of associated CPE units for broadband wire- less access applications is controlled by a central BS. In the downstream direction data are scheduled over consecutive MAC slots, while in the up- stream direction the channel capacity is shared by the CPE units based on a demand-assigned multiple access (DAMA) scheduling scheme. The concept of a connection plays an important part in the 802.22 MAC. The mapping of all services to connections, as performed in the convergence sublayer, facil- itates bandwidth allocation, QoS and traffic parameter association, and data delivery between the corresponding convergence sublayers. A 48-bit univer- sal MAC address that serves as the station identification is used for each 802.22 station, and an 802.22 system has the 12-bit connection identifica- tions (CIDs) for data transmissions.

A superframe structure is employed in the 802.22 MAC with the purpose of efficiently managing data communication and facilitating a number of cognitive functions for licensed incumbent protection, WRAN synchronization, and self-coexistence. In Figure 6.13, a BS transmits a superframe on its operating channel beginning with a special preamble, and contains a superframe control header (SCH) and 16 MAC frames.

The MAC frame size is 10 ms, and each MAC frame comprises a downstream subframe and an upstream subframe with an adaptive boundary in between. The

Table 6.2   IEEE 802.22 Features Compared to IEEE 802.16

	IEEE 802.22	IEEE 802.16e
Air Interface	OFDMA	OFDMA, OFDM, Single Carriers
Fast Fourier transform	Single mode (2048)	Multiple modes (2048, 1024, 512, 128)
OFDMA channel profile (MHz)	6, 7, or 8(according to regulatory domain)	28, 20, 17.5, 14, 10, 8.75, 7, 3.5, 1.25
Burst allocation	Linear	Two-dimensional
Subcarrier permutation	Distributed with enhanced interleaver	Adjacent or distributed
Multiple-antenna techniques	Not supported	Support multiplexing, space time coding, and beamforming
Superframe/frame structure	Support a superframe structure based on groups of 16 frames; frame size: 10 ms	Superframe is not supported; supported frame sizes: 2, 5, 10, or 20 ms
Coexistence with incumbents	Spectrum sensing management, geolocation management, incumbent database query, and channel management	Not supported
Self-coexistence	Dynamic spectrum sharing	Master frame assignment
Internetwork communications	Over-the-air coexistence beacon or over-the-IP-network	Over-the-IP-network (primarily)

Source: Stevenson, Carl R., et al. IEEE 802.22: the first cognitive radio wireless regional area network standard. *IEEE Communications Magazine*, 47.1 (2009): 130–138.

downstream subframe only contains a single PHY protocol data unit (PDU), the upstream subframe may have a number of PHY PDUs which are scheduled from different CPE units, as well as contention intervals for initialization, bandwidth request, UCS notification, and self-coexistence. The reason is that the downstream traffic for CPE located far from the BS can be scheduled early in the downstream subframe; such a data layout allows the MAC to absorb the round-trip delay for large distances. So as to absorb the propagation delay considering a distance of up to 100 km, the self-coexistence window should include a time buffer of one symbol. Similarly, in

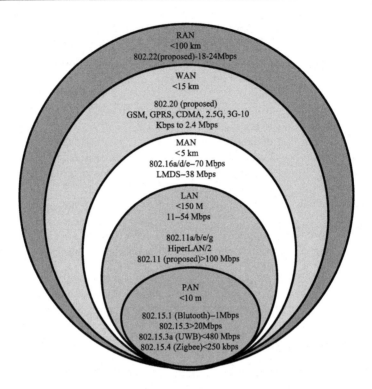

Figure 6.12: 802.22 wireless RAN classification compared to other popular wireless standards. (From Cordeiro, Carlos, et al. IEEE 802.22: an introduction to the first wireless standard based on cognitive radios. *Journal of Communications***, 1.1 (2006): 38–47.)**

order to absorb the round-trip delay in the initial ranging process, a time buffer of two symbols is included before and after the ranging burst.

■ *Network entry and initialization:* Different from other existing wireless access technologies, processes such as synchronization, ranging, capacity negotiation, authorization, registration, and connection setup are not only defined in the network entry and initialization procedures of the 802.22 MAC, but also explicitly specify the operations of geolocation, channel database access, initial spectrum sensing, internetwork synchronization, and discovery.

Satellite-based geolocation technology can provide a global time source, and thus it would be used by BSs and CPE to facilitate synchronization among neighboring networks. BSs and CPE perform spectrum sensing to obtain the list of available TV channels by referring to an up-to-date TV channel usage database and augmenting.

■ *Self-coexistence:* Multiple 802.22 systems may operate in the same vicinity in a typical deployment scenario. Due to co-channel operation, mutual inter-

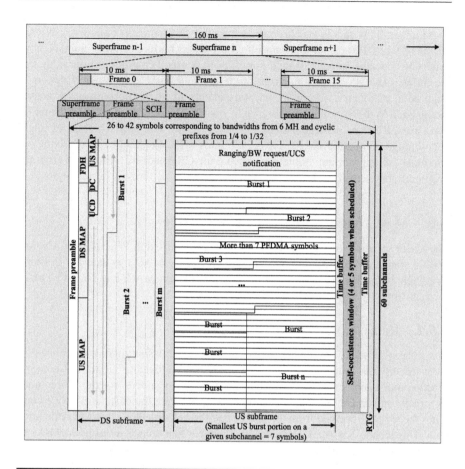

Figure 6.13: Superframe and frame structure. (From Stevenson, Carl R., et al. IEEE 802.22: the first cognitive radio wireless regional area network standard. *IEEE Communications Magazine***, 47.1 (2009): 130–138.)**

ference among these collocated WRAN systems could degrade the system performance significantly. In order to address this issue, a self-coexistence mechanism based on the coexistence beacon protocol (CBP) is specified by the 802.22 MAC, consisting of spectrum-sharing schemes that address different coexistence needs in a coherent manner.

To convey all necessary information across TV channels to facilitate network discovery, coordination, and spectrum sharing, a CBP is proposed that is a communication protocol based on beacon transmissions among the coexisting WRAN cells. A CBP packet is delivered in the operating channel through the beacon transmission in a dedicated self-coexistence window (SCW) at the end of some frames, and comprises a preamble, an SCH, and a CBP MAC PDU.

During a SCW that is synchronized across the TV channels of interest, a WRAN station (BS or CPE) can either transmit CBP packets on its operating channel or receive CBP packets on any channel. For efficient intercell communications, although the SCWs can also be scheduled on an on-demand basis, each WRAN system is required to maintain a minimum repeating pattern of SCWs in transmit (or active) mode. Each WRAN system can reserve its own SCWs on the operating channel for exclusive CBP transmission or share the active SCWs with other co-channel neighbors through contention-based access. A WRAN system can schedule receiving operation at the appropriate moment to capture the CBP packets transmitted from the neighboring systems of interest, by knowing the SCW patterns of its neighbors.

6.5 MAC Mechanisms in Literature

The above sections focus on the MAC protocols proposed in WLAN, WMAN, and WPAN standards. In this section, we highlight some MAC mechanisms in literature, including relay/cooperative MAC, energy-efficient MAC, and cognitive MAC mechanisms.

6.5.1 Relay/Cooperative MAC

In wireless networks, the channel capacity and transmission quality have been affected mainly due to the deteriorated multipath fading channel. Cooperative communication makes use of the antennas of the idle nodes (helpers) for the source node to forward information to the destination node, and then the destination node combines the signals received from the source and the helpers, thus creating cooperation gain. This technology shares antennas in a manner that creates a virtual MIMO system and obtains cooperation gain, which can improve networks' quality and system performance obviously.

An example scenario for relay/cooperative communications is shown in Figure 6.14, in which Sh is the helper for the transmission from source node Ss to the destination node Sd.

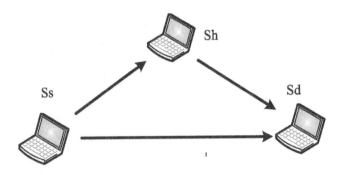

Figure 6.14: A relay/cooperative communication example.

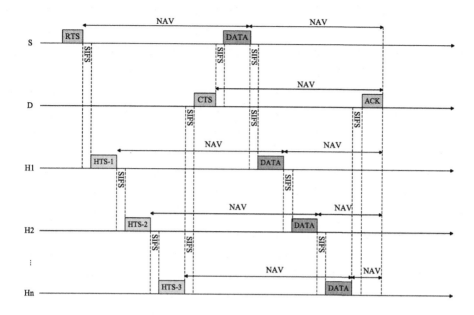

Figure 6.15: Data transmission in DCMAC. (From DCMAC Globecom.)

In the relay/cooperative communication, because of cooperative helpers, the transfer mode changes two phases: one is the transmission from the Ss to Sh and Sd, and the other is from Sh to Sd. In this case, more nodes are involved with the high frequency of collisions, resulting in new hidden/exposed terminal problems. Existing nonrelay/cooperative MAC protocols cannot solve the above problems. Therefore, new relay/cooperative MAC protocols are required to select suitable helpers, reserve resources, and reduce conflict in relay/cooperative wireless networks.

In [36, 37], the authors proposed a cooperative MAC for wireless LANs, which uses a good multihop channel instead of bad single-hop channel. However, this mechanism fails to make cooperation gain. In [38], the authors proposed a distributed method for the distributed MAC protocol with automatic relay selection, but it is costly and requires dual-channel support, and can only choose one cooperative helper.

A typical data transmission chart for relay/cooperative MAC is shown in Figure 6.15, in which more than one helper nodes is included in the relay/cooperative communications.

6.5.2 Energy-Efficient MAC

One main task of the MAC protocol is scheduling the transmission of different nodes to avoid collisions. Most MAC protocols determine whether there is transmission or not by listening to the wireless channel; even a node has no packets ready for trans-

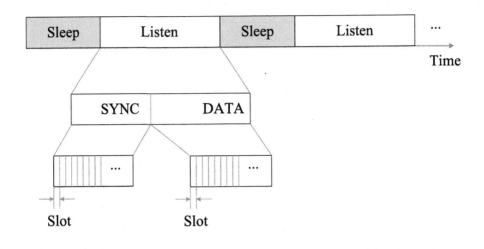

Figure 6.16: Periodic listen and sleep.

mission. This always-on listening wastes the energy of wireless nodes, especially when the traffic is light.

In some networks, such as mobile ad hoc network and sensor network, the energy is supplied by battery. Saving the energy and in turn prolonging the network lifetime is an important issue. An energy-efficient MAC mechanism can turn off or sleep a wireless node to save energy when it has no packets waiting for transmission. In the following, we introduce S-MAC, a typical energy-efficient MAC mechanism designed for wireless sensor networks.

The main purpose of S-MAC is to reduce energy consumption, while supporting collision avoidance. In most wireless sensor networks, the data rate d is very low when no abnormal events happen, so it is not necessary to keep nodes listening all the time. S-MAC reduces the listening time by letting a node go in to periodic sleep mode.

Figure 6.16 shows the basic periodic listen and sleep scheme. When a node has no packets for transmission, it goes to a sleep state for some time, and then wakes up and listens to see if any other node wants to talk to it. In the listen state, a node transmits data to other nodes or receives data from other nodes. During sleep, the node turns off its radio and sets a timer to wake itself later.

The listen/sleep scheme in S-MAC requires synchronizing different neighboring nodes, which means neighboring nodes need to periodically update each other on their schedules to prevent long-time clock. For synchronization, the listen period is further divided into SYNC part and DATA part. Each part includes a few slots for carrier sense. The receiver starts carrier sense when the sleep state is ended, and stops the carrier sense in a randomly selected slot. If it does not detect any transmission during the carrier sense, it sends the SYNC packet. In the DATA part, a node sends a DATA packet according to the same procedure as in the SYNC part.

The sent SYNC packet includes the address of the sender and the time of its next sleep. Receivers will adjust their timers immediately after they receive the SYNC packet. A node will go to sleep when the timer times out.

6.5.3   Cognitive MAC

Different from the classical MAC mechanisms, a cognitive MAC must be adaptive to the environment. It must be intelligent to sensing the spectrum to avoid the interference to the primary user (PU) and other cognitive users (CUs). So in most cases, the cognitive MAC couples with the physical layer. The physical layer can help the MAC layer realize carrier sense and identify the origin of the radiated power by spectrum analysis.

The cognitive MAC mechanism can be coarsely categorized into the centralized and the distributed. Taking IEEE 802.22 as an example, the former is based on some fixed infrastructure, in which a centralized coordinator manages the spectrum allocation and sharing among the CU. The CRs may participate in the spectrum sensing function and provide channel information to the central coordinator. Although the centralized cognitive MAC can optimize the spectrum sharing and avoid collision among different CUs, it is not suitable for some special network scenarios, such as mobile ad hoc network, in which all nodes are mobile and it is difficult to have a fixed coordinator to synchronize CUs and allocate a sensed idle spectrum. In these non-infrastructure and mobile scenarios, distributed cognitive MAC is implemented. For the distributed cognitive MAC mechanism, all mobile nodes sense the idle spectrum dependently, share the sensed information, and access the idle spectrum coordinately.

As IEEE 802.22, an example of centralized cognitive MAC, has been illustrated in Section 6.4, we focus on distributed cognitive MAC in this section.

A typical CSMA-based distributed cognitive MAC includes the following procedure: dynamical spectrum sensing, available channel negotiation, and data transmission, which are shown in Figure 6.17.

Spectrum sensing is used to explore idle spectrum opportunities in cognitive networks. Based on the detection of the weak signal from a primary transmitter, PUs can be detected by local CUs. The main concerns of spectrum sensing are increasing the detection accuracy [40] and optimizing the detection time [41–43].

After detecting the available spectrum, the CU sender and the CU receiver will negotiate the data channel through a determined control channel. Figure 6.17 shows an ACS negotiation procedure through RTS-CTS-RES fames. The CU sender transmits a RTS frame that encapsulates an available transmission channel list (ATCL) for transmission to the CU receiver. Then the CU receiver selects the final data channels (FDCs) from the common parts of ATCL and its available receiving channel list (ARCL), and tell the CU sender the FDC through the CTS frame. The CU sender sends the RES frame to reserve the FDC. Finally, the CU sender and CU receiver exchange data packets through FDC.

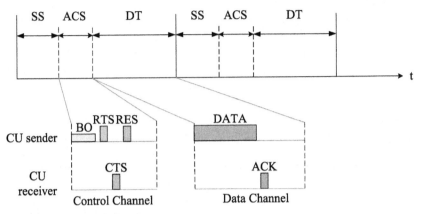

SS: Spectrum sensing
ACS: Available channel selection
DT: Data transmission
BO: Backoff

Figure 6.17: An example for cognitive MAC.

References

1. IEEE Computer Society LAN MAN Standards Committee. Wireless LAN medium access control (MAC) and physical layer (PHY) specifications. 1997.

2. Hiertz, Guido R., et al. The IEEE 802.11 universe. *IEEE Communications Magazine*, 48.1 (2010): 62–70.

3. Crow, Brian P., et al. IEEE 802.11 wireless local area networks. *IEEE Communications Magazine*, 35.9 (1997): 116–126.

4. Cao, Bin, Gang Feng, Yun Li, and Chonggang Wang. Cooperative media access control with optimal relay selection in error-prone Wireless networks, *IEEE Transaction on Vehicular Technology*, 63.1 (2014): 252–265.

5. Bianchi, Giuseppe. Performance analysis of the IEEE 802.11 distributed coordination function. *IEEE Journal on Selected Areas in Communications*, 18.3 (2000): 535–547.

6. Gutierrez, Jose A., Edgar H. Callaway, and Raymond L. Barrett. *Low-rate wireless personal area networks: enabling wireless sensors with IEEE 802.15.4*. IEEE Standards Association, 2004.

7. http://www.ieee802.org/15/pub/TG6.html. 2012.

8. Callaway, Ed, et al. Home networking with IEEE 802.15.4: a developing standard for low-rate wireless personal area networks. *IEEE Communications Magazine*, 40.8 (2002): 70–77.

9. Gutierrez, Jose A., et al. IEEE 802.15. 4: a developing standard for low-power low-cost wireless personal area networks. *IEEE Network*, 15.5 (2001): 12–19.

10. Zheng, Jianliang, and Myung J. Lee. *A comprehensive performance study of IEEE* 802.15. 4. IEEE Press, 2004. New Jersey, USA.

11. Cao, Bin, Yu Ge, Chee Wee Kim, Gang Feng, Hwee Pink Tan, and Yun Li. An experimental study for inter-user interference mitigation in wireless body sensor networks. *IEEE Sensors Journal*, 13.10 2013: 3585–3595.

12. Ramachandran, Iyappan, Arindam K. Das, and Sumit Roy. Analysis of the contention access period of IEEE 802.15. 4 MAC. *ACM Transactions on Sensor Networks (TOSN)*, 3.1 (2007): 4.

13. IEEE P802.15.6, Wireless medium access control (MAC) and physical layer (PHY) specifications for wireless personal area networks (WPANs) used in or around a body 2012.

14. Kwak, Kyung Sup, Sana Ullah, and Niamat Ullah. An overview of IEEE 802.15. 6 standard. Applied Sciences in Biomedical and Communication Technologies (ISABEL). 2010 3rd International Symposium on IEEE, 2010.

15. Ullah, Sana, Manar Mohaisen, and Mohammed A. Alnuem. A review of ieee 802.15.6 MAC, PHY, and security specifications. *International Journal of Distributed Sensor Networks*, 2013 (2013).

16. IEEE Std 802.16-2004. IEEE Standard for Local and Metropolitan Area Networks-Part 16: Air Interface for Fixed BroadbandWireless Access Systems.October 2004. http://ieeexplore.ieee.org/servlet/opac?punumber=9349.

17. Kas, Miray, et al. A survey on scheduling in IEEE 802.16 mesh mode. *IEEE Communications Surveys Tutorials*, Vol. 12, No. 2, pp. 205–221, 2010.

18. Marks, R. B., B. G., Kiernan, C. J., Bushue. IEEE sta. 802.16-2001. *Local and metropolitan area network*, Part 16: air interface for fixed broadband wireless access systems.

19. Eklud, C., R. Marks, S. Ponnuswamy, K. Stanwood, N. Van Waes. *Wireless MAN: inside IEEE 802.16 standard for wireless metropolitan area network.* 1st ed. IEEE Press, 2006. New Jersey, USA.

20. IEEE LAN/MAN Standards Committee. *IEEE standard for local and metropolitan area networks. Part 16: air interface for fixed and mobile broadband wireless access systems amendment 2: physical and medium access control layers for combined fixed and mobile operation in licensed bands and corrigendum 1.* IEEE Std. 802.16-2004/Cor 1-2005 (2006).

21. Morales A, Villapol M. Reviewing the Service Specification of the IEEE 802.16 MAC Layer Connection Management: A Formal Approach[J]. *CLEI Electronic Journal*, 2013, 16(2): 2–2.

22. Andrews, Jeffrey G., Arunabha Ghosh, and Rias Muhamed. Fundamentals of WIMAX. (2007): 113–145. London,UK.

23. IEEE 802.16 Working Group. *IEEE standard for local and metropolitan area networks. Part 16: air interface for fixed broadband wireless access systems.* IEEE Std. 802. 16-2004.

24. Attanasio, Andrea, et al. Auction algorithms for decentralized parallel machine scheduling. *Parallel Computing*, 32.9 (2006): 701–709.

25. Zhu, Hua, and Kejie Lu. *Performance of IEEE 802.16 Mesh Coordinated Distributed Scheduling under Realistic Non-Quasi-Interference Channel.* ICWN.2006 - ww1.ucmss.com.

26. M. Calabrese and J. Snider. Up in the Air. *The Atlantic Monthly*, New America Foundation, 2003, 292(2):46–49.

27. Federal Communications Commission (FCC). Spectrum Policy Task Force, ET Docket 02-135. November 15, 2002.

28. Kolodzy, P. Spectrum Policy Task Force: findings and recommendations. International Symposium on Advanced Radio Technologies (ISART), March 2003.

29. M. McHenry. *Report on Spectrum Occupancy Measurements.* Shared Spectrum Company. VA Vienna, USA.

30. J. Mitola et al. Cognitive Radios: Making Software Radios more Personal. *IEEEPersonal Communications*, vol. 6, no. 4, August 1999.

31. Mitola, J. Cognitive radio: an integrated agent architecture for software defined radio. PhD dissertation, Royal Institute of Technology (KTH), Stockholm, Sweden, 2000.

32. Haykin, Simon. Cognitive radio: brain-empowered wireless communications. *IEEE Journal on Selected Areas in Communications*, 23.2 (2005): 201–220.

33. IEEE 802.22 Working Group on Wireless Regional Area Networks. http://www.ieee802.org/22/.

34. Cordeiro, Carlos, et al. IEEE 802.22: an introduction to the first wireless standard based on cognitive radios. *Journal of Communications*, 1.1 (2006): 38–47.

35. Stevenson, Carl R., et al. IEEE 802.22: the first cognitive radio wireless regional area network standard. *IEEE Communications Magazine*, 47.1 (2009): 130–138.

36. Liu, P., Z. Tao, S. Narayanan, T. Korakis, and S. S. Panwar. CoopMAC: a cooperative MAC for wireless LANs. *IEEE Journal on Selected Areas in Communications*, 25.2, 2007: 340–354.

37. Korakis, T., Z. Tao, Y. Slutskiy, and S. Panwar. A cooperative MAC protocol for ad hoc wireless networks. Proceedings of the Fifth Annual IEEE International Conference on Pervasive Computing and Communications Workshops, 2007.

38. Chou, C.-T., J. Yang, and D. Wang. Cooperative MAC protocol with automatic relay selection in distributed wireless networks. Proceedings of the Fifth Annual IEEE International Conference on Pervasive Computing and Communications Workshops, 2007.

39. Li Y, Cao B, Wang CG, You XH. Dynamical cooperative MAC based on optimal selection of multiple helpers. In: *IEEE Global Telecommunications Conf.* 2009.

40. Kim, H. and K. G. Shin. *Adaptive MAC-layer sensing of spectrum availability in cognitive radio networks*, Technical Report CSE-TR-518-06. University of Michigan, 2006.

41. Kim, H. and K. G. Shin. Efficient discovery of spectrum opportunities with MAClayer sensing in cognitive radio networks. *IEEE Transaction on Mobile Components*, 7 (2008), 533–545.

42. Lee, W. Y. and I. F. Akyildiz. Optimal spectrum sensing framework for cognitive radio networks. *IEEE Transaction on Wireless Communications*, 7.10 (2008): 845–3857.

43. Wang, P., L. Xiao, S. Zhou, and J. Wang. Optimization of detection time for channel efficiency in cognitive radio systems. *Proceedings of Wireless Communications and Networking Conference*, March 2008, pp. 111–115.

Chapter 7

Resource Allocation in Cognitive Radio Networks

Shaowei Wang

Nanjing University

CONTENTS

7.1 Introduction to Cognitive Radio Networks

7.1.1 Definition of Cognitive Radio

The concept of cognitive radio (CR) was first proposed as "the point in which wireless personal digital assistants (PDAs) and ubiquitous networks are sufficiently computationally intelligent about radio resources and related computer-to-computer communications to: (a) detect user communications needs as a function of use context, and (b) provide radio resources and wireless services most appropriate to those needs" [1]. However, the concept of CR is not limited to wireless devices such as PDAs. There is disagreement on how to clearly define CR. As a result, many different definitions have been given by various organizations or even outstanding individuals [2]. The working definition of CR, constructed by the Software Define Radio (SDR) Forum Cognitive Radio Working Group [3], allows for future evolution of CR as well as strictly defining the concept:

> A Cognitive Radio is a software-defined radio that possesses the attributes of being RF and spatially aware with the ability to autonomously adjust to its environment accordingly (frequency, power and modulation).

In a CR system, secondary users (SUs) are allowed to sense the spectrum registered by primary users (PUs) and use the idle part of the spectrum in an opportunistic manner [4]; that is, if an SU detects the presence of a PU in the channel that the SU is using, it releases the channel and switches to a vacant one, or stops transmitting data if no vacant channel is available. However, owing to the inherent feedback delays, estimation errors, and quantization errors in practical wireless systems, there are inevitable sensing errors, which can lead to heavy interference to the PUs. To avoid unacceptable performance degradation of the PUs, the interference generated by the SUs should be regularly controlled, and the physical layer of CR systems should be very flexible to meet these requirements.

Orthogonal frequency division multiplexing (OFDM), which offers a high flexibility for radio resource allocation (RA) in wireless environment, is widely recognized as a promising air interface of a CR system [5]. An OFDM-based system divides a broadband channel into N narrowband subchannels to combat selective fading efficiently. The OFDM-based CR network is investigated in this chapter.

7.1.2 Advantages and Challenges of Cognitive Radio

CR is one critical enabling technology for future communications and networking that can utilize the limited radio resources in an efficient and flexible way. It differs from traditional communication paradigms in that the radios/devices in a CR system can adapt their operating parameters, such as transmission power, frequency, modulation type, etc., to the variations of their surrounding radio environment [6]. Before CRs adjust their operating modes to adapt environment variations, they have to get necessary information of the radio environment. This kind of characteristic is referred to as cognitive capability [7], which enables CR devices to be aware of the transmitted waveform, radio frequency (RF) spectrum, communication network type/protocol, geographical information, locally available resources and services, user needs, security policy, and so on. After CR devices gather necessary information from their radio environment, they can dynamically change their transmission parameters according to the sensed environment variations and achieve optimal performance, which is referred to as reconfigurability [7].

A typical duty cycle of CR includes detecting spectrum white space, selecting the best frequency bands, coordinating spectrum access with other users, and vacating the frequency when a primary user appears. Such a cognitive cycle is supported by the following functions:

- Spectrum sensing and analysis

- Spectrum management and handoff

- Spectrum allocation and sharing

Through spectrum sensing and analysis, CR can detect the spectrum white space, i.e., a portion of frequency band that is not being used by the PUs, and utilize the spectrum. On the other hand, when PUs start using the licensed spectrum again, CR can detect their activity through sensing, so that no harmful interference is generated due to secondary users' transmission.

With recognizing the spectrum white space by sensing, the spectrum management and handoff function of CR enables SUs to choose the best frequency band and hop among multiple bands according to the time-varying channel characteristics to meet various quality of service (QoS) requirements [8]. For instance, when a PU reclaims its frequency band, the SU that is using the licensed spectrum can change to other available frequencies, according to the channel capacity constrained by the noise and interference levels, path loss, channel error rate, holding time, etc.

In dynamic spectrum access, an SU may share the spectrum resources with primary users, other secondary users, or both. Hence, spectrum allocation and sharing mechanism is critical to achieve high spectrum efficiency. Since primary users own the licensed spectrum, when secondary users co-exist in a licensed band with primary users, the interference level due to secondary spectrum usage should be limited to a certain threshold. When multiple secondary users share a frequency band, their access should be coordinated to alleviate collisions and interference.

Among a number of potential advantages of CR, there are three essential ones that make CR so promising for applications:

- Improving link performance

- Improving spectrum utilization

- Potential cost reduction

The idea of improving link reliability arises in the situation where a user loses its connection with the service provider in rural areas or vicinities with poor coverage. CR could essentially play a major role in the improvement of link performance by being able to adapt to poor channels or connections. The technology also has the capability of improving the transmission rate over a channel with high gain, where the CR user looks at its available alternatives and learns the best course of action for the situation. Another aspect that CR will help improve, is indirectly related to link reliability regarding the issue of interoperability. Interoperability refers to the ability of different types of wireless devices and applications to work together effectively, without prior communication, in order to exchange information in a useful and meaningful manner. As an example, suppose a national emergency occurs and public safety personnel from different parts of the country respond to the disaster; all jurisdictions will be able to communicate seamlessly without interference. CR will become a must-have technology for situations as these with its frequency agility and flexibility, the ability to enhance interoperability between different radio standards, and the capability to sense the presence of interferers [9]. With spectrum-sharing capabilities, cognitive radios can prove their effectiveness by utilizing some of the existing spectrum that is not widely used while maintaining call priority and response time.

With the ever-increasing wireless applications, radio spectrum becomes more and more crowded, especially in the band below 6 GHz. It has been disclosed by many studies, however that large portions of spectrum are highly underutilized due to inefficient conventional regulatory policies [10]. CR can aid in solving this problem by filling in unused spectrum and shifting away from overoccupied spectrum. A concept that can be utilized by CR to enhance spectrum utilization is to have radios (or networks) identify spectrum opportunities at run-time and transparently (to legacy systems) fill in the gaps (time, frequency, space) [2].

It is hoped that the concept of CR will lead to the deployment of cheaper radio modules. For CR the goal is to incorporate inexpensive RF components and cheaper processing [2]. The emergence of CR will change the face of current commercial wireless products, such as cellular phones, PDAs, high-speed Internet, commercial radios, video conferences, GPS, etc. The major areas where CR will be most beneficial are military applications, government and regulatory areas, and public safety. Cognitive radio technology is an important innovation for the future of communications and likely to be a part of the new wireless standards, becoming almost a necessity for situations with large traffic and interoperability concerns [9].

For any emerging technology there are drawbacks that can potentially affect the development of the system. Challenges that arise from CR include:

■ Loss of control

■ Regulatory concerns

■ Commercialization issues

Moreover, with any software-defined radio-based technology there are always complexity drawbacks. Since the core of CR resides with software-defined radio,

CR inherits most of the same problems associated with SDR. Some of the drawbacks associated with SDR are:

- Security

- Software reliability

- Keeping up with higher data rates

The process of CRs' sense, learn, and adapt cycle changes the outside world for other cognitive radios with each cycle. This can potentially result in loss of control if a group of radios interact with each other causing chaotic networks [2]. As far as regulatory concerns apply to CR, the current FCC regulations are not specified or designed for CR use. The fear is that regulations will retard the development of cognitive radio or actually reduce available spectrum [2]. The concept of CR is still evolving, and there are many research areas that need to be addressed properly. These include information collection and modeling, decision processes, learning processes, and hardware support [11].

7.2 Dynamic Resource Allocation in OFDM-Based CR Networks

7.2.1 Resource Allocation for Spectral Efficiency

As one of the most important issues in OFDM systems, adaptive resource allocation (RA) has been studied extensively during the past two decades. A comprehensive survey can be found in [12] and references therein. RA algorithms in conventional OFDM systems can be classified into two categories: margin adaptive and rate adaptive. The optimization objective of the former is generally to minimize the required power under given rate requirements of users, while algorithms for rate adaptive usually try to maximize the system throughput of an OFDM system under transmission power limitation.

For the arising OFDM-based CR networks, dynamic RA is very important because it is the prerequisite to achieve high system performance, such as capacity and quality of service (QoS). RA in an OFDM-based CR network, however, is more complex than that in a conventional OFDM system because the PUs may not adopt OFDM modulation, leading to the interference between the two systems. Moreover, the unavoidable sensing errors in the CR network can aggravate the interference, and the interference introduced to the PUs must be carefully controlled below a threshold to prevent the degeneration of the performances of the PUs. Thus, many existing RA algorithms are no longer suitable for OFDM-based CR networks.

7.2.2 General System Model of CR Networks

Consider the downlink of an OFDM-based CR system with K SUs, denoted by $\mathscr{K} = \{1, 2, \ldots, K\}$, coexisting with L active PUs served by licensed system. Figure

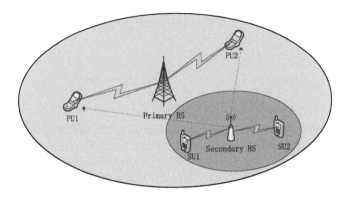

Figure 7.1: System model: Coexistence of primary and secondary users.

7.1 shows the system model. The available bandwidth W is divided into N OFDM subchannels in the CR system, denoted by $\mathcal{N} = \{1, 2, \ldots, N\}$. The bandwidth of the nth subchannel spans from $f_0 + (n-1)W/N$ to $f_0 + nW/N$, where f_0 is the starting frequency. The lth PU's nominal band ranges from f_l to $f_l + B_l$, where f_l and B_l are the starting frequency and the lth PU's occupied bandwidth, respectively.

We assume that perfect channel state information is available at the transceivers of the SUs and the PUs. The interference introduced to the lth PU by SUs' access on the nth subchannel with unit transmission power can be represented as follows [13]:

$$I_{n,l}^{SP} = \int_{f_l - f_0 - (n-1/2)W/N}^{f_l + B_l - f_0 - (n-1/2)W/N} g_{n,l}^{SP} \phi_n^{SU}(f) df \qquad (7.1)$$

where $g_{n,l}^{SP}$ is the power gain from the SU's transmitter to the lth PU's receiver on the nth subchannel. $\phi_n^{SU}(f)$ is the power spectrum density (PSD) of the OFDM subchannel used by an SU, which can be expressed as $\phi_n^{SU}(f) = T_s \left(\frac{\sin \pi f T_s}{\pi f T_s}\right)^2$, where T_s is OFDM symbol duration. On the other hand, the interference generated by the lth PU into the nth subchannel used by the kth SU is

$$I_{k,n,l}^{PS} = \int_{f_0 + (n-1)W/N - f_l - B_l/2}^{f_0 + nW/N - f_l - B_l/2} g_{k,n,l}^{PS} \phi_l^{PU}(f) df \qquad (7.2)$$

where $g_{k,n,l}^{PS}$ is the power gain from the lth PU's transmitter to the kth SU's receiver on the nth subchannel and $\phi_l^{PU}(f)$ is the PSD of the lth PU's signal. Note that we do not assume that the PUs also adopt OFDM modulation.

Define the signal-to-noise ratio (SNR) of the kth SU on the nth subchannel as

$$H_{k,n} = \frac{g_{k,n}^{SS}}{\Gamma(N_0 W/N + \sum_{l=1}^{L} I_{k,n,l}^{PS})} \qquad (7.3)$$

where $g_{k,n}^{SS}$ is the power gain of the kth SU on the nth subchannel, N_0 is the PSD of additive white Gaussian noise, and Γ is the SNR gap and can be represented as $\Gamma = -\frac{\ln(5BER)}{1.5}$ for an uncoded MQAM with a specified BER [14]. The transmission rate of the nth subchannel used by the kth SU is

$$r_{k,n} = \rho_{k,n} \log(1 + p_{k,n} H_{k,n}) \tag{7.4}$$

where $p_{k,n}$ is the kth SU's transmission power on the nth subchannel, and $\rho_{k,n}$ can only be either 1 or 0, indicating whether the nth subchannel is used by the kth SU or not. The data rate of the kth SU, denoted by R_k, can be represented as

$$R_k = \sum_{n=1}^{N} r_{k,n} \tag{7.5}$$

Next, we will introduce several specific RA problems in OFDM-based CR networks based on this general model.

7.2.3 Problem 1: Resource allocation for CR system with Single SU

In this section, we investigate a simple model with only one SU in the CR system. An efficient barrier method is proposed to maximize the system throughput.

For a single SU case, $K = 1$. The interference introduced to the nth OFDM subchannel by the lth PU is

$$I_{n,l}^{PS} = \int_{f_0+(n-1)W/N-f_l-B_l/2}^{f_0+nW/N-f_l-B_l/2} g_{n,l}^{PS} \phi_l^{PU}(f) df \tag{7.6}$$

where $g_{n,l}^{PS}$ is the power gain from the lth PU's transmitter to the SU's receiver on the nth subchannel. Denote the SNR of the nth OFDM subchannel with unit power as

$$h_n = \frac{g_n^{SU}}{\Gamma(N_0 W/N + \sum_{l=1}^{L} I_{n,l}^{PS})} \tag{7.7}$$

where g_n^{SU} is the power gain between the base station and the SU's receiver. The concerned problem is formulated as

$$\max_{p_n} \sum_{n=1}^{N} \frac{W}{N} \log(1 + p_n h_n)$$

$$s.t. \quad C1 : \sum_{n=1}^{N} p_n \le P_t$$

$$C2 : p_n \ge 0, \forall n \tag{7.8}$$

$$C3 : \sum_{n=1}^{N} p_n I_{n,l}^{SP} \le I_l^{th}, \forall l$$

where p_n is the power of the nth subchannel. C1 and C2 are the transmission power constraints of the base station of the CR system. C3 contains the interference constraints of the PUs.

It is easy to prove that (7.8) defines a convex optimization problem with N variables and $N+L+1$ constraints [15]. Generally, barrier method is treated as a standard technique to solve convex optimization problems with inequality constraints. The objective problem is converted into a sequence of unconstrained minimization problems by defining a logarithmic barrier function with parameter t, which decides the accuracy of the approximation. Particularly, each unconstrained minimization problem determined by a parameter t can be solved by Newton method, and the solution to this problem is called a central point in the central path related to the original problem. As t increases, the central point will be more and more accurate to approximate the optimal solution of the original problem. The barrier method consists of two stages: centering step and Newton step. The former is the outer iteration to compute the central point, while Newton step is the inner iteration executed in the centering step.

The complexity of the barrier method is generally $O(N^3)$, where N is the number of the OFDM subchannels. Since there are always thousands of subchannels in OFDM systems, the computation cost is unacceptable because RA should be processed online. Thus, we propose a fast barrier method to work out the optimal solution of the optimization problem by exploiting its special structure. The high computational load is dramatically reduced. For a given number of PUs, the complexity of our proposed fast barrier method is approximately linear to the number of subchannels.

The first step of the barrier method is to reformulate the optimization problem (7.8) into a set of unconstrained optimization problems, making all inequality constraints implicit in an objective function. The barrier function of (7.8) is

$$\phi(P) = \quad -\sum_{n=1}^{N} \log(p_n) - \log(P_t - \sum_{n=1}^{N} p_n)$$
$$-\sum_{l=1}^{L} \log(I_l^{th} - \sum_{n=1}^{N} p_n I_{n,l}^{SP}) \tag{7.9}$$

where $P = (p_1, p_2, \cdots, p_N)$. Denote

$$f(P) = \sum_{n=1}^{N} \frac{W}{N} \log(1 + p_n h_n) \tag{7.10}$$

The optimal solution of (7.8) can be approximated by solving the following unconstrained minimization problem:

$$min \, \psi_t(P) = -tf(P) + \phi(P) \tag{7.11}$$

where $t > 0$. This is an unconstrained minimization problem that can be solved efficiently by the Newton method [15]. As t increases, the approximation becomes more and more accurate.

The computational load of the barrier method mainly lies in the computation of the Newton step ΔP_{nt} at P, that is, solving the equation

$$\nabla^2 \psi_t(P) \Delta P_{nt} = -\nabla \psi_t(P) \tag{7.12}$$

where $\nabla^2 \psi_t(P)$ is the Hessian and $\nabla \psi_t(P)$ is the gradient of $\psi_t(P)$, respectively.

Generally, solving (7.12) has a cost of $O(N^3)$. The Hessian of the problem has the following form:

$$\bigtriangledown^2 \psi_t(P) = H + \sum_{l=0}^{L} g_l g_l^T \qquad (7.13)$$

where

$$H = \begin{bmatrix} \lambda_1 & & & \\ & \lambda_2 & & \\ & & \ddots & \\ & & & \lambda_N \end{bmatrix} \qquad (7.14)$$

$\lambda_i = t \frac{W}{N} \frac{h_i^2}{(\Gamma + h_i p_i)^2} + \frac{1}{p_i^2}$, and g_i's are vectors with length N,

$g_0 = (\frac{1}{P_t - \sum_{n=1}^{N} p_n}, \frac{1}{P_t - \sum_{n=1}^{N} p_n}, \cdots, \frac{1}{P_t - \sum_{n=1}^{N} p_n})^T$,

$g_l = \frac{1}{I_l^{th} - \sum_{n=1}^{N} p_n I_{n,l}^{SP}} (I_{1,l}^{SP}, I_{2,l}^{SP}, \cdots, I_{N,l}^{SP})$. Define $H_i = H + \sum_{l=0}^{i} g_l g_l^T, i = 0, 1, 2, \cdots, L$.

As H is diagonal and $\lambda_i > 0$, so H is obviously positive definite. $g_0 g_0^T$ is positive semidefinite, and then $H_0 = H + g_0 g_0^T$ is positive definite. Since $g_l g_l^T$ is always positive semidefinite, H_is are positive define sequentially.

It follows that the matrix of (7.12) is invertible. Since the Hessian H_L can be treated as the sum of a diagonal matrix and $L + 1$ number of rank 1 matrices, we can use this special structure to calculate the Newton step ΔP_{nt} with approximate linear complexity.

Rewrite the KKT system (7.12) as follows:

$$H_L u^0 = -\bigtriangledown \psi_t(P) \qquad (7.15)$$

where $u^0 = \Delta P_{nt}$. Recall $H_L = H_{L-1} + g_L g_L^T$; (7.15) can be written as

$$(H_{L-1} + g_L g_L^T) u^0 = -\bigtriangledown \psi_t(P) \qquad (7.16)$$

Since H_is are positive define and invertible, then

$$u^0 = (H_{L-1} + g_L g_L^T)^{-1} (-\bigtriangledown \psi_t(P)) \qquad (7.17)$$

Using the matrix inversion lemma [17], we have

$$u^0 = H_{L-1}^{-1}(-\bigtriangledown \psi_t(P)) - \frac{g_L^T H_{L-1}^{-1}(-\bigtriangledown \psi_t(P))}{1 + g_L^T H_{L-1}^{-1} g_L} H_{L-1}^{-1} g_L \qquad (7.18)$$

Step 1: Denote two intermediate variables $u_1^1, u_2^1 \in R^n$ as the solutions of the following two sets of linear equations:

$$\begin{aligned} H_{L-1} u_1^1 &= -\bigtriangledown \psi_t(P) \\ H_{L-1} u_2^1 &= g_L \end{aligned} \qquad (7.19)$$

Then (7.18) can be written as

$$u^0 = u_1^1 - \frac{g_L^T u_1^1}{1 + g_L^T u_2^1} u_2^1 \qquad (7.20)$$

It means that u^0 can be worked out if u_1^1 and u_2^1 have been calculated.

Step 2: Similarly, u_1^1 and u_2^1 can be obtained by solving the following three sets of linear equations

$$
\begin{aligned}
H_{L-2}u_1^2 &= -\nabla\psi_t(P) \\
H_{L-2}u_2^2 &= g_L \\
H_{L-2}u_3^2 &= g_{L-1}
\end{aligned}
\tag{7.21}
$$

where $u_1^2, u_2^2, u_3^2 \in R^n$ are other intermediate variables.

Continue this process to *step L + 1*, $L+1$ variables $u_1^L, u_2^L, \cdots, u_{L+1}^L \in R^n$ are obtained by solving $L+2$ sets of linear equations:

$$
\begin{aligned}
Hu_1^{L+1} &= -\nabla\psi_t(P) \\
Hu_2^{L+1} &= g_L \\
&\vdots \\
Hu_{L+2}^{L+1} &= g_0
\end{aligned}
\tag{7.22}
$$

Since H is diagonal, each set of equations in (7.22) can be solved at a cost of $O(N)$. The computation cost of solving (7.22) is $O(LN)$. Using (7.20), we calculate all u_i^L, $i = 1, 2, \cdots, M+1$, with $O(LN)$ complexity. Carrying out the iteration process inversely, we can calculate all the intermediate variables $u_1^{i-1}, u_2^{i-1}, \cdots, u_i^{i-1}$ with a cost of at most $O(LN)$ until u^0 is worked out. The total cost is $O(L^2N)$. Notice that all H_is are positive definite, and the condition of using the matrix inversion lemma is always satisfied during the computations.

Recall that L is the number of PUs and generally $L \ll N$ in CR systems, so the complexity of the algorithm is almost linearly related to N.

7.2.4 Problem 2: Resource Allocation for Multiuser OFDM-Based Cognitive Radio Networks with Heterogeneous Services

In this section, we investigate a general spectrum-sharing case for a CR system: the SUs can access the regulated portion of licensed spectrum as long as the interference to the PUs is kept below their tolerable thresholds. Heterogeneous services are taken into consideration.

The CR system adopts OFDM modulation and operates in a centralized manner; that is, an access point serves all SUs in the CR network, just as a conventional base station does. We do not assume that the licensed system serving the PUs also operates in an OFDM manner. Heterogeneous services, for both real-time (RT) SUs and non-real-time (NRT) SUs, exist in the CR network. Different from the conventional OFDM system model [17], to prohibit the performance degradation of the PUs, the interference introduced to the PUs must be considered carefully.

We assume there are $K1$ RT SUs, and each of them has minimal rate requirement $R_{k,min}$, $k = 1, 2, \cdots, K1$. We try to maximize the sum rate of the NRT SUs while guaranteeing the minimal rate requirements for the RT SUs, with total transmission power budget of the CR system and the interference constraints of the PUs. The optimization problem can be described as follows:

$$OP1 \quad \max_{p_{k,n}, \rho_{k,n}} \sum_{k=K_1+1}^{K} \sum_{n=1}^{N} \rho_{k,n} r_{k,n}$$

$$s.t. \quad C1 : \sum_{n=1}^{N} \rho_{k,n} r_{k,n} \geq R_{k,min}, k = 1, \cdots, K_1$$

$$C2 : p_{k,n} \geq 0, \forall k, n$$

$$C3 : \sum_{n=1}^{N} \sum_{k=1}^{K} \rho_{k,n} p_{k,n} \leq P_t \qquad (7.23)$$

$$C4 : \sum_{n=1}^{N} \sum_{k=1}^{K} \rho_{k,n} p_{k,n} I_{n,l}^{SP} \leq I_l^{th}, \forall l$$

$$C5 : \rho_{k,n} \in \{0,1\}, \forall k, n$$

$$C6 : \sum_{k=1}^{K} \rho_{k,n} = 1, \forall n$$

C1 guarantees the target rate requirements of the RT SUs. C2 and C3 are the transmission power constraints, while C4 is the interference constraints of the PUs. C5 and C6 declare that each subchannel is kept from being shared among the SUs.

Both binary variables $\rho_{k,n}$ and real variables $p_{k,n}$ are involved in the optimization problem. The main difficulty of solving the problem lies in the integer constraints. An intuitive way to tackle them is the time-sharing method, which relaxes the integer variables into continuous ones. Redefine $\rho_{k,n} \in [0,1]$ as the faction of the nth subchannel allocated to kth SU, temporarily permitting that each OFDM subchannel can be shared by multiple SUs, and $s_{k,n} = p_{k,n}\rho_{k,n}$. The original problem can be converted into

$$OP2 \quad \max_{s_{k,n}, \rho_{k,n}} \sum_{k=K_1+1}^{K} \sum_{n=1}^{N} \rho_{k,n} \log\left(1 + \frac{s_{k,n} H_{k,n}}{\rho_{k,n}}\right)$$

$$s.t. \quad C1 : \sum_{n=1}^{N} \rho_{k,n} \log\left(1 + \frac{s_{k,n} H_{k,n}}{\rho_{k,n}}\right) \geq R_{k,min}, k = 1, \cdots, K_1$$

$$C2 : s_{k,n} \geq 0, \forall k, n$$

$$C3 : \sum_{n=1}^{N} \sum_{k=1}^{K} s_{k,n} \leq P_t \qquad (7.24)$$

$$C4 : \sum_{n=1}^{N} \sum_{k=1}^{K} s_{k,n} I_{n,l}^{SP} \leq I_l^{th}, \forall l$$

$$C5 : \rho_{k,n} \in [0,1], \forall k, n$$

$$C6 : \sum_{k=1}^{K} \rho_{k,n} = 1, \forall n$$

The $s_{k,n}$ can also be characterized as the actual power consumption of the kth SU on the nth subchannel in a time frame interval. It is easy to prove that OP2 is a convex optimization problem [15] that can be solved by standard convex optimization techniques.

The solution to OP2 contains sharing factor $\rho_{k,n}$,s, which are not binary variables. Rounding technique is necessary to obtain a feasible solution to the original problem OP1. Intuitively, larger $\rho_{k,n}$ implies the kth SU obtains a higher rate over the nth subchannel. It is straightforward to allocate the nth subchannel to the k^*th SU that has the largest $\rho_{k,n}$, that is,

$$\rho_{k,n}^* = \begin{cases} 1 & \text{for } k^* = \underset{k}{\arg\max}\,\rho_{k,n} \\ 0 & \text{otherwise} \end{cases} \tag{7.25}$$

However, such a rounding method can result in unreasonable subchannel allocation assignments when K is large because the difference between the largest $\rho_{k,n}$ and second largest one is negligible in many cases. Moreover, the power distribution among subchannels has to be reallocated after the rounding procedure in order to maximize the sum capacity of the CR system. Furthermore, solving OP2 by using standard convex optimization techniques generally has a complexity of $O(2KN + N)^3$, which is too high for practical wireless systems.

So the solution to OP2 is an upper bound of OP1. The optimal solution of OP2 is generally infeasible for OP1, and the upper bound can only be approximated by other feasible solutions. In this work, we consider a two-stage approach to tackle OP1. At the first stage, subchannel allocation is implemented with an efficient heuristic method, which removes the integer constraints in OP1. At the second stage, power distribution across subchannels is carried out to maximize the sum rate of the NRT SUs while also keeping the rate requirements of the RT SUs satisfied.

7.2.4.1 Subchannel Allocation Scheme

We propose an efficient subchannel allocation scheme to figure out the binary variables $\rho_{k,n}$, specifying a subchannel allocation assignment. The intuition of our method follows. In an OFDM-based CR system, an OFDM subchannel with high SNR for a CR user may also generate more interference to the PUs. That is, the water-filling-like method [18] for the conventional OFDM systems is no longer suitable for the CR scenario because the interference threshold constraint also gives an upper bound of the transmission power for each subchannel. The SNR of a subchannel and the interference introduced to PUs should be jointly considered during the subchannel allocation process, as well as the power distribution among subchannels. From the viewpoint of channel capacity, the possible transmission rate of an OFDM subchannel is dependent on not only its SNR, but also the transmission power, which is determined by both the interference thresholds of PUs and the transmission power limit of the CR system. Taking into account these factors, we can evaluate the possible achievable rate of the nth subchannel employed by the kth SU as follows:

$$r_{k,n}^M = \log(1 + p_{k,n}^M H_{k,n}), \tag{7.26}$$

where $p_{k,n}^M$ is the maximum possible power for SU k on subchannel n,

$$p_{k,n}^M = \min\left(P_t, \min_{l\in\mathcal{L}}\left(\frac{I_l^{th}}{I_{n,l}^{SP}}\right)\right). \tag{7.27}$$

Obviously, $p_{k,n}^M$ is related to the power limit and the interference introduced to PUs.

Algorithm 7.1 : Suboptimal Subchannel Allocation

---

1. **Initialization:**
2. $\mathcal{N}_f = \mathcal{N}$, $\Omega_k = \emptyset, \forall k$
3. Set the RT SUs' rates to zero: $R_{t_i} = 0$, for $i = 1, 2, ..., K_1$
4. **For the RT SUs:**
5. *While* $\mathcal{N}_f \neq \emptyset$ and $R_{t_k} < R_{k,min}$ for any $1 \leq k \leq K_1$
6.   Find k^* satisfies $R_{t_{k^*}} - R_{k^*,min} \leq R_{t_k} - R_{k,min}, 1 \leq k \leq K_1$
7.   For k^*, find n^* satisfies $r^M_{k^*,n^*} \geq r^M_{k^*,n}, \forall n \in \mathcal{N}_f$
8.   Update $R_{t_{k^*}} = R_{t_{k^*}} + \log_2(1 + p_{k^*,n^*} H_{k^*,n^*})$
9.   Update $\Omega_{k^*} = \Omega_{k^*} \cup n^*, \mathcal{N}_f = \mathcal{N}_f \setminus n^*$
10. *Endwhile*
11. **For the NRT SUs:**
12. *For* $i = 1$ to length($\mathcal{N}_f$)
13.   For $n^* = \mathcal{N}_{f_i}$, find k^* satisfies $r^M_{k^*,n^*} \geq r^M_{k,n^*}$
14.   Update $\Omega_{k^*} = \Omega_{k^*} \cup n^*$
15. *Endfor*

The channel capacity is normalized by (7.26) in this way. Let Ω_k denote the subchannel set employed by SU k, and $\mathcal{N}$ is the set of subchannels. The outline of our subchannel allocation scheme is describe in Algorithm 7.1. The subchannel allocation algorithm includes two steps. First, the RT SUs are allocated subchannels to meet their requirements. Then the remaining subchannels are allocated to the NRT SUs.

The principle of our subchannel allocation algorithm for the RT SUs is that the SU whose current rate is the farthest away from the target one has the priority to get a new subchannel among the available candidates. This procedure continues until all RT SUs' rate requirements are satisfied. Preferably, the subchannel with the highest achievable rate associated with an RT SU will be chosen at this step. To simplify computation, the power distribution is temperately set as $p_{k,n} = \min(P_t/N, \min_{l \in \mathcal{L}}(I_l^{th}/I_{n,l}^{SP}))$ in order to always satisfy the power and the interference constraints. By doing so, we need not to consider the RT SUs when allocating subchannels to the NRT SUs. We allocate each of the remaining subchannels to the NRT SU that has the highest achievable rate over this channel to maximize the sum capacity of the CR system, as shown in Algorithm 7.1.

7.2.4.2 Optimal Power Allocation

When subchannel assignment is given, the binary variables $\rho_{k,n}$,s are fixed to 1 or 0, and the constraints C5 and C6 in OP1 vanish. Now we need to figure out the power allocation among subchannels to maximize the sum capacity while keeping all constraints satisfied. Recall that Ω_k is the subchannel set of the kth SU, and the optimal power allocation is to solve the following optimization problem:

$$OP3 \quad \max_{p_{k,n}} \sum_{k=K_1+1}^{K} \sum_{n\in\Omega_k} \log(1+p_{k,n}H_{k,n})$$

$$s.t. \quad C_1: \sum_{n\in\Omega_k} r_{k,n} \geq R_{k,min}, 1,\ldots,K_1$$

$$C_2: \sum_{k=1}^{K} \sum_{n\in\Omega_k} p_{k,n} \leq P_t \tag{7.28}$$

$$C_3: \sum_{k=1}^{K} \sum_{n\in\Omega_k} p_{k,n}I_{n,l}^{SP} \leq I_l^{th}, l=1,\ldots,L$$

$$C_4: p_{k,n} \geq 0, \forall k,n$$

Obviously, (7.28) defines a convex optimization problem because the objective function and constraints are all convex. OP3 can be solved by the barrier method, and the complexity of the barrier method is generally $O(N^3)$, where N is the number of subchannels. Since there are always thousands of subchannels in OFDM systems, the complexity is unacceptable because the RA is a real-time problem. So OP3 is still intractable in practice for CR networks. Similar to problem 1, we propose a fast barrier method to work out the optimal solution of OP3 by exploiting its special structure. The high computational load is dramatically reduced. For a given number of PUs, the complexity of the fast barrier method is approximately linear to the number of subchannels.

The first step of the barrier method is to reformulate the OP3 into a set of unconstrained optimization problems, making all inequality constraints implicit in an objective function. The logarithmic barrier function of the OP3 is as follows [15]:

$$\phi(x) = \quad -\sum_{k=1}^{K_1} \log\left(\sum_{n\in\Omega_k} \log(1+p_{k,n}H_{k,n}) - R_{k,min}\right)$$

$$-\sum_{k=1}^{K} \sum_{n\in\Omega_k} \log p_{k,n} - \log\left(P_t - \sum_{k=1}^{K} \sum_{n\in\Omega_k} p_{k,n}\right) \tag{7.29}$$

$$-\sum_{l=1}^{L} \log\left(I_l^{th} - \sum_{k=1}^{K} \sum_{n\in\Omega_k} p_{k,n}I_{n,l}^{SP}\right)$$

where $x = (p_1, p_2, \ldots, p_N)$. Notice that the subscript k can be omitted now as it has been determined for a given subchannel allocation assignment. Denote

$$f(x) = \sum_{k=K_1+1}^{K} \sum_{n\in\Omega_k} \log(1+p_{k,n}H_{k,n}). \tag{7.30}$$

The optimal solution of OP3 can be approximated by solving the following unconstrained minimization problem:

$$\min \ \Psi_t(x) = -tf(x) + \phi(x). \tag{7.31}$$

Since function $\Psi_t(x)$ is convex and twice continuously differentiable, (7.31) has a unique optimal solution. In each centering step of the barrier method, we adopt the Newton method to compute the central point for a given parameter t. The Newton step at x, which is denoted by Δx_{nt}, is given by

$$\nabla^2 \Psi_t(x) \Delta x_{nt} = -\nabla \Psi_t(x), \tag{7.32}$$

where $\nabla^2 \Psi_t(x)$ and $\nabla \Psi_t(x)$ are the Hessian and the gradient of $\Psi_t(x)$, respectively.

The outline of the barrier method is summarized in Algorithm 7.2. ε and ε_n are the tolerances of the barrier method and the Newton step, respectively. α and β are two constants used in the backtracking line search in the Newton step, where $\alpha \in (0, 0.5)$ and $\beta \in (0, 1)$, respectively. The step size of the backtracking line search is s with $s > 0$. t and μ are the parameters that are associated with a trade-off between the number of outer and inner iterations.

Algorithm 7.2 : The Barrier Method

---

1. **Initialization for Barrier method**
2.   Find strictly feasible point x, $t := t^{(0)} > 0$, tolerance $\varepsilon > 0, \mu > 1$
3. **Outer Loop for Barrier method**
4.   Centering step: Compute $x^*(t)$ derived by (7.31) via Newton method
5.   **Initialization for Newton method**
6.   Starting point x, tolerance $\varepsilon_n > 0, \alpha \in (0, 1/2), \beta \in (0, 1)$
7.   **Inner Loop for Newton method**
8.   Compute Δx_{nt} and $\lambda^2 := -\nabla \Psi_t(x) \Delta x_{nt}$;
9.   Quit if $\lambda^2/2 \leq \varepsilon_n$
10.   Backtracking line search on $\Psi_t(x)$, $s := 1$
11.   *while* $\Psi_t(x + s\Delta x_{nt}) > \Psi_t(x) - \alpha s \lambda^2$
12. $s := \beta s$
13.   *endwhile*
14.   Update:$x := x + s\Delta x_{nt}$
15.   Update:$x^*(t) = x$.
16.   Stopping criterion:$(N + K_1 + L + 1)/t < \varepsilon$.
17.   Increase:$t := \mu t$.

If computing the Δx_{nt} in (7.32) directly by matrix inversion, the computation complexity is $O(N^3)$. It is too high to apply as discussed above. For the considered optimization problem (7.31), we analyze its structure and develop a fast algorithm to calculate the Newton step with lower complexity by exploiting the special structure. For notation brevity, we denote

$$
\begin{aligned}
f_0 &= P_t - \sum_{k=1}^{K} \sum_{n \in \Omega_k} p_{k,n} \\
f_k &= \sum_{n \in \Omega_k} r_{k,n} - R_{k,min}, \qquad k = 1, \ldots, K_1 \\
g_l &= I_l^{th} - \sum_{k=1}^{K} \sum_{n \in \Omega_k} p_{k,n} I_{n,l}^{SP}, \quad l = 1, \ldots, L
\end{aligned}
\tag{7.33}
$$

The gradient of $\Psi_t(x)$ is given by

$$\frac{\partial \Psi_t(x)}{\partial p_{k,n}} = -\frac{\chi_k H_{k,n}}{1 + p_{k,n} H_{k,n}} - \frac{1}{p_{k,n}} + \frac{1}{f_0} + \sum_{l=1}^{L} \frac{I_{n,l}^{SP}}{g_l} \tag{7.34}$$

where χ_k yields

$$\chi_k = \begin{cases} \frac{1}{f_k} & k = 1, \ldots, K_1 \\ t & k = K_1 + 1, \ldots, K \end{cases} \tag{7.35}$$

The Hessian of $\Psi_t(x)$ follows:

$$\frac{\partial^2 \Psi_t(x)}{\partial p_{k,n}^2} = \begin{bmatrix} D_1 & & & \\ & D_2 & & \\ & & \ddots & \\ & & & D_N \end{bmatrix} + \frac{\nabla f_0 \nabla f_0^T}{f_0^2}$$

$$+ \sum_{k=1}^{K_1} \frac{\nabla f_k \nabla f_k^T}{f_k^2} + \sum_{l=1}^{L} \frac{\nabla g_l \nabla g_l^T}{g_l^2} \tag{7.36}$$

$$= D + \sum_{m=1}^{M} F_i F_i^T$$

where $D = diag(D_1, D_2, \ldots, D_N) \in \mathscr{R}^{N \times N}$ and $M = K_1 + L + 1$ with

$$D_n = \frac{1}{p_{k,n}^2} + \chi_k \frac{H_{k,n}^2}{(1 + p_{k,n} H_{k,n})^2} \tag{7.37}$$

And F_is are all vectors with N elements,

$$F_i(x) = \begin{cases} \frac{\nabla f_0}{f_0}, & i = 1 \\ \frac{\nabla f_k}{f_k} & k = 1, \ldots, K_1, i = k+1 \\ \frac{\nabla g_l}{g_l} & l = 1, \ldots, L, i = K_1 + l + 1 \end{cases} \tag{7.38}$$

Since it is easy to prove that the matrix D and all $F_i F_i^T$s are positive definite, it follows that $\nabla^2 \Psi_t(x)$ is also positive definite. Hence, the Karush-Kuhn-Tucker (KKT) matrix $\nabla^2 \Psi_t(x)$ on the left in (7.32) is invertible.

It can be provend that Equation (7.32) can be solved with the complexity of $O(NM^2)$.

Recall that $M = K_1 + L + 1$ and generally $M \ll N$ in practical wireless systems. The computational complexity is significantly reduced by the proposed fast algorithm compared with matrix inversion, which needs $O(N^3)$ computations.

In the initialization step, it requires a strictly feasible starting point for the barrier method. A preparatory procedure is necessary to compute and prove that the feasible points exist or not. As discussed in [15], finding feasible solutions is equivalent to solving minimization problem by introducing a crucial indicator parameter z. The optimization problem for the warm start procedure can be formulated as follows:

$$OP4 \quad \min_{z,p_{k,n}} \quad z$$

$$s.t. \quad C_1 : \sum_{n \in \Omega_k} r_{k,n} + z \geq R_{k,min}, k = 1, \ldots, K_1$$

$$C_2 : \sum_{k=1}^{K} \sum_{n \in \Omega_k} p_{k,n} \leq P_t \tag{7.39}$$

$$C_3 : \sum_{k=1}^{K} \sum_{n \in \Omega_k} p_{k,n} I_{n,l}^{SP} \leq I_l^{th} + z, l = 1, \ldots, L$$

$$C_4 : p_{k,n} \geq 0, \forall k, n$$

where z can be interpreted as a bound on the maximum infeasibility of the inequalities C1 and C3, and our goal is to drive the maximum infeasibility below zero.

Note that OP4 is also a convex optimization problem. Since we can easily find $p_{k,n}$s while maintaining C2 and C4 (i.e., $p_{k,n} = P_t/N$) satisfied, after which we can choose any z satisfying C1 and C3, feasible solutions of OP4 always exist. Due to its similar structure with OP3, we can also apply the fast barrier method to solve OP4.

By solving OP4, a strictly feasible point may be computed, or there are no feasible points. If the optimal solution to OP4 satisfies $z \leq 0$, the strictly feasible power allocation variables $p_{k,n}$ can be then used as the starting point of the barrier method for the OP3. Otherwise, no feasible point exists for OP3, and we regard such a case as system outage.

The fast barrier algorithm of solving the OP3 consumes M decomposition, while each decomposition yields an additional equation. We need to solve the $M+1$ matrix system in the first step of the reverse substitution and the computational complexity for each system is measured by $O(N)$. After M reverse substitution steps, the total computational cost for the optimal power allocation is $O(NM^2)$, which is much lower than $O(N^3)$ if using matrix inversion directly.

Since we also can apply the proposed fast algorithm to the warm start procedure, the computational complexity of this procedure is equal to that of the OP3 because of the same structure. Therefore, we conclude that the complexity of the optimal power allocation is $O(NM^2)$. Notice that the number of PUs is always much smaller than that of the subchannels in wireless systems, that is, $M \ll N$; the complexity is reduced dramatically.

7.2.4.3 Efficient Approximations and Algorithms for Power Distribution

We also give a heuristic power allocation algorithm to approximate the optimal solution with low complexity. The heuristic method can produce solutions close to the optimal. Its complexity is even lower than that of the fast barrier method.

In [19], an index function is defined to measure the cost of allocating each possible bit, which poses an efficient bit loading algorithm. Similarly, we define a normalized cost function indicating the cost of allocating a certain rate on each subchannel,

$$F_c(r_{k,n}) = \frac{e^{r_{k,n}} - 1}{e^{r_{k,n}^M} - 1} \tag{7.40}$$

The channel SNR and the interference to PUs are jointly considered in the normalized cost function, reflecting the ability of carrying bits for each subchannel. With the aid of the cost function, we can convert the transmission power constraint C2 and the interference threshold constraints C3 in OP3 into a normalized capacity. Given a subchannel allocation assignment, instead of solving OP3, we transform OP3 into the following form:

$$
\begin{aligned}
OP5 \quad & \max_{r_{k,n}} \sum_{k=K_1+1}^{K} \sum_{n\in\Omega_k} r_{k,n} \\
s.t. \quad & C_1 : \sum_{n\in\Omega_k} r_{k,n} \geq R_{k,min}, k = 1,\ldots,K_1 \\
& C_2 : r_{k,n} \geq 0, \forall k,n \\
& C_3 : \sum_{k=1}^{K} \sum_{n\in\Omega_k} F_c(r_{k,n}) \leq C
\end{aligned}
\tag{7.41}
$$

The parameter C in OP5 is a constant determined by the transmission power and interference constraints, informing the maximum sum capacity of all subchannels. The key role of this transformation is that we unify a series of constraints of OP3, such as C2 and C3, into a single constraint C3 in OP5. OP5 can be regarded as an approximation to OP3. Notice that we do not know the value of C in advance, and it can be only worked out when all $r_{k,n}$s are known. We will show that it is not necessary to know the value of C when solving OP5 in the following.

Without consideration of C2 in OP5, the Lagrangian of OP5 is given by

$$
\begin{aligned}
L(r_{k,n}, \lambda_0, \mu_k) = \quad & -\sum_{k=K_1+1}^{K} \sum_{n\in\Omega_k} r_{k,n} \\
& +\lambda_0 \Big(\sum_{k=1}^{K} \sum_{n\in\Omega_k} F_c(r_{k,n}) - C \Big) \\
& + \sum_{k=1}^{K_1} \mu_k \Big(R_{k,min} - \sum_{n\in\Omega_k} r_{k,n} \Big)
\end{aligned}
\tag{7.42}
$$

where λ_0 and $\mu_k(k = 1,\ldots,K_1)$ are the Lagrange multipliers with $\lambda_0 \geq 0$ and $\mu_k \geq 0$. Applying KKT conditions [15], the optimal solution $r_{k,n}^*$ satisfies the following equations when concerning C2 in OP5:

$$
\frac{\partial L}{\partial r_{k,n}^*} \begin{cases} = 0, & r_{k,n}^* \geq 0 \\ > 0, & r_{k,n}^* = 0 \end{cases}
\tag{7.43}
$$

$$
\lambda_0 \Big(\sum_{k=1}^{K} \sum_{n\in\Omega_k} F_c(r_{k,n}^*) - C \Big) = 0
\tag{7.44}
$$

$$
\mu_k \Big(R_{k,min} - \sum_{n\in\Omega_k} r_{k,n}^* \Big) = 0, \quad k = 1,\ldots,K_1
\tag{7.45}
$$

With simple transformations of (7.43), we get the optimal solution for the RT SUs,

$$r_{k,n}^* = [\log(\frac{\mu_k(e^{r_{k,n}^M}-1)}{\lambda_0})]^+, \quad k = 1, \ldots, K_1 \tag{7.46}$$

where $[x]^+ := \max(x, 0)$. And the optimal solutions for the NRT SUs are

$$r_{k,n}^* = [\log(\frac{e^{r_{k,n}^M}-1}{\lambda_0})]^+, \quad k = K_1 + 1, \ldots, K \tag{7.47}$$

For user k, denote Ω_k^* as the subchannels set in which $r_{k,n}^* > 0, \forall n \in \Omega_k^*$. Sort the $r_{k,n}^*$s in ascending order. Then for $m, n \in \Omega_k^*$, we have

$$r_{k,m}^* - r_{k,n}^* = \log(\frac{p_{k,m}^M H_{k,m}}{p_{k,n}^M H_{k,n}}), \quad k = 1, \ldots, K \tag{7.48}$$

Denote $w_{m,n}^k = \frac{p_{k,m}^M H_{k,m}}{p_{k,n}^M H_{k,n}}$, $N_k = |\Omega_k^*|$. The rate of user k is given by

$$R_k = N_k r_{k,1^*} + \sum_{n \in \Omega_k^*} \log(w_{n,1^*}^k), \quad k = 1, \ldots, K \tag{7.49}$$

where $r_{k,1^*}^*$ represents the minimal element in Ω_k^*. Substituting (7.48) and (7.49) into (7.45) for the case of $\mu_k \neq 0$, we can obtain the closed-form expression of $r_{k,n}$ for the RT SUs,

$$r_{k,n} = \frac{1}{N_k}(R_{k,min} - \sum_{n \in \Omega_k^*} \log(w_{n,1^*}^k)) + \log(w_{n,1^*}^k) \tag{7.50}$$

The details of power allocation for the RT users are illustrated in Algorithm 7.3. Denote the power allocated to SU k and the interference to PU l introduced by SU k as P_k and I_k^l, respectively. It is easy to get the following equations for each SU:

$$P_k = \sum_{n \in \Omega_k} \frac{e^{r_{k,n}}-1}{H_{k,n}} \qquad \forall k$$

$$I_k^l = \sum_{n \in \Omega_k} \frac{(e^{r_{k,n}}-1)I_{n,l}^{SP}}{H_{k,n}} \qquad \forall k, l = 1, \ldots, L \tag{7.51}$$

Algorithm 7.3 : Power Distribution for the RT SUs

---

1. **Initialization:**
2. For the RT SUs, denote $\Omega_k^* = \Omega_k, k = 1, \ldots, K_1$
3. **Power Distribution among the RT SUs**
4. *For $k = 1$ to K_1*
5.   *While* true
6.   Calculate $r_{k,n}$ by (7.50) $n \in \Omega_k^*$ and sort $r_{k,n}$s in ascending order
7.   *If $r_{k,1^*} \geq 0$*
8.   break
9.   *Else*
10. $r_{k,1^*} = 0, \Omega_k^* = \Omega_k^*/1^*$
11.   *Endif*
12.   *Endwhile*
13. *Endfor*
14. Power loading: $p_{k,n} = (e^{r_{k,n}}-1)/H_{k,n}, n \in \Omega_k, k = 1, \ldots, K_1$

Consider the power limitation and the interference constraints for the $K - K_1$ NRT SUs; we have

$$\sum_{k=K_1+1}^{K} P_k = \sum_{k=K_1+1}^{K} \sum_{n\in\Omega_k} \frac{e^{r_{k,n}} - 1}{H_{k,n}} \leq P_t - P_r$$

$$\sum_{k=K_1+1}^{K} I_k^l = \sum_{k=K_1+1}^{K} \sum_{n\in\Omega_k} \frac{(e^{r_{k,n}} - 1)I_{n,l}^{SP}}{H_{k,n}} \leq I_l^P - I_l^r \qquad (7.52)$$

where $P_r = \sum_{k=1}^{K_1} P_k$ and $I_l^r = \sum_{k=1}^{K_1} I_k^l$ are the actual power consumption and interference introduced to PU l by all RT SUs, respectively.

Substituting (7.47) into (7.52), the minimum value of Lagrange multiplier λ_0 is determined by solving the following inequalities:

$$\sum_{k=K_1+1}^{K} \sum_{n\in\Omega_k} \left[\frac{(e^{r_{k,n}^M} - 1)/\lambda_0 - 1}{H_{k,n}} \right]^+ \leq P_t - P_r$$

$$\sum_{k=K_1+1}^{K} \sum_{n\in\Omega_k} \left[\frac{I_{n,l}^{SP}(e^{r_{k,n}^M} - 1)/\lambda_0 - 1}{H_{k,n}} \right]^+ \leq I_l^P - I_l^r \qquad (7.53)$$

From (7.47), we know the minimal λ_0 can maximize the sum capacity of NRT SUs. With the value of λ_0 obtained from (7.53), we get the optimal rate allocation for NRT SUs by (7.47), which is equivalent to achieving the power distribution among NRT SUs.

The complexity of solving OP5 can be counted roughly as follows. In the procedure of power loading for the RT SUs, the complexity is $O(N_k)$ for each RT SU, where N_k is the number of subchannels allocated to SU k. On the other hand, the rate allocation for the NRT SUs can be obtained in $O(L|\mathcal{N}_n|)$, where $|\mathcal{N}_n|$ denotes the number of subchannels allocated to NRT SUs. To sum up, the complexity of the power allocation scheme is bounded by $O(LN)$.

7.2.4.4 Simulation Results

We evaluate the performance of the proposed algorithms with a series of experiments. Consider a multiuser OFDM-based CR system, where all users are randomly located in a 3×3 km area, and each receiver is uniformly distributed in the circle within 0.5 km its transmitter. The path loss exponent is 4, the variance of shadowing effect is 10 dB, and the amplitude of multipath fading is Rayleigh. We assume that each PU's bandwidth is randomly generated by uniform distribution and the maximum value is $2W/3L$. The noise power is 10^{-13} W, and the interference thresholds of all PUs are set to 5×10^{-12} W.

We consider the following algorithms: time-sharing (TS) method defined by the OP2, subchannel assignment followed by the optimal power allocation method (SA-OP), subchannel assignment followed by the approximation power allocation method (SA-AOP), maximum SNR priority subchannel allocation (MS), and minimum interference priority subchannel allocation (MI). For the MS, we always allocate a subchannel to an SU that acquires the highest SNR over this channel, while the MI

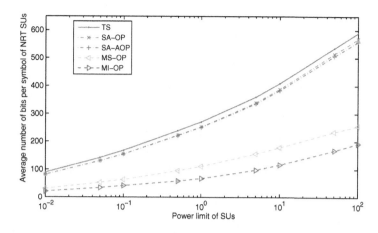

Figure 7.2: Average number of bits per OFDM symbol of NRT SUs as a function of the transmission power. $R_{k,min}$ = 20 bits/symbol, $N = 64$, $K = 4$, $K_1 = 2$, $L = 2$.

scheme allocates a subchannel to an SU that generates the minimum interference to PUs. Both the MS and the MI adopt the optimal power allocation strategy, marked as MS-OP and MI-OP in Figure 7.2. As discussed above, the solution of the TS method is an upper bound of the original optimization problem OP1.

Figure 7.2 shows the average number of bits per symbol of NRT SUs as a function of transmission power limit. The number of PUs and SUs are 2 and 4, respectively, including two RT SUs. There are 64 subchannels in the OFDM-based CR system, and the rate requirements of RT SUs are uniformly set to 20 bits/symbol. We can observe that the sum capacity of NRT SUs increases as the transmission power limit increases. Both of our proposed RA schemes outperform the MS-OP and the MI-OP, obviously. The average capacity gap between the SA-OP and the TS is less than 6%, while the SA-AOP can achieve more than 92% of the upper bound. Since the MS and the MI schemes don't consider the power limit and the interference jointly, they can only achieve about 40 and 30% capacity of the upper bound, respectively.

Figure 7.3 illustrates the average number of bits per symbol of NRT SUs as a function of the number of subchannels. The number of SUs is 4, including two RT SUs. The transmission power limit of SUs is set to 1 W, and the minimal rate requirement for each RT SU is 20 bits/symbol. As the number of subchannels increasesthe sum rate of NRT SUs increases and the rate difference per subchannel between our proposed two schemes and the optimal becomes smaller. The reason is that the SUs can benefit from channel diversity in a wireless environment. And our two schemes (SA-OP and SA-AOP) significantly outperform the MS-OP and the MI-OP, both of which fail to flexibly meet the target rates of the RT SUs while maximizing the sum rate of NRT SUs. Specifically, when $N = 256$, the difference between the SA-OP (SA-AOP) and the upper bound decreases to 3% (4%), while the MS-OP and the MI-OP just achieve about 50 and 30% capacity of the TS, respectively. This obser-vation suggests that the performance loss due to the heuristic subchannel allocation

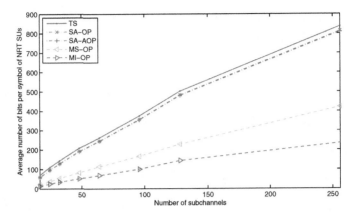

Figure 7.3: Average number of bits per OFDM symbol of NRT SUs as a function of the number of subchannels. $R_{k,min}$ = 20 bits/symbol, $P_t = 1$ W, $K = 4$, $K_1 = 2$, $L = 2$.

method is neglectful when N is sufficiently large, which is suitable for application in practical systems because there are always thousands of subchannels.

We also demonstrate the multiuser diversity effect for the mentioned algorithms in Figure 7.4. The number of RT SUs is $K_1 = 4$, and the number of NRT SUs varies from 1 to 16. Figure 7.4 shows the average number of bits per symbol as a function of the number of NRT SUs, $R_{k,min} = 10$ bits/symbol and $P_t = 1$ W. The number of subchannels is 128. As can be seen in Figure 7.4, the capacity of the CR system increases with the increasing number of NRT SUs. The difference between the TS and the SA-OP is roughly stays at 6%, while the performance of the SA-AOP is

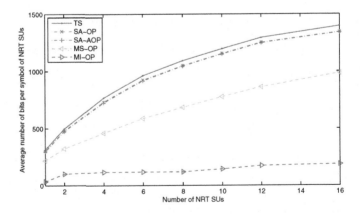

Figure 7.4: Average number of bits per OFDM symbol of NRT SUs as a function of the number of NRT SUs. $R_{k,min}$ = 10 bits/symbol, $N = 128$, $K_1 = 4$, $L = 2$.

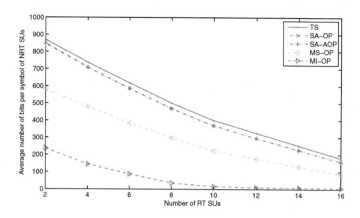

Figure 7.5: Average number of bits per OFDM symbol of NRT SUs as a function of the number of RT SUs. $R_{k,min}$ **= 10 bits/symbol,** $N = 128$, $K - K_1 = 4$, $L = 2$.

almost the same as that of the SA-OP, both of which outperform the MS-OP and the MI-OP. We can explain this phenomenon as a result of multiuser diversity. When there are more SUs, there are more chances for a subchannel to be allocated to a user with relatively high channel gain.

Figure 7.5 depicts the average number of bits per symbol as a function of the number of RT SUs. The minimal rate requirement $R_{k,min}$ of each RT SU is 10 bits/symbol, while the number of RT SUs varies from 2 to 16. The number of NRT SUs is fixed to 4. The number of subchannels is 128 and the transmission power limit is $P_t = 1$ W. Figure 7.5 shows that the sum capacity of NRT SUs decreases as the number of RT SUs increases. This phenomenon can be explained intuitively. When there are more RT SUs trying to access the CR network, more subchannels or power is consumed by RT SUs to meet their rate requirements. Moreover, the radio resource will be more frequently exhausted when the number of RT SUs becomes larger, which intensifies the capacity loss of the NRT SUs. The average capacity gap between the SA-OP and the TS is about 6%, while the SA-AOP can achieve more than 99% capacity of the SA-OP. It also shows that our proposed algorithms perform much better than the MS-OP and the MI-OP.

Finally, we investigate the convergence of the proposed fast barrier method. The computational load of the fast barrier method mainly lies in the number of Newton iterations. From Figure 7.6(a) and (b), we observe that the average numbers of Newton iterations required for a guaranteed duality gap of 10^{-3} are 110 and 40 for the OP2 and the OP3, respectively. Furthermore, the number of iterations varies in a narrow range. We conservatively conclude the fast barrier method is effective and efficient.

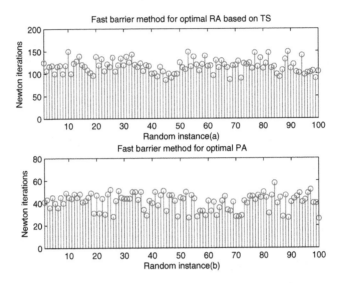

Figure 7.6: Number of Newton iterations required for convergence for 100 channel realizations. $N = 64$, $K = 4$, $K_1 = 2$, $L = 2$. **(a) Fast barrier method for RA based on time sharing. (b) Fast barrier method for the optimal power allocation with given subchannel allocation assignments.**

7.2.5 Problem 3: Resource Allocation for Heterogeneous Cognitive Radio Networks with Imperfect Spectrum Sensing

In this section, the mutual interference and the spectrum sensing errors are taken into consideration in the CR system model. The SUs opportunistically use the spectrum licensed by the PUs via an access point (AP). To support diverse services, we also model a heterogeneous CR network that serves for both real-time (RT) SUs and non-real-time (NRT) SUs. We try to maximize the sum rate of all SUs while guaranteeing the required rates of the RT users and a set of proportional rate constraints among the NRT users to make the resource allocation much fairer.

In practical systems, there are typically two kinds of sensing errors [20]. The first is misdetection, which occurs when the CR system fails to detect the PUs' signals. The band of a subchannel is identified to be vacant, but it is truly used by the PU. The other kind of sensing error is false alarm, which means the CR network identifies the band of a subchannel is unavailable, but it is vacant actually. Generally, the AP in the CR system collects the sensed information of all SUs and makes a decision on which subchannel can be used by SUs. Then the set of sensed available subchannels $\mathcal{M}_v^l$ in the subband of the lth PU is determined, as well as the set of sensed unavailable subchannels $\mathcal{M}_o^l$.

Table 7.1 Probability Information from Imperfect Spectrum Sensing

	Actual State	Sensing Result	Probability Information
1	Active (O_n)	Occupied $(\tilde{O}_n)$	$P\{\tilde{O}_n\|O_n\} = 1 - q_n^m$
2	Active (O_n)	Vacant $(\tilde{H}_n)$	$P\{\tilde{H}_n\|O_n\} = q_n^m$
3	Idle (H_n)	Vacant $(\tilde{H}_n)$	$P\{\tilde{H}_n\|H_n\} = 1 - q_n^f$
4	Idle (H_n)	Occupied $(\tilde{O}_n)$	$P\{\tilde{O}_n\|H_n\} = q_n^f$

The probabilities of misdetection and false alarm on the nth subchannel are q_n^m and q_n^f, respectively. Again, we assume the values of q_n^m and q_n^f can be obtained by either local spectrum sensing via the transceivers of SUs, or preferably, cooperative spectrum sensing [20] and distributed spectrum clustering [21]. Note that we do not consider the coexistence of both PUs' and SUs' transmitters in this model, which may arise in practical CR systems. For this case, interference probability is a useful metric, and a collaborative sensing scheme for the presence of multiple co-channel transmitters is necessary [22].

Obviously, misdetection results in co-channel interference to the PUs, while false alarm lowers the utilization efficiency of spectrum. There are four possible scenarios for spectrum sensing, as shown in the Table 7.1, where H_n and O_n are the hypotheses of the absence and presence of a certain PU's signal on the nth subchannel, $\tilde{H}_n$ and $\tilde{O}_n$ are the events that the nth subchannel is available or unavailable based on the sensed information, respectively. Denoting $P_{1,n}$ as the probability that the nth subchannel is truly used by a PU while the CR network makes a correct judgment, we have

$$
\begin{aligned}
P_{1,n} &= P\{O_n|\tilde{O}_n\} \\
&= \frac{P\{\tilde{O}_n|O_n\}P\{O_n\}}{P\{\tilde{O}_n|O_n\}P\{O_n\} + P\{\tilde{O}_n|H_n\}P\{H_n\}} \\
&= \frac{(1 - q_n^m)q_n^L}{(1 - q_n^m)q_n^L + q_n^f(1 - q_n^L)}
\end{aligned}
\tag{7.54}
$$

where q_n^L is the a priori probability that the subband of the nth subchannel is used by PUs. Similarly, let $P_{2,j}$ denote the probability that the jth subchannel is truly occupied when the CR system deems it as vacant.

Then the interference introduced to the lth PU by the access of an SU on the nth subchannel with unit transmission power is

$$
I_{n,l}^{SP} = \sum_{j \in \mathcal{M}_o^l} P_{1,j} I_{j,l}^n + \sum_{j \in \mathcal{M}_v^l} P_{2,j} I_{j,l}^n
\tag{7.55}
$$

where $I_{j,l}^n$ is the interference introduced to the jth subchannel in the subband of the lth PU,

$$I_{j,l}^n = \int_{(j-1)W/N-(n-1/2)W/N}^{jW/N-(n-1/2)W/N} g_{n,l}^{SP} \phi_n^{SU}(f) df \tag{7.56}$$

The optimization problem follows:

$$\max_{p_{k,n}, \rho_{k,n}} \sum_{k=1}^{K} \sum_{n=1}^{N} \rho_{k,n} \log(1 + p_{k,n} H_{k,n})$$

$$
\begin{aligned}
s.t. \quad & C1: p_{k,n} \geq 0, \forall k, n \\
& C2: \sum_{n=1}^{N} \sum_{k=1}^{K} \rho_{k,n} p_{k,n} \leq P_t \\
& C3: \sum_{n=1}^{N} \sum_{k=1}^{K} \rho_{k,n} p_{k,n} I_{n,l}^{SP} \leq I_l^{th}, \forall l \\
& C4: R_1 : R_2 : \cdots : R_{K_0} = \gamma_1 : \gamma_2 : \cdots : \gamma_{K_0} \\
& C5: R_k = R_k^{req}, k = K_0 + 1, \cdots, K \\
& C5: \rho_{k,n} \in \{0,1\}, \forall k, n \\
& C6: \sum_{k=1}^{K} \rho_{k,n} = 1, \forall n
\end{aligned}
\tag{7.57}
$$

where P_t is the transmission power limit of the AP. C3 means that the interference to the lth PU cannot exceed its threshold I_l^{th}. C4 is the proportional rate constraints of the NRT users. C5 is the fixed-rate requirements of the RT users. C6 and C7 indicate that each subchannel is not shared by SUs.

7.2.5.1 Optimal Power Allocation Using Fast Barrier Method

We can use the method proposed in problem 3 to allocate subchannels to SUs. Given a subchannel assignment, the binary variables $\rho_{k,n}$ in (7.57) are fixed to 0 or 1, the integer constraints vanish, and power distribution across subchannels follows. Recalling that Ω_k is the set of subchannels allocated to the kth SU, the power distribution problem can be written as follows:

$$
\begin{aligned}
\max_{r_{k,n}} \quad & \sum_{k=1}^{K} \sum_{n \in \Omega_k} r_{k,n} \\
s.t. \quad C1 \quad & p_{k,n} \geq 0, \forall n \in \mathcal{N}, \forall k \\
C2 \quad & \sum_{k=1}^{K} \sum_{n \in \Omega_k} p_{k,n} \leq P_t \\
C3 \quad & \sum_{k=1}^{K} \sum_{n \in \Omega_k} p_{k,n} I_{n,l} \leq I_l^{th}, l = 1, \ldots, L \\
C4 \quad & \sum_{n \in \Omega_k} r_{k,n} = \beta_k \sum_{n \in \Omega_1} r_{1,n}, k = 2, \ldots, K_0 \\
C5 \quad & \sum_{n \in \Omega_k} r_{k,n} = R_k^{req}, k = K_0 + 1, \ldots, K
\end{aligned}
\tag{7.58}
$$

where $\beta_k = \gamma_k/\gamma_1, k = 1, \ldots, K_0$. It is easy to prove that (7.58) defines a convex problem that can be solved by the barrier method.

First, we convert all inequality constraints into a logarithmic barrier function $\phi(r)$,

$$\phi(r) = -\sum_{k=1}^{K} \sum_{n \in \Omega_k} \log r_{k,n} - \log(P_t - \sum_{k=1}^{K} \sum_{n \in \Omega_k} p_{k,n}) \\ - \sum_{l=1}^{L} \log(I_l^{th} - \sum_{k=1}^{K} \sum_{n \in \Omega_k} p_{k,n} I_{n,l}) \tag{7.59}$$

where all variables $r_{k,n}$ are collected into a unified vector r, and $r = \{r_n\}_{n=1}^{N}$. Notice that the subscript k can be omitted as it is fixed for a given subchannel assignment. Denoting $f(r) = \sum_{k=1}^{K} \sum_{n \in \Omega_k} r_{k,n}$, the minimization problem with a certain parameter t is

$$\begin{aligned} \min \quad & \psi_t(r) = -tf(r) + \phi(r) \\ s.t. \quad & Ar = b \end{aligned} \tag{7.60}$$

where A is a $(K-1) \times N$ matrix and $b \in \mathscr{R}^{K-1}$ with

$$A_{k,n} = \begin{cases} -\beta_{k+1} & k = 1, \ldots, K_0 - 1, n \in \Omega_1 \\ 1 & k = 2, \ldots, K-1, n \in \Omega_k \\ 0 & otherwise. \end{cases}, \quad b_k = \begin{cases} 0 & k = 1, \ldots, K_0 - 1 \\ R_k^{req} & k = K_0, \ldots, K-1 \end{cases} \tag{7.61}$$

The optimal solution to (7.60) is an approximation of the original problem. As t increases, the approximation becomes more and more close to the optimal solution.

At the centering step of the barrier method, the Newton method is employed to compute the central point. With a given parameter t, Newton step Δr and the associated dual variables v are given by the following Karush-Kuhn-Tucker (KKT) systems:

$$\begin{bmatrix} \nabla^2 \psi_t(r) & A^T \\ A & 0_n \end{bmatrix} \begin{bmatrix} \Delta r \\ v \end{bmatrix} = \begin{bmatrix} -\nabla \psi_t(r) \\ 0_v \end{bmatrix} \tag{7.62}$$

where $0_n \in \mathscr{R}^{(K-1)\times(K-1)}$ and vector $0_v \in \mathscr{R}^{(K-1)\times 1}$. $\nabla^2 \psi_t(r)$ and $\nabla \psi_t(r)$ are the Hessian and the gradient of $\psi_t(r)$, respectively.

Similar to problem 2, if computing the Newton step by solving (7.62) directly, it generates a complexity of $O((N+K)^3)$, which is too high to apply in practical systems. We propose a fast computation of the Newton step by exploiting the structure of problem (7.58). Denoting

$$\begin{aligned} f_0 &= P_t - \sum_{k=1}^{K} \sum_{n \in \Omega_k} p_{k,n} \\ f_l &= I_l^{th} - \sum_{k=1}^{K} \sum_{n \in \Omega_k} p_{k,n} I_{n,l}, l = 1, \ldots, L \end{aligned} \tag{7.63}$$

the gradient and the Hessian of $\psi_t(r)$ are given by

$$\nabla \psi_t(r) = -t - \frac{1}{r_n} + \frac{e^{r_n}}{f_0 H_n} + \sum_{l=1}^{L} \frac{I_{n,l} e^{r_n}}{f_l H_n} \tag{7.64}$$

$$\nabla^2 \psi_t(r) = \begin{bmatrix} 1/r_1^2 & & \\ & \ddots & \\ & & 1/r_N^2 \end{bmatrix} + \frac{\nabla f_0 \nabla f_0^T}{f_0^2}$$

$$+ \frac{\nabla^2 f_0}{f_0} + \sum_{l=1}^{L} \frac{\nabla f_l \nabla f_l^T}{f_l^2} + \sum_{l=1}^{L} \frac{\nabla^2 f_l}{f_l} \qquad (7.65)$$

$$= \begin{bmatrix} D_1 & & \\ & \ddots & \\ & & D_N \end{bmatrix} + g_0 g_0^T + \sum_{l=1}^{L} g_l g_l^T$$

where $D_n = 1/r_n^2 + e^{r_n}/(H_n f_0) + \sum_{l=1}^{L} e^{r_n} I_{n,l}/(H_n f_l), n \in \mathcal{N}$ and $g_i = \nabla f_i/f_i, i = 0, 1, \ldots, L$.

The Hessian is positive definite because the diagonal matrix $D = diag(D_1, \ldots, D_N)$, and $g_0 g_0^T$ and $g_l g_l^T$ are all positive definite matrices. Moreover, since A is a full-row rank matrix, the KKT matrix at the left side of (7.62) is invertible. However, if we compute the inversion of the KKT matrix directly, it has a complexity of $O((N+K)^3)$, which is too high to apply because there are thousands of OFDM subchannels in practical wireless systems. By exploiting the structure of the problem, it can be proven that Equation (7.62) can be solved with the complexity of $O(L^2 N)$. As can be observed from (7.65), the Hessian $\nabla^2 \psi_t(r)$ is a sum of diagonal matrix D and $L+1$ rank 1 matrices $g_l g_l^T$, which leads to a fast computation of the matrix inversion of $\nabla^2 \psi_t(r)$ [19]. Thus, the inversion of the KKT matrix can be computed in a recursive manner with complexity of $O(L^2 N)$.

7.2.5.2 Simulation Results

Experiments are conducted to evaluate the performance of our proposed algorithms. Consider a multiuser OFDM-based CR system, where each receiver is uniformly distributed in a circle within 0.5 km of its transmitter. The path loss exponent is 4, the variance of the shadowing effect is 10 dB, and the amplitude of multipath fading is Rayleigh. The noise power is 10^{-13} W. The probabilities of PUs' activity (q_n^L), misdetection (q_n^m), and false alarm (q_n^f) are uniformly distributed over $[0,1]$, $[0.01, 0.05]$, and $[0.05, 0.1]$, respectively. The bandwidth of each OFDN subchannel is 62.5 kHz. The AP in the CR network identifies available subchannels randomly. We also assume that each PU's bandwidth is randomly generated by uniform distribution among the licensed spectrum.

To evaluate sum capacity, we compare our proposed algorithms, including integer subchannel assignment with optimal power allocation (INT-OP) and integer subchannel assignment with rate loading scheme (SA-OP), with the other three algorithms: EPC, IFPC and MS. The EPC and IFPC are introduced in [23]. The EPC assumes that equal power is distributed among subchannels, while IFPC allocates power inversely proportional to the interference level. The MS always allocates a subchannel to the user that acquires the highest SNR over this channel. All schemes adopt our proposed

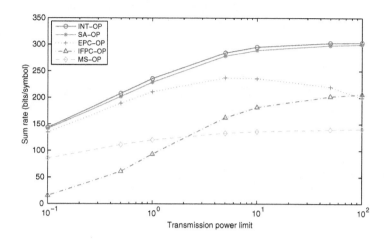

Figure 7.7: The sum rate of all SUs as a function of transmission power limit. $N_t = 64$, $K = 4$, $K_0 = 2$, $L = 2$.

optimal power allocation. There are 64 subchannels, $K = 4$, $K_0 = 2$, $\gamma_1 : \gamma_2 = 1{:}1$, and $R_{k,min} = 10$ bits/symbol. Figure 7.7 shows the sum capacity as a function of transmission power limit P_T.

From Figure 7.7 we can observe that the sum capacity of all SUs grows with the increase of power budget. Our proposed algorithms, the INT-OP and the SA-OP, perform better than the others. When power budget is small, the EPC works quite well. Conversely, the IFPC obtains solutions close to our proposed schemes when P_T is large. The reason is that the power limit and the interference level are jointly considered in our proposed subchannel allocation schemes, while the EPC and the IFPC take only one of them into consideration. It is worth noticing that the gap between the INT-OP and the SA-OP is small, suggesting the suboptimal power allocation algorithm provides a good approximation to the optimal.

Figure 7.8 illustrates the sum rate of all SUs versus the transmission power limit for different numbers of subchannels. There are four SUs, including two RT users, whose rate requirements are uniformly set to 20 bits/symbol. The interference threshold for each PU is 5×10^{-13} W. It is obvious that the sum rate increases when the number of subchannels becomes larger. We can clarify this phenomenon as a result of the channel diversity in wireless environment. Anyway, the SA-OP is always capable of achieving more than 97% of the INT-OP in the different cases, which means that the performance loss due to the suboptimal power allocation is negligible.

We depict the sum rate of all SUs versus the interference threshold for different numbers of PUs in Figure 7.9 There are 64 subcarriers in the considered network and the power limit is 1 W. There are two NRT users with $\gamma_1 : \gamma_2 = 1{:}1$ and two RT users with $R_k^{req} = 10$ bits/symbol. All PUs have the same interference threshold. The sum rate increases with the growth of the interference threshold. The performance of the EPC and IFPC differs with different interference threshold. We can explain this

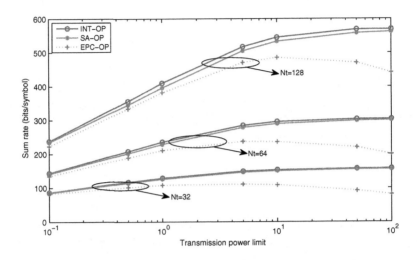

Figure 7.8: Sum capacity as a function of transmission power limit. $K = 4$, $K_0 = 2$, $L = 2$.

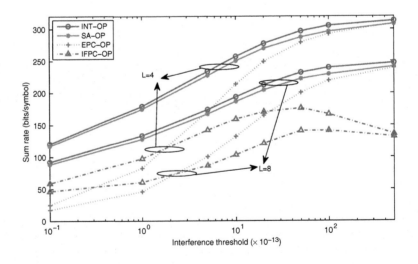

Figure 7.9: Sum capacity as a function of the interference threshold. $N_t = 64$, $K = 4$, $K_0 = 2$.

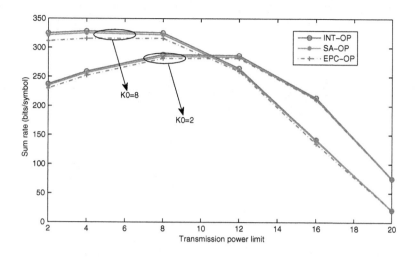

Figure 7.10: Sum capacity as a function of number of RT users. $N_t = 64$, $L = 2$, $R_k^{req} = 20$ **bits/symbol.**

phenomenon intuitively. When the interference threshold is relatively smaller, most of the subchannels are interference limited, which makes the performance of IFPC better than that of EPC. When the interference threshold increases, more and more subchannels are power limited, and the EPC performs better than IFPC. It is worth noting that the EPC and our proposed algorithms perform almost the same when the interference threshold is large enough. The reason is that all subchannels become power limited on the conditions of large interference tolerance. If the interference threshold is small, more subchannels are interference limited, and the performance gap between the EPC and our proposed methods becomes larger as the threshold decreases.

We also verify the effect of multiuser diversity for the CR network. Figure 7.10 shows the sum capacity versus the number of the RT users for two cases, where the numbers of NRT users are $K_0 = 2$ and $K_0 = 8$, respectively. The number of the RT users varies from 2 to 20 with fixed-rate requirements of $R_k^{req} = 20$ bits/symbol. We assume equal fairness among the NRT users. The number of subchannels is $N_t = 64$ and $P_T = 1$ W. When the number of the RT users is relatively small, the sum rate increases slightly with the growth of the number of the RT users. However, the rate loss occurs at a cutoff of the number of RT users, where a sharp decrease of the sum rate can be found when the number of RT users becomes larger. This phenomenon can be explained intuitively.

At the beginning, the CR network benefits from multiuser diversity because a subchannel is more likely to be allocated to an SU that has good channel gain over it. So we can find the sum capacity of the CR network increases with the growth of the number of the RT users. However, when there are more RT users trying to access the CR network, most of the subchannels and more power are consumed by the RT

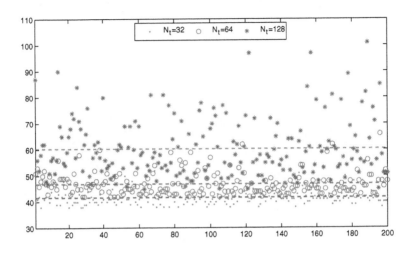

Figure 7.11: Number of Newton iterations required for convergence during 200 channel realizations. $K = 4$, $K_0 = 2$.

users to meet their rate requirements. And the radio resource will be more frequently exhausted when the number of the RT users is larger than the cutoff, which leads to capacity loss of the NRT users. So does the CR network. We can also see from Figure 7.10 that the SA-OP performs almost as well as the INT-OP, both of which are slightly better than the EPC-OP.

Finally, we investigate the convergence of our proposed barrier method. The computational load of the barrier method mainly lies in the computation of the Newton step. If the number of Newton iterations is large or varies in a wide range, the algorithm would be difficult to apply in practical wireless systems. Figure 7.11 and 7.12 show that this not the case for our proposed method for all concerned settings. Figure 7.11 shows the number of Newton iterations for the barrier method to converge in 200 random instances. The average number of Newton iterations is shown in Figure 7.11 by the corresponding dashed line for a certain N_t. Figure 7.12 gives the cumulative distribution function (CDF) of the number of Newton iterations. Both Figure 7.11 and 7.12 show that the number of Newton iterations varies in a narrow range with a given N_t. Our proposed algorithm is effective and efficient.

7.3 Conclusion

In this chapter, we discussed the resource allocation problem in OFDM-based CR networks. First, we introduced the concept of cognitive radio and pointed out the advantages and challenges raised by the CR networks, and gave a general resource allocation model of the OFDM-based CR networks. Then we analyzed several representative scenarios that lead to different optimization problems and proposed effective and efficient algorithms to address them. Some insightful numerical results

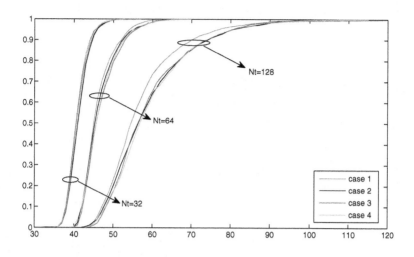

Figure 7.12: CDF of the number of Newton iterations required for convergence over 1000 channel realizations. $K = 4$, $K_0 = 2$, $P_T = 1$ **W.** $\gamma_1 : \gamma_2 = 1{:}1$ **(case 1),** $\gamma_1 : \gamma_2 = 1{:}2$ **(case 2),** $\gamma_1 : \gamma_2 = 1{:}3$ **(case 3),** $\gamma_1 : \gamma_2 = 1{:}4$ **(case 4).**

validate our proposal, indicating that the developed resource allocation algorithms are promising for applications. Readers who are interesting in the details of the proposed algorithms can refer to [24–32].

References

1. J. Mitola, Cognitive radio an integrated agent architecture for software defined radio, PhD thesis, KTH Royal Institute of Technology, Stockholm, Sweden, 2000.

2. J. Reed, Cognitive radio, Virginia Tech ECE5674, Fall Semester 2005.

3. SDR Forum Cognitive Radio Working Group, cognitive radio definitions and nomenclature, January 2006

4. S. Huang, X. Liu, and Z. Ding, Opportunistic spectrum access in cognitive radio networks, in *Proceedings of IEEE INFOCOM 2008*, Phoenix, AZ, 2008, pp. 1427–1435.

5. T. A. Weiss and F. K. Jondral, Spectrum pooling: an innovative strategy for the enhancement of spectrum efficiency, *IEEE Commun. Mag.*, vol. 42, no. 3, pp. 8–14, 2004.

6. Federal Communications Commission, *Spectrum policy task force report*, FCC Report, ET Docket 02-135, November 2002.

7. S. Haykin, Cognitive radio: brain-empowered wireless communications, *IEEE J. Sel. Areas Commun.*, vol. 23, no. 2, pp. 201–220, 2005.

8. I. F. Akyildiz, W.-Y. Lee, M. C. Vuran, and S. Mohanty, Next generation/dynamic spectrum access/cognitive radio wireless networks: a survey, *Comput. Netw.*, vol. 50, pp. 2127–2159, 2006.

9. D. Maldonado, Bin Le, A. Hugine, T. W. Rondeau, C. W. Bostian, Cognitive radio applications to dynamic spectrum allocation: a discussion and an illustrative example, *IEEE DySPAN*, pp. 597–600, 2005.

10. Federal Communications Commission, *Facilitating opportunities for flexible, efficient, and reliable spectrum use employing cognitive radio technologies*, FCC Report, ET Docket 03-322, December 2003.

11. C. Bostian and J. Reed, Understanding the issues in software defined cognitive radio, tutorial presented at IEEE DySPAN, Baltimore, MD, November 2005.

12. S. Sadr, A. Anpalagan, and K. Raahemifar, Radio resource allocation algorithms for the downlink of multiuser OFDM communication systems, *IEEE Commun. Surv. Tutor.*, vol. 11, no. 3, pp. 92–106, 2009.

13. G. Bansal, M. J. Hossain, and V. K. Bhargava, Optimal and suboptimal power allocation schemes for OFDM-based cognitive radio systems, *IEEE Trans. Commun.*, vol. 7, no. 11, pp. 4710–4718, 2008.

14. A. J. Goldsmith and S. Chua, Variable-rate variable-power MQAM for fading channels, *IEEE Trans. Commun.*, vol. 45, no. 10, pp. 1218–1230, 1997.

15. S. Boyd and L. Vandenberghe, Convex optimization, Cambridge University Press, Cambridge, 2004.

16. C. D. Meyer, *Matrix analysis and applied linear algebra*, SIAM Press, Philadelphia, 2000.

17. M. Tao, Y.-C. Liang, and F. Zhang, Resource allocation for delay differentiated traffic in multiuser OFDM systems, *IEEE Trans. Wireless Commun.*, vol. 7, no. 6, pp. 2190–2201, 2008.

18. C. Y. Wong, R. S. Cheng, K. B. Lataief, and R. D. Murch, Multiuser OFDM with adaptive subcarrier, bit, and power allocation, *IEEE J. Sel. Areas Commun.*, vol. 17, no. 10, pp. 1747–1758, 1999.

19. S. Wang, Efficient resource allocation algorithm for cognitive OFDM systems, *IEEE Commun. Lett.*, vol. 14, no. 8, pp. 725–727, 2010.

20. X. Zhou, G. Li, D. Li, D. Wang, and A. Soong, Probabilistic resource allocation for opportunistic spectrum access, *IEEE Trans. Wireless Commun.*, vol. 9, no. 9, pp. 2870–2879, 2010.

21. H. Zhang, Z. Zhang, H. Dai, R. Yin, and X. Chen, Distributed spectrum-aware clustering in cognitive radio sensor networks, in *Proceedings of IEEE GLOBECOM 2011*, Houston, TX, 2011, pp. 1–6.

22. A. Nasif and B. Mark, Opportunistic spectrum sharing with multiple cochannel primary transmitters, *IEEE Trans. Wireless Commun.*, vol. 8, no. 11, pp. 5702–5710, 2009.

23. S. M. Almalfouh and G. L. Stuber, Interference-aware radio resource allocation in OFDMA-based cognitive radio networks, *IEEE Trans. Veh. Technol.*, vol. 60, no. 4, pp. 1699–1713, 2011.

24. S. Wang, M. Ge, and C. Wang, Efficient resource allocation for cognitive radio networks with cooperative relays, *IEEE J. Sel. Areas Commun.*, vol. 31, no. 11, pp. 2432–2441, 2013.

25. S. Wang, Z.-H. Zhou, M. Ge, and C. Wang, Resource allocation for heterogeneous cognitive radio networks with imperfect spectrum sensing, *IEEE J. Sel. Areas Commun.*, vol. 31, no. 3, pp. 464–475, 2013.

26. S. Wang, M. Ge, and W. Zhao, Energy-efficient Rresource allocation for OFDM-based cognitive radio networks, *IEEE Trans. Commun.*, vol. 61, no. 8, pp. 3181–3191, 2013.

27. M. Ge and S. Wang, Fast optimal resource allocation is possible for multiuser OFDM-based cognitive radio networks with heterogeneous services, *IEEE Trans. Wireless Commun.*, vol. 11, no. 4, pp. 1500–1509, 2012.

28. S. Wang, F. Huang and Z.-H. Zhou, Fast power allocation algorithm for cognitive radio networks, *IEEE Commun. Lett.*, vol. 15, no. 8, pp. 845–847, 2011.

29. S. Du, F. Huang and S. Wang, Power Allocation for OFDM-based Cognitive Radio Networks with Cooperative Relays. *IET Communications*, vol.8, no.6, pp. 921–929, 2014.

30. S. Wang, Z.-H. Zhou, M. Ge and C. Wang, Resource Allocation for Heterogeneous Multiuser OFDM-based Cognitive Radio Networks with Imperfect Spectrum Sensing. In: *Proceedings of IEEE INFOCOM'12*, pp. 2264–2272, Orlando, FL, 2012.

31. S. Wang, F. Huang, M. Ge and C. Wang, Optimal Power Allocation for OFDM-based Cooperative Relay Cognitive Radio Networks. In: *Proceedings of IEEE ICC'12*, pp. 1676–1680, Ottawa, Canada, 2012.

32. S. Wang, Q. Yang, W. Shi and C. Wang, Interference Mitigation and Resource Allocation in Cognitive Radio-Enabled Heterogeneous Networks. In: *Proceedings of the IEEE GLOBECOM'13*, pp. 5060–5065, Atlanta, GA, 2013.

Chapter 8

Advanced Multicast Transmissions for Future Wireless Networks

Yiqing Zhou, Hang Liu, Lin Tian, and Jinling Shi

Wireless Research Center, Institute of Computing Technology, Chinese Academy of Sciences and Beijing Key Laboratory of Mobile Computing and Pervasive Devices, Beijing, China

Zhengang Pan

Green Communication Research Center (GCRC) of China Mobile Research Institute (CMRI), Beijing, China

CONTENTS

8.1   Development of Multicast Services

With the rapid development of wireless broadband communication systems and consuming electronics, the demand for mobile multimedia services becomes stronger and stronger. It has been reported that the number of mobile broadband subscriptions has grown steadily to around 1.7 billion during the first quarter of 2013, and these data are expected to be more than 9 billion by the end of 2018. Meanwhile, mobile data traffic doubled between 2012 and 2013, and 12 times growth is expected between 2012 and 2018 [1]. Moreover, it is predicted that handset data traffic will grow over 300% by 2017 to 21 exabytes [2]. The exponential increase of mobile data traffic presents big challenges to future mobile networks. There is a consensus that by 2020, the capacity of mobile communication networks should be boosted thousands of times to support the huge data traffic. Various possible solutions have been proposed, such as striving for more spectrums, developing enhanced communication technologies with high spectral efficiency, and adopting heterogeneous cellular network architectures with dense wireless access nodes.

As a point-to-multipoint transmission scheme, multicast can provide the same information to numerous subscribers simultaneously using the same radio resource. Compared to the point-to-point unicast transmission, multicast is highly spectral efficient and should be one of the promising technologies for future mobile networks. As

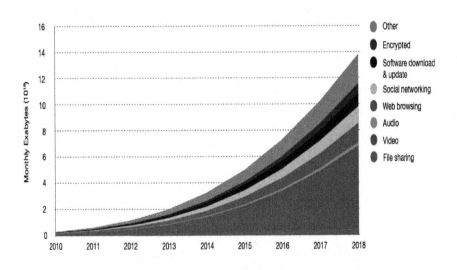

Figure 8.1: Mobile data traffic by applications type. (From Ericson Mobility Report, June 2013, http://www.ericsson.com/mobility-report.)

shown in Figure 8.1, mobile video has been and will still be the leading driver of data traffic on mobile networks, given that it has exceeded 50% of the total mobile data traffic in 2011 [3] and is expected to increase with compound annual traffic growth of 60% until 2018 [1]. Moreover, it is observed that many video resources such as hot music television programs like Gangnam Style are required by many subscribers at the same time. If multicast transmission is employed to convey this kind of video traffic instead of unicast, not only will the network traffic load be reduced, but also the system spectral efficiency will be improved.

Multicast transmission has been extensively investigated and standardized as a key technique for mobile communication systems by various international organizations. For example, multimedia multicast and broadcast services (MBS) are supported by mobile worldwide interoperability for microwave access (WiMAX) systems [4, 5], Multimedia Broadcast Multicast Services (MBMS) are defined in the Third Generation Partnership Project (3GPP) Release 6 for the third-generation mobile communication systems (3G) [6], and the enhanced MBMS is a key feature of 3GPP Long Term Evolution (LTE) systems [7]. Using 3G cellular networks, mobile TV services have been launched worldwide, e.g., by T-Mobile in UK, Verizon in the United States, and Reliance in India. According to RNCOS's research report [8], the number of mobile TV subscribers is projected to grow at a compound annual growth rate of over 45% between 2009 and 2013 to reach around 450 million by the end of 2013.

Besides the mobile communication standards supporting multicast services, there are also global standards specially designed for mobile multicast transmission such as the Digital Video Broadcast-handheld (DVB-H) developed by DVB Project and China Mobile Multimedia Broadcasting (CMMB). Usually dedicated terminals instead of mobile phones should be purchased to enjoy multicast services provided by DVB-H or CMMB. In China, the development of CMMB is not Promising, with only 5% subscribers among all mobile users, due to the lack of high-performance terminals and high-quality programs [9]. The development of DVB-H is also not as florescent as expected. The main downside was the cost, requiring fairly substantial up-front investment, and it will take many years to see any returns on this investment. Therefore, operators in developed countries prefer to employ the broadband connection including the internet and a wireless network to deliver the data streaming. Another problem of DVB-H is the lack of compatible devices, which is the main reason why Swisscom canceled DVB-H services in 2010 [10].

Providing multicast services via cellular communication systems instead of broadcast systems has advantages, like no need to construct dedicated networks and no need to deploy specific receivers at mobile phones, since they are naturally integrated in the communication system. However, there are still a lot of challenges, both non-technical and technical. For example, one big nontechnical challenge is how to encourage the development of high-quality programs to attract subscribers. Moreover, in China, another challenge is that the cellular communication network and the multicast programs are supervised by different government departments. How to design a scheme to make these departments work efficiently is critical for the success of mobile multicast services. On the other hand, advanced multicast technologies are also needed to solve technical challenges, such as enhancing the spectral and energy efficiency. Note that in 3GPP Release 6, the spectrum efficiency of multicast services is only 0.020.2 bps/Hz. Then in Release 8, LTE employs the multicast-broadcast single-frequency network (MBSFN) technique, where the same multicast signal can be transmitted by multiple synchronized BSs simultaneously so that the spectrum efficiency is improved to 1 bps/Hz. In future mobile networks, much higher spectrum efficiency, such as 5 bps/Hz is expected for multicast services. How to achieve this target still remains a big challenge. Moreover, power consumption of wireless networks has attracted a lot of attention recently, as energy costs make up a vast portion of operational expenditure (OPEX). For example, China Mobile spends more than 40% OPEX on electric energy, among which about 72% is consumed by the BS [11]. Improving power efficiency of wireless networks has become a hot topic in both academia and industry [12]. How to design energy-efficient multicast schemes is also a main challenge.

In the following context, focusing on the multicast transmission in cellular networks, various schemes will be introduced in Section 8.2, starting with the conventional multicast, followed by MBSFN, layered multicast, digital fountain-coded multicast, and cooperative multicast. Section 8.3 focuses on the energy- and spectrum-efficient two-stage cooperative multicast, illustrating its performance in various scenarios. Finally, conclusions are drawn in the last section.

8.2 Recent Advances in Multicast Transmission

8.2.1 Conventional Multicast

The conventional multicast scheme is simple and easy to implement. The system allocates a time slot T and a bandwidth of B for multicast services. Then the BS transmits multicast data with a rate of R_{one} in the given slot to all M subscribers in the cell. According to Shannon theory, the maximum data rate a mobile station (MS) i ($i \in [1,M]$) can support is limited by the signal-to-noise ratio (SNR) of the received signal, given by $R_{one,i} = B \log_2(1 + SNR_i)$. It is well known that the received SNR at MSs varies significantly with the MS locations due to various channel effects, such as shadowing, small-scale fading, and especially path loss, which is a function of the distance between the MS and BS and decides the average signal strength. As a result, the supported data rate varies when the MS's location changes. For example, in LTE, the data rate supported by MSs close to the BS could be as large as 10 Mbps, while this number is only several kbps for the MS at the cell edge. Therefore, to ensure that all subscribers could receive the multicast data, the transmission rate of BS is limited by the worst SNR experienced by the subscribers, i.e., $R_{one} = \min_{i \in [1,M]} [B \log_2(1 + SNR_i)]$. In research works, it is always assumed that all N subscribers should receive the data; i.e., 100% coverage is achieved. In this case, the total system throughput $N \cdot R_{one}$ gets saturated as N increases [13]. This is because when N gets larger, the worst SNR becomes smaller, and so does the BS transmission rate R_{one}. As a whole, $N \cdot R_{one}$ approaches a constant value when N is sufficiently

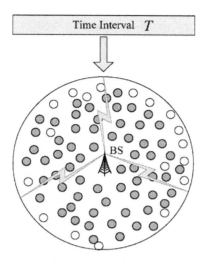

SMS: MS that successfully receives the data   ○ UMS: MS that fails to receive the data

Figure 8.2: Conventional multicast transmission.

large. However, in practice, it is difficult and costly to ensure a 100% coverage for a given data rate R_{one}. As shown in [14], given $R_{one} = 128$ kbps, the BS transmission power should be as high as 80 dBm if all MSs in a cell with a radius of 1500 m should receive the data. Usually R_{one} should be properly set so that most of (e.g., 95%) but not all the subscribers could be covered by the multicast services. The relationship between R_{one} and the coverage is derived in the following context.

As shown in Figure 8.2, considering an orthogonal frequency division multiplexing (OFDM) system with single antenna transceivers, the BS transmits multicast data at a rate of R_{one} with a power of P_{BS} in the multicast time interval T. Assume that a block of continuous subcarriers is allocated for multicast transmission, and they experience similar channel fading [15, 16]. Therefore, in the following context, the received signal on one subcarrier is concerned. Given a slow fading channel, after an ideal FFT, the signal received at a subcarrier of MS k could be given by

$$Y_k = \sqrt{P_{BS}}\alpha_{BS,k}H_k d + \eta_k \tag{8.1}$$

where $\alpha_{BS,k}^2 = A_1 \cdot D_{BS,k}^{-\gamma}$ represents the path loss from the BS to MS k, where A_1 is a constant, $D_{BS,k}$ is the distance between BS and MS k, and γ is the path loss parameter. H_k stands for the Rayleigh channel fading experienced on the subcarrier with an average power of 1. Note that shadowing is not considered in the channel model since it plays a less important role than the path loss and microscopic fading [17]. d represents the transmitted symbol with unit power and η_k is the zero mean Gaussian noise with a variance of $\sigma^2 = N_0 B$, where N_0 is the power spectrum density of the noise. Then the signal-to-noise ratio (SNR) of MS k could be given by $SNR_k = P_{BS}\alpha_{BS,k}^2|H_k|^2/\sigma^2$. Given the multicast data R_{one}, one can successfully receive the message from BS only when the SNR at the destination is no less than the SNR threshold SNR_{one}, which could be derived using Shannon theory and given by $SNR_{one} = 2^{(R_{one}/B)} - 1$.

For one MS with a distance of z from the BS, the success probability $P(SNR_k \geq SNR_{one})$ could be expressed as an exponential distribution with parameter $SNR_{one} \cdot \sigma^2/P_{BS}A_1 z^{-\gamma}$, when H_k follows a Rayleigh distribution with unit power. Moreover, assume that all MSs are uniformly distributed in the cell with a radius of R; then the total coverage of the cell could be written as

$$C_{th} = \int_0^R P(SNR_k \geq SNR_{one}) \cdot \frac{2z}{R^2}dz = \frac{2(A_1 \cdot P_{BS})^{2/\gamma}}{\gamma R^2 \sigma_{noi}^{4/\gamma}} SNR_{one}^{-2/\gamma}\Gamma\left(\frac{2}{\gamma}, \frac{R^\gamma SNR_{one}\sigma_{noi}^2}{A_1 \cdot P_{BS}}\right)$$
(8.2)

Therefore, the relationship between the transmission data rate R_{one}, BS power P_{BS}, and total coverage C_{th} is set up. For a given C_{th} and R_{one}, the required P_{BS} can be calculated from (8.2). On the other hand, for a given P_{BS} and C_{th}, the affordable data transmission rate R_{one} can also be obtained.

8.2.2 Multicast Broadcast Single-Frequency Network

Conventional multicast transmission is introduced in the third generation of mobile communication systems (3G) [18] to provide Multimedia Broadcast Multicast Services (MBMS). However, it is difficult to ensure the quality of service (QoS) of the cell edge MSs. Targeting improvement of the QoS of cell edge users and motivated by Coordinated Multipoints (CoMP) Transmission, the multimedia broadcast service single-frequency network (MBSFN) was proposed and applied in LTE-A systems to improve the performance of conventional multicast and provide enhanced MBMS (eMBMS).

In MBSFN, SFN regions are defined that are composed of multiple adjacent BSs, similar to the clusters in CoMP. The BSs in one SFN region broadcast/multicast the same message simultaneously using the same radio resources. Figure 8.3 plots an example SFN region with three neighboring BSs. It can be seen that the MS within the region could receive three copies of the multicast signal. The received signal can be taken as a multipath signal. Since LTE/LTE-A employs OFDM in the downlink transmission, as long as the length of the cyclic prefix (CP) is longer than the maximum path delay, the same multicast signals transmitted from the BSs in one SFN region could be easily combined as useful signals instead of interference as in conventional multicast. Note that in LTE/LTE-A, the length of CP is normally designed according to the maximum delay spread observed in one cell. So for unicast services, a downlink transmission slot (0.5 ms) is composed of seven OFDM symbols, each with a CP of length 4.69 us. To facilitate the employment of MBSFN, an extended CP scheme is introduced where a frame is divided into six OFDM symbols, each with a CP of length 16.67 us, so that a larger delay spread can be supported. It should also be noted that although the performance of MBMS could be improved with MBSFN, it also brings great challenge in synchronization and coordination between BSs.

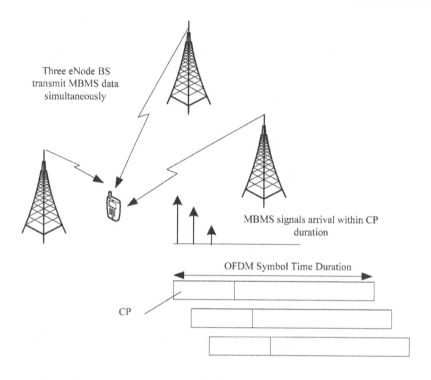

Figure 8.3: MBSFN transmission.

8.2.3 Layered Multicast

In conventional multicast, the BS transmits at a fixed data rate R_{one} to all subscribers, which is limited by the worst channels experienced by the subscribers. So for the MSs near the cell center, their good channel conditions could not be exploited to improve the user experiences. In order to improve the system performance as well as the QoS of users with good channel conditions, layered multicast was proposed for wireless multicast transmission [19].

As early as the 1970s, it was established that for the broadcast/multicast environment where a BS communicates with multiple MSs with varying signal strengths, the optimal solution is multiresolution [20]. Based on this theorem, it was proposed in [19] that the source information should be represented in a hierarchy of resolutions and a multiresolution transmission scheme should be designed. For example, scalable video coding can be employed to encode the video source into one base layer and multiple enhanced layers according to time, space, and video quality. The base layer only ensures basic video quality with a low frame rate or low resolution. When the channel conditions are good, enhanced layers could be received to increase the frame rate or resolution so that high video quality is obtained. To efficiently convey the layered source information to multiple MSs, a corresponding multiresolution

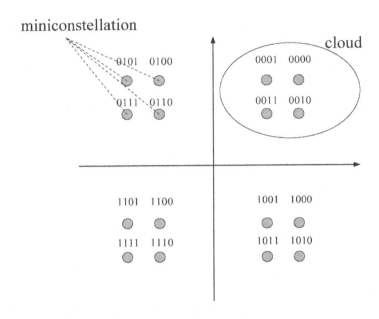

Figure 8.4: Signal constellation of multiresolution 16QAM.

transmission scheme is needed. One possible solution is to use multiresolution modulations (also known as hierarchical modulations) [19].

Figure 8.4 plots the constellation of a multiresolution 16QAM. The basic idea is that the constellation is composed of "clouds" of miniconstellations, where the base layer information is carried by the clouds while the enhanced layer is represented by the miniconstellations. For the example multiresolution 16QAM, the first two bits are for clouds and the last two bits carry miniconstellation information. The receiver should first decide the clouds, subtract the cloud value from the received signal, and then decode the information carried by the miniconstellations.

As shown in Figure 8.5, for a cell edge MS with poor channel conditions, it may be able to distinguish the clouds but cannot decode the miniconstellations correctly. So it can only receive the base layer and be served with basic qualities. On the other hand, for a MS with good channels, both the clouds and miniconstellation information can be received correctly. Thus, it can obtain high video quality with both base and enhanced layers.

A lot of research has been carried out on layered multicast. An adaptive layered multicast is proposed in [21], where the constellation size, the priority parameters of the hierarchical signal constellations, and the mapping of bit positions to bits from different layers could be changed according to the channel conditions. Considering the simultaneous transmission of voice and multiclass data over Nakagami-m fading channels, the adaptive layered multicast using hierarchical BPSK/MQAM modulation is spectrally more efficient for data transmission than the scheme employing

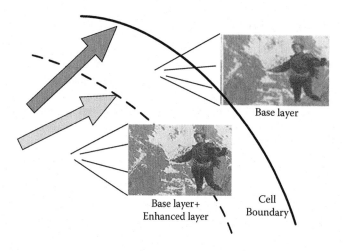

Figure 8.5: Layered multicast transmission.

BPSK/M-AM, while both schemes provide the same outage probability for voice and data. It should be noted that the benefit of the adaptive scheme is obtained at the cost of providing channel state information (CSI) to the transmitter, which is not needed in conventional layered multicast.

Optimization of layered multicast is another research focus. Defining a utility function of the decoded rates of the users, the optimal transmission power allocation and decoding threshold of each layer is investigated in [22], where the Lagrangian dual theorem is employed to solve the optimization problem efficiently. Moreover, aiming to reduce the energy consumption, an energy-efficient layered broadcast/multicast (ELBM) mechanism is proposed [23], which includes the formulation of the energy efficiency evaluation function and a layered modulation solution selection algorithm to determine optimal modulation parameters. Layered multicast is also extended to wiretap channels [24]. First, superposition coding is employed to encode information into a number of layers, and then each layer is encoded by stochastic coding to keep the corresponding information secret from an eavesdropper. Both Gaussian wiretap channels and block fading wiretap channels are considered. The secrecy rate and the optimal power allocation over the layers are derived for the proposed layered multicast scheme.

It should be noted that in layered multicast, the layered source coding has been widely supported by various video standards, such as H.264 and MPEG-4. Moreover, layered transmission can also be naturally supported in 4G wireless networks [25-26]. Although layered multicast could improve the spectral efficiency compared to that of traditional ones, it brings unfairness among the MSs who pay the same for the multicast service but are served with different qualities.

8.2.4 Reliable Multicast Based on Digital Fountain Coding

The concept of digital fountain was proposed by John Byers and Michael Luby, etc., in 1998 [27], aiming to solve the reliable transmission problem of Internet TCP protocol, where the source will repeat sending the same packet to the destination until an acknowledgment (ACK) signal is received, which is not resource efficient. The main idea is to first divide the source information into k different blocks, and then encode these blocks to endless information packets. At the receiver, one can successfully decode the source message with high probability if any n(n > k) of the information packets are received, just the same as one can fill his or her cup (successfully receive the message) from a water fountain when enough drops of water (packets) are obtained at any time.

Digital fountain coding is a rateless coding, whose main advantage is that the transmitter encodes the source message and sends it without considering the rate while the destination just receives the packets continually until the whole message can be decoded. It is suitable for the reliable transmission of multicast data, which usually could not employ conventional schemes such as hybrid ARQ (HARQ) to ensure that the data are correctly received by each subscriber. Using digital fountain coding, all of the subscribers of multicast services could successfully receive the source message, no matter whether the channel condition is good or bad, since they all can decoded sufficient number of packets as the transmission goes on. It can be seen that digital fountain coding could adapt to channel conditions, and all the users will be served with the same QoS.

The Luby transform (LT) code is the first practical digital fountain code for binary erasure channel [28]. The encoding of LT code is the same as the linear random coding. As shown in Figure 8.6, an integer sequence $\{d_1, d_2, \ldots d_n, \ldots\}$ should be generated first, following a designed distribution $f(d)$, where d_i is known as degree. Then d_i information symbols should be selected randomly from the K_{LT} ($K_{LT} = 4$ in Figure 8.6) information symbols and be summed up with binary arithmetic to obtain the ith encoded symbol. The graph which combines the source and encoded symbols is used to decode the message. The decoding starts from the point with degree 1, and the edge connected to this point will be deleted from the graph after updating the decoding symbols iteratively. Therefore, all the source symbols can be obtained one by one.

The degree distribution function is critical to the performance of the LT code. One of the most widely used distribution functions is the ideal soliton distribution $f_1(d)$, given by

$$f_1(d) = \begin{cases} 1/K_{LT} & d = 1 \\ 1/(d(d-1)) & 2 \le d \le K_{LT} \end{cases} \tag{8.3}$$

LT code can recover the message with high probability on the condition that a high degree is used to connect almost all the information symbols. However, this will significantly reduce the resource efficiency and increase the complexity of encoding and decoding progress. Therefore, the raptor code was proposed [29], which is a combination of LT code and linear block code called precode. As shown in Figure 8.7, instead of being encoded with LT code directly, the source information is

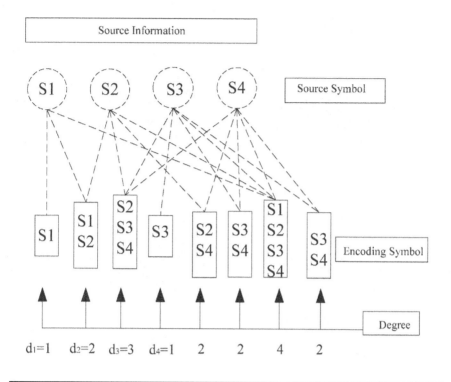

Figure 8.6: LT code.

encoded with a precode first. Hence, the LT code is employed to encode most of the message but not the entire one, while the precode is responsible for correcting the other error symbols. This two-stage coding significantly reduces the complexity and also improves the performance of LT code. Raptor codes had been adopted in 3GPP for mobile cellular wireless broadcast and multicast, and DVB-H standards for IP data cast [30].

8.2.5 Cooperative Multicast

To improve the performance of cell edge MSs, cooperative multicast (CM) is proposed, which introduces MS-based cooperative relay in multicast transmission [31]. As an example, Figure 8.8 illustrates the transmission scheme of two-stage CM, where the time slot allocated for multicast services T is divided into two time intervals, T_1 and T_2. At the first stage, the BS transmits multicast data at a data rate $R_{two,1}$ with a power of $P_{BS,CM}$ so that only part of the MSs, i.e., the MSs with good channel conditions, could successfully receive the data. Then, at the second stage, with a time duration of T_2, some successful MSs (SMSs) are selected as mobile relays (MRs) to convey the multicast data at a rate of $R_{two,2}$ to the unsuccessful MSs

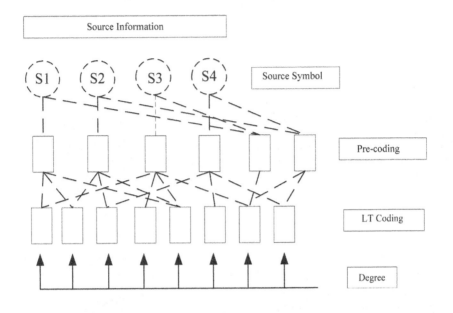

Figure 8.7: Raptor code.

(UMSs). To provide the same service to all MSs, the same amount of data should be transmitted during the first and second stages, i.e., $R_{two,1}T_1 = R_{two,2}T_2$.

Compared to traditional one-stage multicast (TM), the advantages of the two-stage CM include:

- Path-loss gain: As shown in Figure 8.8, the distance between the MRs and UMSs is much shorter than that between the BS and UMSs. To convey the multicast data successfully, the transmission power of MRs is much lower than that of the BS.

- Spatial diversity: At the second stage of CM, multiple distributed SMSs are employed as MRs to forward the same multicast data to UMSs. These signals can be combined at each UMS to provide better performance. For example, in OFDM-based systems such as LTE and LTE-A, when the CP is sufficiently long, signals from all MRs could fall within a CP duration and a stronger signal may be constructed (CP combining: CPC).

- Time diversity: If the same data are transmitted at the first and second stages, which is possible when $T_1 = T_2$, each UMS could further combine the signals received at the first and second stages with techniques such as maximum ratio combining (MRC). Thus, a better received SNR can be obtained.

- User fairness: In cellular systems, it is always difficult to ensure the service quality of cell edge users. In multicast transmission, the number of UMSs gets larger as their distances to the BS increase. Fortunately, SMSs exist even

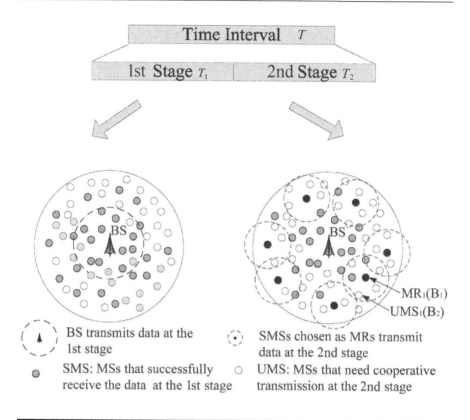

Figure 8.8: Two-stage cooperative multicast transmission.

for locations close to cell edges. Employing those SMSs as MRs, there is a high possibility that the UMSs could receive multicast data successfully at the second stage. Since the same amount of data is transmitted during the first and second stage, the cell edge and cell center users are served with the same quality, so that user fairness can be ensured.

With the path loss gain provided by shorter communication distances, spatial diversity provided by distributed MRs, and time diversity provided by two-stage transmission, CM should be more spectral efficient than the traditional ones given the same power consumption, or more energy efficient to provide the same spectral efficiency, equivalently. It has been shown that by employing all SMSs as MRs, a much higher spectral efficiency can be achieved compared to that of TM [31, 32]. However, the total power consumption also increases significantly with the number of MRs. Since energy efficiency has been widely taken as a key feature of future wireless systems [33], it is necessary to investigate spectral and energy-efficient CM schemes, which will be presented in Section 8.3.

Although CM could enhance the spectral and energy efficiency of the traditional one, it is obtained at the cost of increased complexity. First, additional signaling is needed in CM if the BS is to control the second stage transmission. For example, SMSs may send signals to the BS to indicate that they are willing to act as relays. The BS should then inform the selected SMSs to relay data to other UMSs. Moreover, the location information of MSs may be needed in CM. Thus, after the first stage transmission, the BS could know the positions of SMSs and UMSs and efficient MR selection could be performed. With the rapid development location-based services, the location of MSs might be obtained using terminal-based or network-based positioning techniques [34]. In addition, SMSs in CM spend extra power to relay the data, which is undesirable considering the limited battery life at mobile devices. There are two possible ways to mitigate the problem. One is that the number of SMSs selected as MRs should be as small as possible. The other is to introduce proper stimulating schemes to encourage SMSs to participate in the CM [35].

8.3 Spectral and Energy Efficient Cooperative Multicast

Recently, power consumption of wireless networks has attracted a lot of attention as energy costs make up a vast portion of operational expenditure (OPEX). For example, China Mobile spends more than 40% OPEX on electric energy, among which about 72% is consumed by the BS [11]. Therefore, improving power efficiency of wireless networks has become a hot topic in both academia and industry. In cooperative multicast transmission, a few studies have been conducted to save energy. For example, to employ the cooperative multicast transmission with reasonable power consumption, the nearest neighbor protocol is proposed [32], where only the closest SMSs to UMSs are chosen as MRs. Compared to [31], the power consumption can be improved a lot at the cost of providing the CSI of all MSs. Meanwhile, assuming multiple antennas at the BS, transmit beamforming is studied in [36], targeting minimization of the BS transmission power in two-stage CM. Furthermore, to design energy-efficient schemes, a location-based service technology is employed in [37, 38] to select proper MRs at the second stage, which can reduce the energy consumption by 10 to 18%. Reference [38] modifies the scheme in [31] by letting cooperative transmission happen only if a SMS receives requests from UMSs. Thus, power consumption can be reduced by eliminating unnecessary transmissions. However, the coverage performance cannot be guaranteed by this scheme.

This section provides a detailed introduction on the spectral and energy-efficient two-stage CM transmission, which aims to guarantee a practical coverage ratio (such as 95%) with minimized total power consumption P_{tot}. It should be noted that P_{tot} is related to a number of system parameters, such as the transmission power of BS at the first stage $P_{BS,C}$, the transmission power of MR at the second stage P_{MS}, the MR arrangement schemes, the density of MSs in the cell, the signal processing scheme at the receiver, and so on. As illustrated before, most of the energy in current mobile networks is consumed by BS. Therefore, $P_{BS,C}$ is taken as a key parameter when minimizing P_{tot} for the spectral and energy-efficient two-stage CM and P_{MS} is fixed. Moreover, assume that totally N_{MR} MRs are needed at the second stage; then $P_{tot} = P_{BS,C} \frac{T_1}{T} + N_{MR} P_{MS} \frac{T_2}{T}$. Therefore, besides $P_{BS,C}$, the MR arrangement scheme is another key to minimizing P_{tot}, which decides the value of N_{MR}.

In this section, the traditional one-stage multicast (TM) is employed as a comparing scheme. For fair comparison, the CM and TM schemes should provide the same multicast service and the same coverage performance. Denote the transmission data rates of TM and CM at the first and second stages as R_{one}, $R_{two,1}$, and $R_{two,2}$, respectively, and the received SNRs as SNR_{one}, $SNR_{two,1}$, and $SNR_{two,2}$, respectively. Assume that the transmission data rate can be chosen according to the channel capacity, the relationship between the data rate and the received SNR can be obtained as $R_{one} = B \log_2(1 + SNR_{one})$, $R_{two,1} = B \log_2(1 + SNR_{two,1})$, and $R_{two,2} = B \log_2(1 + SNR_{two,2})$. Hence, to provide the same multicast service with TM and CM, it is required that the same amount of data is transmitted during each multicast time interval, i.e., $T_1 \cdot B \log_2(1 + SNR_{two,1}) = T \cdot B \log_2(1 + SNR_{one})$ and $T_2 \cdot B \log_2(1 + SNR_{two,2}) = T \cdot B \log_2(1 + SNR_{one})$. The corresponding SNR thresholds at the first and second stages are then given by $SNR_{two,1} = (1 + SNR_{one})^{T/T_1} - 1$ and $SNR_{two,2} = (1 + SNR_{one})^{T/T_2} - 1$, respectively. In this section, it is assumed that $T_1 = T_2$ and $SNR_{two,1} = SNR_{two,2} = SNR_{two}$.

8.3.1 Optimized Two-Stage CM for High User Density

Assuming high user density, the optimized two-stage CM is designed that targets minimization of the total power consumption P_{tot} while guaranteeing a coverage ratio of C_{th}. Note that the coverage performance of TM has been given in (8.2). For the two-stage CM, the coverage at the first stage can also be obtained from (8.2). At the second stage, more than one SMS can be employed to forward data to UMSs. In real systems, when the CP is sufficiently long, signals from all MRs could fall within a CP duration and a stronger signal may be constructed (CP combining: CPC). Although CPC is practical, it would be difficult to decide the coverage area of a MR with CPC since any location is covered by multiple MRs. Therefore, to facilitate the performance analysis, a selective combining scheme based on average received signal strength (SCA) is adopted where UMS only receives the signal from the nearest MR, while the signals from other MRs are ignored. Then, the coverage ratio of MRs at the second stage could be obtained similarly to (8.2) [39].

8.3.2 MR Arrangement Based on Sector Ring Structures

Next, the MR arrangement scheme should be designed. Since the user density is sufficiently high, it can be ensured that any location in the cell can be covered by at least one MR at the second stage. Aiming to provide uniform coverage performance for different locations of the macro cell, a MR arrangement based on sector ring structures is employed for the second stage transmission as follows. As shown in Figure 8.9, given the desired coverage ratio C_{th}, for any BS transmission power $P_{BS,C}$, a radius $R_{BS,C}$ can be obtained using (8.2), within which the coverage ratio is C_{th}.

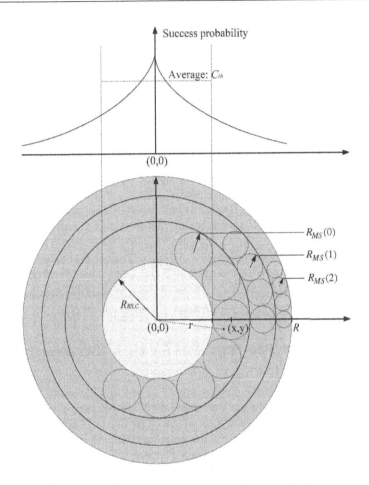

Figure 8.9: MR arrangement based on sector ring structures at the second stage.

Hence, there is no need to further employ MRs in this area and MRs should be located in the annular region with radius from $R_{BS,C}$ to R. This annular region is further divided into several rings, where the coverage ratio at the first stage is below C_{th} and decreases as the distance from the BS increases. So each ring needs MRs to improve the overall coverage ratio to C_{th} after the two-stage transmission, and the required coverage ratio of the MRs increases with their distances from the BS. Note that MRs have a fixed transmission power. So their coverage radius decreases as their distances from the BS increase and is denoted as $R_{MS}(0)$, $R_{MS}(1)$, and so on. The farther away the ring is from the BS, the more MRs that should be employed. The following issues should be noted for this MR arrangement scheme.

First, after two-stage CM transmission, to guarantee a coverage ratio of C_{th} for the macro cell, it is not necessary for the coverage ratio of each ring to be C_{th}. It can be lower than C_{th} if another ring provides a coverage ratio higher than C_{th}. However, from the fairness point of view, it is desirable that each ring of the macro cell can achieve the same coverage ratio of C_{th}.

Second, although the transmission power of MR is fixed, the required coverage capability varies according to its position. When it is closer to the BS, the success probability of the first transmission is higher. Therefore, to achieve C_{th} after two-stage transmission, the required success probability of the second transmission can be lower. Accordingly, the required coverage ratio of the MR is lower and the corresponding coverage radius is larger.

Third, assume that the cellular coverage area of MR is equivalent to the sector ring that contains it. If there are totally M rings, denote the coverage radius of MR in the ith ring as $R_{MS}(i)(i=0,\ldots,M-1)$, so the corresponding area is given by $S_{MS,i} = \pi R_{MS}^2(i)$. Moreover, the area of the ith ring is $S_{Ring,i} = \pi \left(R_{BS,C} + 2\sum_{j=0}^{i} R_{MS}(i) \right)^2 - \pi \left(R_{BS,C} + 2\sum_{j=0}^{i-1} R_{MS}(i) \right)^2$. Therefore, to ensure the coverage ratio of the ith ring, the number of MRs needed can be approximated by

$$N(i) \approx \frac{S_{Ring,i}}{S_{MR,i}} = \frac{\left(R_{BS,C} + 2\sum_{j=0}^{i} R_{MS}(i) \right)^2 - \left(R_{BS,C} + 2\sum_{j=0}^{i-1} R_{MS}(i) \right)^2}{R_{MS}^2(i)} \tag{8.4}$$

Then, the total power consumption is given by

$$P_{tot} = P_{BS,C}\frac{T_1}{T} + \sum_{i=0}^{M-1} N(i)P_{MS}\frac{T_2}{T} \tag{8.5}$$

which is decided by $P_{BS,C}$ and $N(i)$. Therefore, to carry out power consumption analysis, the relationship between $P_{BS,C}$ and $N(i)$ (or $R_{MS}(i)$, equivalently) should be obtained.

8.3.3   Relationship between $P_{BS,C}$ and $N(i)$

Since $P_{BS,C}$ decides the coverage ratio at the first stage, the relationship between $P_{BS,C}$ and $N(i)$ could be set up by ensuring the coverage ratio of the ith ring to be no less than C_{th}, which could be approximated by [39]

$$C_{th} \approx C_{BS}(i) + C_{MS}(i) - C_{BS}(i)C_{MS}(i) \tag{8.6}$$

where $C_{BS}(i)$ and $C_{MS}(i)$ are the coverage ratios of the cellular area of radius $R_{MS}(i)$ provided by the BS and MRs at the first and second stages, respectively. $C_{BS}(i)$ can be further approximated by the coverage ratio of the sector ring between $R_{BS,C} + 2\sum_{j=0}^{i-1} R_{MS}(j)$ and $R_{BS,C} + 2\sum_{j=0}^{i} R_{MS}(j)$, which is decided by $P_{BS,C}$ and $\sum_{j=0}^{i} R_{MS}(j)$. Moreover, $C_{MS}(i)$ can be obtained from (8.2), which depends on

$R_{MS}(i)$. Consider the zeroth ring where $\sum_{j=0}^{i} R_{MS}(j) = R_{MS}(0)$. The relationship between $P_{BS,C}$ and $R_{MS}(0)$ can be obtained, which can be denoted as $R_{MS}(0, P_{BS,C})$. Then for the first ring, the relationship between $P_{BS,C}$ and $R_{MS}(1)$ can be expressed as $R_{MS}(1, P_{BS,C}, R_{MS}(0)) = R_{MS}(1, P_{BS,C})$. Similarly, $R_{MS}(i)$ is related to $P_{BS,C}$ by $R_{MS}(i, P_{BS,C})$. Then, $N(i)$ can be expressed as $N(i, P_{BS,C})$ and the total power consumption can be minimized by

$$P_{tot} = \min_{P_{BS,C}} \left\{ P_{BS,C} \frac{T_1}{T} + \sum_{j=0}^{M-1} N(j, P_{BS,C}) P_{MS} \frac{T_2}{T} \right\} \tag{8.7}$$

where the total number of rings M is decided by increasing the number of rings until the coverage area of the BS and all the rings is no less than the macro cell coverage area, which can be given by

$$M = \min_{i} \left\{ R \leq R_{BS,C} + 2 \sum_{j=0}^{i} R_{MS}(j) \right\} \tag{8.8}$$

Note that for a given cell radius of R, it is possible that the chosen M results in $R_{BS,C} + 2 \sum_{j=0}^{M-2} R_{MS}(i) < R < R_{BS,C} + 2 \sum_{j=0}^{M-1} R_{MS}(i)$. That means, the macro cell cannot be covered by integer number of rings. In this case, to ensure a coverage ratio of C_{th} for the cell, part of the transmission power of MRs is wasted. To be energy efficient, the corresponding $P_{BS,C}$ should be reduced or increased to ensure that the cell can be covered by an integer number of rings. So it can be expected that to minimize the total transmission power, the optimal $P_{BS,C}$ should make the macro cell be exactly covered by the BS and integer number of rings.

8.3.4 *Numerical Search for Optimal $P_{BS,C}$*

According to the previous analysis, numerical computations are needed to get the optimal $P_{BS,C}$ that achieves the minimum total power consumption. The algorithm is described as follows:

Step 1: Set system parameters like P_{MS}, R, C_{th}, SNR_{one}, T, T_1, and T_2 and calculate SNR_{two}.

Step 2: Search for the optimal $P_{BS,C}$ from a high value (e.g., the transmission power of BS in TM P_{BS}) with a search step Δ_p. Set $n = 0$ as the index for the value of $P_{BS,C}$ and $P_{BS,C}(n) = P_{BS}$.

Step 2.1: According to (8.2), $R_{BS,C}$ can be calculated with the given values of $P_{BS,C}(n)$, C_{th}, and SNR_{two}. Set $i = 0$ as the index for the ring next to the coverage area of the BS.

Step 2.2: For the MR in the ith ring, calculate the required coverage ratio $C_{BS}(i)$.

Step 2.3: Similar to step 2.1, calculate $R_{MS}(i)$ with the given values of P_{MS}, $C_{BS}(i)$, and SNR_{two}. Then the number of MRs needed in this ring $N(i, P_{BS,C})$ can be obtained.

Step 2.4: If $R_{BS,C} + 2\sum_{j=0}^{i} R_{MS}(j) < R$, the macro cell cannot be covered by the BS and $(i+1)$ rings. Set $i = i + 1$ and go to step 2.2. If $R_{BS,C} + 2\sum_{j=0}^{i} R_{MS}(j) \geq R$, the whole cell has been covered and there is no need for more rings.

Step 2.5: Calculate the total transmission power $P_{tot}(n)$ corresponding to $P_{BS,C}(n)$. Set $P_{BS,C}(n+1) = P_{BS,C}(n) - \Delta_P$. If $P_{BS,C}(n+1)$ is still larger than a predefined lower bound, set $n = n + 1$ and go to step 2.1.

Step 3: Find the minimum total transmission power from the set of values $\{P_{tot}(n)\}$ and the corresponding $P_{BS,C}$ is the desired optimal BS transmission power $P_{BS,C}^*$.

Although the numerical search for $P_{BS,C}^*$ needs a lot of computation, it is done offline and will not increase the complexity of the system. In practice, the two-stage CM with optimized power consumption and the MR arrangement based on sector ring structures could be carried out as follows. At the first stage, the BS transmits multicast data using a power of $P_{BS,C}^*$. Since the number of rings, the number of MRs in each ring, and their coverage radiuses are all calculated when searching for $P_{BS,C}^*$, a map of MR locations suggested by the proposed scheme can be obtained and stored in the BS. After the first stage transmission, each SMS should send a message to the BS, indicating that it has successfully received the data. Location information of each SMS is needed at the BS, which might be obtained using terminal-based or network-based positioning techniques [34]. Then the BS should select the SMS closest to a suggested MR location as the MR and send messages to notify these SMSs. At the second stage, the chosen SMSs act as MRs and transfer multicast data to UMSs. It can be seen that compared to TM, the two-stage CM needs two simple signalings after the first stage transmission to determine which SMSs should be MRs. Moreover, additional location information is needed, which should not be a difficult task with the development of location-based services.

8.3.5 Theoretical Estimation

It can be seen that a lot of computation is needed in the previous numerical search for the optimal $P_{BS,C}$ achieving the minimum total power consumption. In fact, further approximations can be adopted to provide a simple theoretical estimation for the optimal $P_{BS,C}$. Consider any position (x, y) in the ith ring. The success probabilities of the first and second stage transmissions are given by $s_{two,1}(x,y)$ and $s_{two,2}(x,y)$, respectively; i.e., the MS at (x, y) can receive the multicast data from the BS at the first stage with a probability of $s_{two,1}(x,y)$ or from the MR at the second stage with a probability of $s_{two,2}(x,y)$. Then after the two-stage transmission, the final success probability becomes $s_{two}(x,y) = 1 - (1 - s_{two,1}(x,y)) \cdot (1 - s_{two,2}(x,y))$. Here $s_{two}(x,y)$ is set to C_{th} so that the coverage ratio can be guaranteed. Denoting the distance between (x, y)

and the BS as r (see Figure 8.9), it is obtained that

$$s_{two,1}(r) = \exp\left(-\frac{SNR_{two} \cdot \sigma_{noi}^2}{A_1 \cdot P_{BS,C} \cdot r^{-\gamma}}\right) \tag{8.9}$$

Then $s_{two,2}(r)$ should satisfy

$$s_{two,2}(r) = 1 - \frac{1 - C_{th}}{1 - s_{two,1}(r)} = \exp\left(-\frac{SNR_{two} \cdot \sigma_{noi}^2}{A_2 \cdot P_{MS} \cdot R_{MS}^{-\gamma}}\right) \tag{8.10}$$

It can be obtained that

$$R_{MS}(r) = \left(\ln\left(\frac{1 - s_{two,1}(r)}{C_{th} - s_{two,1}(r)}\right) \cdot \frac{A_2 P_{MS}}{\sigma_{noi}^2 \cdot SNR_{two}}\right)^{1/\gamma} \tag{8.11}$$

which is further averaged over $R_{BS,C}$ to R to get a coverage radius $\overline{R_{MS}}$ independent of r. Then, the number of MRs needed at the second stage transmission can be estimated by

$$N_{tot} = \frac{\pi(R^2 - R_{BS,C}^2)}{\pi \overline{R_{MS}}^2} \tag{8.12}$$

As shown in Figure 8.9, the ring from $R_{BS,C}$ to R cannot be seamlessly covered by the MRs with a cellular coverage area. So N_{tot} obtained from (8.12) is a lower bound for the actual number of MRs needed to improve the coverage ratio to C_{th}.

Thus, the total transmission power can be optimized as follows:

$$P_{tot} = \min_{P_{BS,C}}\left\{P_{BS,C}\frac{T_1}{T} + \frac{R^2 - R_{BS,C}^2}{\overline{R_{MS}}^2} \cdot P_{MS}\frac{T_2}{T}\right\} \tag{8.13}$$

Since both $R_{BS,C}$ and $\overline{R_{MS}}$ are functions of $P_{BS,C}$, P_{tot} is only decided by $P_{BS,C}$ and the optimal $P_{BS,C}$ that achieves the minimum total transmission power can be found by numerical methods.

8.3.6 Performance Evaluation

The system parameters are shown in Table 8.I. Referring to practical network settings, the target coverage ratio C_{th} is chosen to be 95%, the transmission power of BS for TM is 33.88 W (45.3 dBm), and the transmission power of MS is 0.20 W (23 dBm). The radius of the macro cell is set to 1500 m. Moreover, two different path loss models are employed for the BS to MS and the MR to MS transmission [40]. It should be noted that the value of path loss depends on a number of elements, including the height of transmitter, the height of receiver, the reference distances, the carrier frequency, and the distance between the transmitter and receiver. Since the BS and MR have different heights and reference distances, the path loss models for the BS to MS and for the MR to MS should be different.

Table 8.1 System Parameters

Carrier frequency	2.5 G
Frequency band B	10 M
Path loss from BS to UE $\overline{PL}_{BS}(dB)$	$17.39 + 37.6\log_{10}(d[m])$
Path loss from UE to UE $\overline{PL}_{MS}(dB)$	$37.78 + 37.6\log_{10}(d[m])$
Path loss parameter γ	3.76
Transmission power of BS for conventional one-stage multicast P_{BS}	33.88 W (45.3 dBm)
Transmission power of MS P_{MS}	0.20 W (23 dBm)
Noise power spectrum N_0	-169 dBm/Hz
Coverage ratio C_{th}	95%
Cell radius R	1500 m
Rate of multicast R_{one}	0.89 bps/Hz

Using these settings, the optimal BS transmission power $P_{BS,C}^*$ achieving the minimum total power consumption is shown in Figure 8.10, which is obtained by numerical searching $P_{BS,C}$ from 4 W to 64 W. Moreover, the corresponding number of MRs needed to guarantee a coverage ratio of 95% is illustrated in Figure 8.11. It can be seen that in general, the total power consumption decreases as $P_{BS,C}$ increases from 4 W and reaches the minimum value of 19.2 W when $P_{BS,C} = 11.32$ W. Corresponding to the optimal $P_{BS,C}$, the macro cell is exactly covered by the BS and three rings, where 31, 47, and 60 MRs are allocated with coverage radiuses of $R_{MS}(0) = 127.9$, $R_{MS}(1) = 104.9$, and $R_{MS}(2) = 94.0$ m, respectively. So the total number of MRs is 138, as shown in Figure 8.11. When $P_{BS,C}$ increases further, the total power consumption becomes higher. Note that the total power consumption of the two-stage CM scheme changes in a staggered way with $P_{BS,C}$. This is because the number of MRs needed to provide the desired coverage ratio does not change smoothly. For example, when $P_{BS,C} = 6$ W, according to the proposed scheme, the whole cell can be exactly covered by the BS and four rings with 190 MRs. As $P_{BS,C}$ increases, its coverage area becomes larger and the sector ring from $R_{BS,C}$ to R gets smaller. But the system still needs four rings to cover the whole cell with a guaranteed coverage ratio of 95%; even some power of MRs is wasted because the actual coverage area is larger than the macro cell. The number of MRs even slightly increases due to the reduced width of rings. When $P_{BS,C}$ is further increased to 11.32 W, the cell can be exactly covered by the BS and three rings with 138 MRs. Thus, the number of MRs and the total transmission power reduce sharply. It can be seen that to be energy efficient, the optimal $P_{BS,C}$ must be obtained when the macro cell can be exactly covered by the BS and integer number of rings. This consists with the analysis of (8.8).

Moreover, the optimal BS transmission power and the number of MRs obtained from the simple theoretical estimation (8.13) are also plotted in Figure 8.10 and Figure 8.11, respectively. It can be seen that the estimation for the optimal BS power is close to the numerical results of the proposed scheme, which verifies the effectiveness of the theoretical estimation (8.13). For comparison, the total transmission power of TM (33.88 W) is also shown in Figure 8.10. Using the two-stage CM trans-

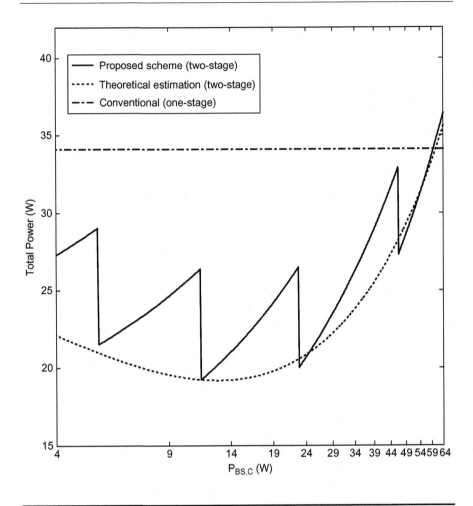

Figure 8.10: Total power consumption as a function of BS transmission power.

mission with a guaranteed coverage ratio of 95%, the total power consumption can be saved by more than 40%. This is because the CM can benefit from path loss gain. Compared to the long communication distance between the BS and MS, the distance between the MR and UMS is much shorter. Although for the same distance the path loss between MR and UMS is larger than that between BS and MS, as shown by the two different path loss models [40], the resultant path loss gain is considerable as a whole. Moreover, the BS power consumption can be saved by more than 80%, thanks to the cooperation provided by MSs. This would be attractive to operators. If customers are properly stimulated and willing to provide cooperation, the power consumption of the BS can be greatly reduced.

Finally, it would be interesting to compare the energy efficiency between the optimized two-stage CM scheme and existing schemes such as the one in [31] (denoted

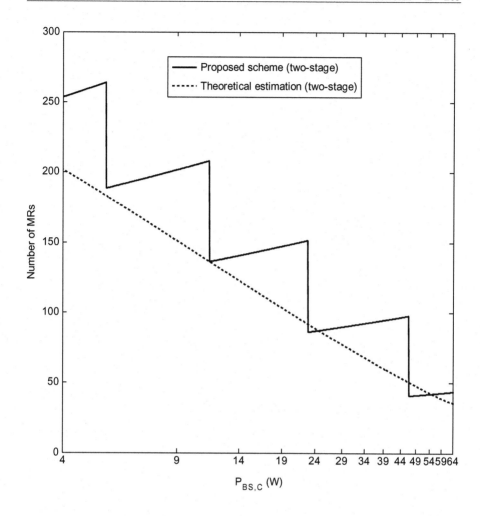

Figure 8.11: Required number of MRs for guaranteed coverage as a function of $P_{BS,C}$**.**

as scheme [31]). Table 8.2 lists the differences between the optimized scheme and Scheme [31]. Energy efficiency is defined as throughput per watt and simulation results are shown in Figure 8.12 for the two schemes. It can be seen that the optimized scheme has two orders higher energy efficiency than scheme [31], and the performance gap increases with user density. This is because according to the optimized scheme, as long as the cell radius and the MS transmission power are given, the number of MRs and their locations are determined and will not change with user densities. When more users are considered, the system throughput increases with the same power consumption. Therefore, the energy efficiency becomes higher and higher as the number of users increases. As a whole, targeting minimization of power

Table 8.2 **Main Differences between Scheme [31] and the Optimized CM Scheme**

		Scheme [31]	Optimized Scheme
Target		Maximize system throughput	Minimize the total transmission power
Coverage ratio		100%	95%
Additional signaling		0	2
Additional information		0	Location information of SMSs needed
1st stage	**BS transmission power**	Fixed (45.3 dBm)	Optimized to achieve the minimum total transmission power
	Coverage ratio	Fixed (50%)	Optimized with the BS power
2nd stage	**How to choose MR**	All SMSs	Selected SMSs closest to the MR position suggested by the proposed scheme

consumption with guaranteed coverage at high density, the optimized scheme outperforms scheme [31] concerning energy consumption and efficiency, at the cost of two additional signalings and the need for the location information of all SMSs.

8.3.7 Discussions and Future Work

The optimized two-stage CM with high user density and SCA has been introduced. Note that SCA is employed to facilitate theoretical analysis. In real systems, CPC should be considered. It has been shown in [41] that given the same conditions, CM with CPC could further reduce the total power consumption by up to 17% compared to that with SCA. Moreover, maximum ratio combining (MRC) could also be used to combine the signal at different stages to further improve the performance.

On the other hand, although CM has been shown to be superior to TM in various scenarios, it is noted that CM may be inferior to TM in some circumstances, such as when the number of MSs in the cell is small (or low user density, equivalently, considering a fixed cell radius). Although it is not explicitly stated in [31], from simulation results, one can expect that CM may perform worse than TM when the number of MSs is less than 10. In practice, it is possible that only a small number of MSs are served, such as at the beginning stage of promoting multicast services, or only a few MSs are willing to relay the data. It remains unknown if CM should always be employed and what kind of MR selection schemes should be adopted in these scenarios.

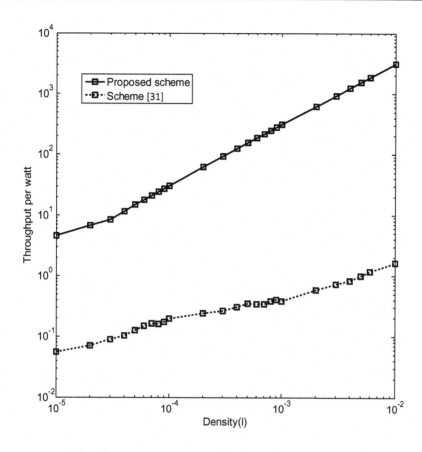

Figure 8.12: Comparison of energy efficiencies between the proposed scheme and scheme [31].

References

1. Ericson Mobility Report, June 2013, http://www.ericsson.com/mobility-report.

2. The Strategy Analytics Forecast, Handset Data Traffic (2001–2017), http://www.prnewswire.com/news-releases/strategy-analytics-handset-data-traffic-to-grow-over-300-by-2017-to-21-exabytes-214113401.html# prettyPhoto.

3. Cisco Visual Networking Index: Global Mobile Data Traffic Forecast Update, 2012–2017, http://www.cisco.com/en/US/solutions/collateral/ns341/ns525/ ns537/ns705/ns827/white_paper_c11-520862.html.

4. *IEEE Standard for Local and Metropolitan Area Networks Part 16: Air Interface for Fixed and Mobile Broadband Wireless Access Systems*, IEEE Standard 802.16 Working Group, 2005.

5. T. Jiang, W. Xiang, H.H. Chen and Q. Ni, "Multicast Broadcast Services Support in OFDMA-Based WiMAX Systems" IEEE Comm. Mag., vol. 45, no. 8, pp. 78–86, Aug. 2007.

6. 3GPP Technical Specifications TS 25.346, *Introduction of the Multimedia Broadcast/Multicast Service (MBMS) in the Radio Access Network (RAN)*, March 2006, http://www.3gpp.org/.

7. 3GPP Technical Specifications TS 36.300, *3rd Generation Partnership Project; Technical Specification Group Radio Access Network; Evolved Universal Terrestrial Radio Access (E-UTRAN) and Evolved Universal Terrestrial Radio Access Network (E-UTRAN); Overall Description*, v10.3.0, Stage 2, March 2011.

8. Mobile/Tablet TV and Video Content, Broadcast and OTT Strategies 2013–2017, Juniper Research, May 2013.

9. What Is the Future of CMMB Mobile TV with Less Than 5% Subscribers, August 2012, http://tech.163.com/12/0821/07/89DORQ2I00094MOK.html.

10. http://www.totaltele.com/view.aspx?ID=454228.

11. China Mobile Research Institute, *C-RAN White Paper: Green Evolution of Wireless Access Network*, v1.0.0 ed., 2010.

12. C. Han, T. Harrold, S. Armour, I. Krikidis, S. Videv, P. M. Grant, H. Haas, J. S. Thompson, I. Ku, C.-X. Wang, T. A. Le, M. R. Nakhai, J. Zhang, and L. Hanzo, Green Radio: Radio Techniques to Enable Energy Efficient Wireless Networks, *IEEE Commun. Mag.*, vol. 49, no. 6, pp. 46–54, June 2011.

13. C. Suh and J. Mo, Resource Allocation for Multicast Services in Multicarrier Wireless Communications, *IEEE Trans. Wireless Commun.*, vol. 7, no. 1, pp. 27–31, 2008.

14. L. Tian, Y. Zhou, Y. Zhang, G. Sun, and J. L. Shi, Resource Allocation for Multicast Services in Distributed Antenna Systems with QoS Guarantees, *IET Commun.*, vol. 6, no. 3, pp. 264–271, 2012.

15. H. Zhu and J. Wang, Chunk-Based Resource Allocation in OFDMA Systems—Part I: Chunk Allocation, *IEEE Trans. Commun.*, vol. 57, no. 9, pp. 2734–2744, 2009.

16. H. Zhu and J. Wang, Chunk-Based Resource Allocation in OFDMA Systems—Part II: Joint Chunk, Power and Bit Allocation, *IEEE Trans. Commun.*, vol. 60, no. 2, pp. 499–509, 2012.

17. N. Guan, Y. Zhou, H. Liu and J.L. Shi, "An Adaptive Multicast Transmission Scheme with Coverage Probability Guarantees," *J. System Simulation*, vol. 25, no. 6, pp. 1235–1240, Jun. 2013.

18. 3GPP TS 25.346, Introduction of the Multimedia Broadcast Multicast Service (MBMS) in the Radio Access Network (RAN), Stage 2, Release 6.

19. K. Ramchandran, A. Ortega, K. M. Uz, and M. Vetterli, Multiresolution Broadcast for Digital HDTV Using Joint Source/Channel Coding, *IEEE J. Sel. Areas Commun.*, vol. 11, no. 1, pp. 6–23, 1993.

20. T. Cover, Broadcast Channels, *IEEE Trans. Inform. Theory*, vol. IT-18, pp. 2–14, 1972.

21. Md. J. Hossain, P. K. Vitthaladevuni, M.-S. Alouini, V.K. Bhargava, and A.J. Goldsmith, Adaptive Hierarchical Modulation for Simultaneous Voice and Multiclass Data Transmission Over Fading Channels, *IEEE Trans. Vehicular Technol.*, vol. 55, no. 4, 2006.

22. M. Shaqfeh, W. Mesbah, and H. Alnuweiri, Utility Maximization for Layered Broadcast over Rayleigh Fading Channels, *IEEE ICC2010*, June 2010, pp. 1–6.

23. J. Q. Mei, H. Ji, and Y. Li, Energy Efficient Layered Broadcast/Multicast Mechanism in Green 4G Wireless Networks, *IEEE Infom2011 Workshop,* pp. 295–300.

24. Y. Liang, L. Lai, H. V. Poor and S. Shamai, The Broadcast Approach over Fading Gaussian Wiretap Channels, Information Theory Workshop, 2009.

25. J. She, X. Yu, P. H. Ho, and E. H. Yang, A Cross-Layer Design Framework for Robust IPTV Services over IEEE 802.16 Networks, *IEEE Sel. Areas Commun.*, vol. 27, no. 2, pp. 235–245, 2009.

26. J. Hong and P. A. Wilford, A Hierarchical Modulation for Upgrading Digital Broadcast Systems, *IEEE Trans. Broadcasting*, vol. 51, no. 2, pp. 223–229, 2005.

27. J. W. Byers, M. Luby, M. Mitzenmacher and A. Rege, A Digital Foutain Approach to Reliable Distribution of Bulk Data, *Proceedings of the ACM SIGCOMM '98*, May 1998, pp. 56–67.

28. M. Luby, LT Codes, *Proceedings of the 43rd Symposium on Foundations of Computer Science*, November 2002, pp. 271–280.

29. A. Shokrollahi, Raptor Codes, *IEEE Trans. Inform. Theory*, vol. 52, no. 6, pp. 2551–2567, 2006.

30. 3GPP Technical Specifications TS 26.346, *Multimedia Broadcast/Multicast Services (MBMS); Protocols and Codecs (Release 6)*, v6.3.0, 2005, http://www.3gpp.org/.

31. F. Hou, L. Cai, P. Han, and X. Shen, A Cooperative Multicast Scheduling Scheme for Multimedia Services in IEEE 802.16 Networks, *IEEE Trans. Wireless Commun.*, vol. 8, pp. 1508–1519, 2009.

32. N. Guan, Y. Zhou, H. Liu, L. Tian, and J. L. Shi, An Energy Efficient Cooperative Multicast Transmission Scheme with Power Control, *IEEE GLOBECOM2011*, December 2011, pp. 1–5.

33. C. Han, T. Harrold, S. Armour, I. Krikidis, S. Videv, P. M. Grant, H. Haas, J. S. Thompson, I. Ku, C.-X. Wang, T. A. Le, M. R. Nakhai, J. Zhang, and L. Hanzo, Green Radio: Radio Techniques to Enable Energy Efficient Wireless Networks, *IEEE Commun. Mag.*, vol. 49, no. 6, pp. 46–54, 2011.

34. I. A. Junglas and R. T. Watson, Location Based Services, *Commun. ACM*, vol. 51, no. 3, pp. 65–69, 2008.

35. H. C. Lu, W. Liao, M. C. Chen, and M. A. Alhussein, Coding-Aware Peer-to-Peer Data Repair in Multi-Rate Wireless Networks: A Game Theoretic Analysis, *IEEE J. Sel. Areas Commun.*, vol. 31, no. 99, pp. 391–398, 2013.

36. T. Han and N. Ansari, Energy Efficient Wireless Multicasting, *IEEE Commun. Lett.*, vol. 15, no. 6, pp. 620–622, 2011.

37. S. M. Elrabiei and M. H. Habaebi, Energy Efficient Cooperative Multicasting for MBS WiMAX Traffic, *IEEE ISWPC2010*, May 2010, pp. 600–605.

38. J. Lee, Y. M. Lim, K. Kim, and S. G. Choi, Energy Efficient Cooperative Multicast Scheme Based on Selective Relay, *IEEE Commun. Lett.*, vol. 16, no. 3, pp. 386–388, 2012.

39. Y. Zhou, H. Liu, Z. G. Pan, L. Tian, J. L. Shi and G. H. Yang, "Two-Stage Cooperative Multicast Transmission with Optimized Power Consumption and Guaranteed Coverage," *IEEE JSAC*, vol. 32, no. 2, pp. 274–284, Feb. 2014.

40. IEEE 802.16m-08/004r5, *IEEE 802.16m Evaluation Methodology Document*, January 2009.

41. H. Liu, Y. Zhou, L. Tian, H. H. Chen, X. Han, and J. L. Shi, Investigation on Energy Efficiency of OFDM-Based Two-Stage Cooperative Multicast with CP Combining, *IEEE WCNC2013*, April 2013, pp. 1575–1580.

Chapter 9

Mobility Management in Femtocell Networks

Mahmoud H. Qutqut and Hossam S. Hassanein

School of Computing, Queen's University, Ontario, Canada

CONTENTS

Abstract

Current wireless broadband networks (WBNs) are facing several limitations and considerations, such as poor indoor coverage, explosive growth in data usage, and massive increase in number of WBN subscribers. Various inventions and solutions are used to enhance the coverage and increase the capacity of wireless networks. Femtocells are seen as a key next step in wireless communication today. Femtocells offer excellent indoor voice and data coverage. As well, femtocells can enhance the capacity and offload traffic from macrocells. There are several issues that must be considered though to enable the successful deployment of femtocells. One of the most important issues is mobility management. Since femtocells will be deployed densely, randomly, and by the millions, providing and supporting seamless mobility and handoff procedures is essential. We present a broad study on mobility management in femtocell networks. Current issues of mobility and handoff management are discussed. Several research works are overviewed and classified. Finally, some open and future research directions are discussed.

List of Abbreviations

3G	Third generation
3GPP	Third generation Partnership Project
4G	Fourth generation
AAS	Advanced antenna system
ACL	Access control/CSG list
AP	Access point
ARPU	Average revenue per user
BS	Base station
CN	Core network
CPE	Consumer premises equipment
CSG	Closed subscriber group
DA	Distributed antenna
DL	Downlink
DSL	Digital subscriber line
eNB	Enhanced NB
E-UTRAN	Evolved UTRAN
FAP	Femto AP
FBS	Femto BS
FDD	Frequency division duplex
Femto-GW	Femto gateway
GPRS	General packet radio service
GPS	Global positioning system
GSM	Global System for Mobile Communications
HCS	Hierarchical cell structure
HLR	Home location register
HO	Handoff
HSDPA	High-speed downlink packet access
HSPA	High speed packet access
HSPA+	Evolved HSPA
HSUPA	High-speed uplink packet access
IEEE	Institute of Electrical and Electronics Engineers
IMT-2000	International Mobile Telecommunications 2000
IP	Internet Protocol
ITU	International Telecommunication Union
Kmph	Kilometer per hour
LAI	Location area identify
LAU	Location area update
LTE	Long Term Evolution
LTE-A	LTE-Advanced
MAC	Media access control
MBN	Mobile broadband network
Mbps	Megabits per second

MHz	Megahertz
MIMO	Multiple-input multiple-output
MOG	Multimedia online gaming
NB	NodeB
NCL	Neighbor cell list
OA&M	Operations, administration, and management
OFDMA	Orthogonal frequency division multiplexing access
OSG	Open subscriber group
PCI	Physical cell identifier
PLMN ID	Public land mobile networks identity
QAM	Quadrature amplitude modulation
QoS	Quality of service
QPSK	Quadrature phase shift keying
RNC	Radio network controller
RSS	Relative/received signal strength
RSSI	Received signal strength indication
SAE	System architecture evolution
SC-FDMA	Single-carrier frequency division multiple access
SeGW	Security gateway
SINR	Signal-to-interference ratio
SOHO	Small office, home office
SON	Self-organization network
TAI	Tracking area identity
TDD	Time division duplex
UE	User equipment
UL	Uplink
UMTS	Universal Mobile Telecommunications System
USIM	Universal Subscriber Identity Model
UTRAN	UMTS Terrestrial Radio Access Network
VLR	Visitor Location Register
VoIP	Voice over IP
WBN	Wireless broadband network
WCDMA	Wideband code division multiple access
Wi-Fi	Wireless Fidelity
WiMAX	Worldwide Interoperability for Microwave Access
WLAN	Wireless local area network
WWAN	Wireless wide area network

9.1 Introduction

There has been a significant interest in wireless broadband technologies over the past two decades. Wireless broadband networks (WBNs) such as third generation (3G) and fourth generation (4G) networks provide high data rates, large coverage areas, and high-quality multimedia services. However, existing 3G and 4G networks share a number of drawbacks, including limited cellular data capacity and poor indoor coverage. The latter poses a major limitation in cellular usage, especially since up to 65%

of voice and 90% of data traffic take place indoors [1]. There is also a tremendous growth of data usage in cell phones, driven by the popular data-hungry applications such as multimedia online gaming (MOG), mobile TV, voice over IP (VoIP), video calling, streaming TV, Web2.0, video on demand, location-based services, and social networks (Facebook, Google+, MySpace). The aforementioned factors mandate a solution that remedies the capacity and coverage constraints. Therefore, different solutions are proposed to solve these issues. The solutions can range from deployment of heterogeneous networks with Wireless Fidelity (Wi-Fi) for dual-mode devices to installation of more cell sites and relay stations, as well as signal boosters. Femtocells were also introduced as a device-compatible and cost-effective solution.

Femtocell networks are seen as a promising solution for enhanced indoor coverage and increased network capacity, as well as offloading traffic from the macro/microcells. Perhaps one of the key requirements for mass deployments and feasibility of femtocells is mobility management. Femtocells have many unique characteristics that make mobility management a critical and difficult process. Such challenges include random deployment in an ad hoc manner, working in a licensed spectrum, overlaying with macro/microcells, and backward compatibility requirements with the existing infrastructures and devices.

This chapter gives an overview of femtocell networks and mobility management. Also, it presents current problems and issues in mobility management. The chapter discusses, reviews, and classifies several recent research efforts on mobility management in femtocell networks.

The remainder of this chapter is organized as follows. An overview of the background topics related to the femtocell networks is presented in Section 9.2. In Section 9.3, mobility management and femtocell network challenges are discussed. The related and common research efforts are studied in Section 9.4, as well as classifications and comparisons of proposed solutions. Finally, Section 9.5 discusses the open research issues and concludes the chapter.

9.2   Femtocells

This section provides an overview of femtocell networks and some related aspects.

9.2.1   What Are Femtocells?

A femtocell is a cell in a cellular network that provides radio coverage and is served by a femto-BS (FBS).[1] FBS, also known as a home-BS or a femto-access point (FAP), is a mini low-power BS installed by end users. FBSs are typically deployed indoors residentially, in small office, home offices (SOHOs), and in enterprises to provide better coverage, especially where access would otherwise be limited or unavailable. FBSs also offer enhanced data capacity and offload traffic from the macro/micro

---

[1] In this chapter, we use FBS to stand for the device itself (BS), and femtocell to refer to the coverage area that is covered by a FBS.

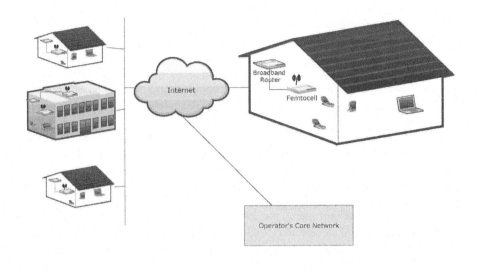

Figure 9.1: Femtocell network overview.

networks. FBSs look like broadband modems and some FBS manufacturers offer a choice of all in one box (DSL modem, Wi-Fi router, and FBS) [2]. Femtocells operate in the licensed spectrum, and have tens of meters of coverage range and can support up to 10 active users in a residential setting. FBSs connect to standard cellular phones and similar devices through their wireless interfaces (LTE, WiMAX, HSDPA+) [3]. Traffic is then routed to the cellular operator's network via a broadband connection (e.g., xDSL, cable), as shown in Figure 9.1.

Femtocells can also be deployed outdoors, and can be used in urban areas and subway stations [4, 5].

9.2.2   Brief History and Current Status

The home base station concept was introduced by Bell Labs of Alcatel-Lucent in 1999 [16]. In 2002, Motorola announced the first 3G home base station [6]. In 2006, femtocell as a term was introduced [17]. In late 2007, the Femto Forum was founded as a nonprofit membership organization to promote and enable femtocell technologies worldwide. The forum changed its name to Smallcell Forum in 2012. The forum supports the adoption of industry-wide standards, regulations, and interoperability of femtocells by telecom operators around the world [7]. Currently, Smallcell Forum includes more than 60 mobile operators and 74 vendors. Furthermore, 27 FBS vendors and more than 45 telecom providers have announced commercial launches of FBSs [7].

9.2.3 Comparison between Femtocells and Other Coverage Solutions

There are many coverage solutions that have been developed to solve the problem of indoor converge. In *microcell and picocell* solutions, the operator installs micro-BSs (with smaller coverage area than macrocells) to improve coverage and capacity in urban or high-density areas with poor reception. In *distributed antenna* (DA) solutions, the operator installs DA elements as signal boosters, which are connected to macro-BS via a dedicated fiber or microwave link. These coverage enhancement solutions are typically expensive and require the operator's involvement. Table 9.1 presents a comparison between these solutions.

As shown in Table 9.1, the cost of a FBS is less than $200, which is a relatively low-cost solution. The cost of a micro-BS is between $200,000 and 400,000, and that of a DA is between $200 and 400. Deploying femtocells will particularly increase the network capacity via using the broadband connections for FBSs' backhaul; deploying more microcells will also increase the network capacity, but not as much as femtocell deployment, due to the limitations of the dedicated backhaul for microcells. DAs will not increase the capacity since they are connected to the macro-BSs. FBSs are installed as end user devices, whereas micro-BSs and DAs are installed by operators. The capital expenditure of femtocells is the hardware cost. However, in microcells, the capital expenditure is in installing a new BS and its cell site, where for DA, installing of antenna elements and backhaul adds to the cost. The operational expenditure for femtocells is in providing backhaul broadband connection.

Table 9.1 **Comparison between Femtocell, Microcell, and Distributed Antenna (DA)**

	Femtocell	Microcell	Distributed Antenna (DA)
Cost	Low	High	Low
Network capacity	Very increased	Increased	Limited to BS capacity
Install	User	Operator	Operator
Capital expenditure	Purchase a FBS	Install a new cell BS	Antenna element and backhaul installation
Operating expenditure	Provide a broadband connection	Electricity, site lease, maintenance, and backhaul	Antenna maintenance and backhaul connection
Indoor coverage problem	Enhance indoor coverage	Does not entirely solve indoor coverage	Does not solve indoor coverage

However, the operational expenditures of microcells are electricity, site lease, maintenance cost, and backhaul, and for DAs they are antenna maintenance and backhaul connection. Femtocells solve the indoor coverage problem; microcells and DAs do not entirely solve indoor coverage.

9.2.4　Benefits of Using FBSs for Users and Operators

There are many potential benefits from the deployment of femtocells. These benefits are summarized below [2, 8–11].

User's benefits:

■ Improved indoor coverage for both data and voice services, since FBS is closed to the users

■ Improved data rate capacity, because FBS uses the user's high data rate broadband connection as its backhaul

■ Reduced indoor cost charged (zone pricing), as the operators will offer attractive pricing plans for indoor calls and data

■ Reduced power consumption for UEs due to the lower transmit power of FBS than of macro/micro-BSs

■ Ability to offer new services, e.g., home gateway, connected home, location-based services

■ Simple deployment, as FBS works as a plug-and-play device

■ No need for new expensive dual-mode UEs, as current UEs work with femtocells

Operator's benefits:

■ Reduced capital expenditures, since no additional expensive macro/micro-BSs are needed

■ Lower operational expenditures, because no new cell site, cell site backhaul, or maintenance costs are needed

■ Increased mobile usage indoors due to the low-cost fare—hence increasing the average revenue per user (ARPU)

■ Reduced customer churn rate because customers will be potentially more satisfied with the offered services through femtocells

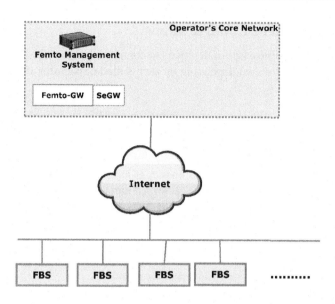

Figure 9.2: Femtocell network architecture.

9.2.5 Femtocell Network Architecture and Functionalities

The following is a description of the main common components of a femtocell network (shown in Figure 9.2).

FBS is a device located at the customer's premises that interfaces with mobile devices over the air radio interface that functions as a BS [12].

Femto gateway (Femto-GW) is an entry element to the operator's core network. It acts as concentrator to aggregate traffic from a large number of FBSs [11]. Also, the femto-GW could operate as a security gateway to provide authentication to allow data to/from authorized FBSs to protect the operator's CN from the public environment of the Internet, and to terminate large numbers of encrypted IP data from hundreds of thousands of FBSs. Also, there could be another element implanted separately called security gateway (SeGW) that does the security functions [13].

Femto management system provides management protocols for plug-and-play operations, administration, and management (OA&M) of FBSs [14]. Broadband Forum TR-069[2] has been selected as the framework for femtocell management and was widely supported by 3GPP2 and 3GPP vendors and carriers as the femtocell management protocol [3, 10, 15].

---

[2] Technical Report 069 (TR-69) is a Broadband Forum technical specification entitled *CPE WAN Management Protocol (CWMP)*. It defines an application layer protocol for remote management of end user devices.

9.2.6 Deployment Configurations

There are many possible cases of deployment configurations for FBSs. The possible configurations are classified depending on access mode, spectrum allocation types, and transmit power.

Access modes: An important characteristic of FBSs is their ability to control access. There are three common access control modes: open and closed, and hybrid [16].

1. *Closed access mode*, also known as closed subscriber group (CSG). In this scenario, a FBS serves a limited number of UEs it defined before in its access control list (ACL) [17]. This can be used in homes or enterprise environments.

2. *Open access mode*, also known as open subscriber group (OSG). In this scenario, any UE can connect to the FBS without restrictions. This can be used in hotspots, malls, and airports.

3. *Hybrid access mode*, is an adaptive access policy between CSG and OSG. In this scenario, a portion of FBS resources are reserved for private use of the CSG and the remaining resources are allocated in an open manner [17].

Spectrum planning: Allocation of the available spectrum in femtocell deployments can be one of the following [9, 18]:

1. *Dedicated spectrum:* In this approach, different frequencies are assigned for femtocells and macro/microcells.

2. *Partial co-channel:* In this approach, macro/microcells and femtocells share some spectrum and the rest of the spectrum is reserved for macro/microcells only.

3. *Shared spectrum:* In this approach, macro/microcells and femtocells share all available spectrum.

Transmit power configuration: The configuration process of downlink and uplink transmit power of FBSs can be fixed maximum or adaptive [19].

9.2.7 Femtocell Challenges and Open Issues

Despite many benefits and advantages of femtocell networks, they also come with their own issues and challenges. These issues and challenges need to be addressed for successful mass deployment of femtocell networks. The most relevant issues include:

■ *Interference:* Unplanned deployment of a large number of FBSs introduces interference issues for the mobile networks. Frequency interference is one of the most crucial issues that impair femtocell deployment. Frequency interference in femtocells includes cross-tier and co-tier interference [20]. In cross-tier interference, a FBS interferes with macro/micro-BS or vice versa. In co-tier interference, a FBS interferes with another neighboring FBS or FBS user.

■ *Security and QoS:* Since FBSs use nondedicated fixed broadband connections (i.e., xDSL) that carry femto and nonfemto traffic, managing and controlling voice/data priority and security over a third party becomes more difficult [21, 22] unless Internet backhaul belongs to the same cellular operator.

■ *Location and synchronization:* FBSs operate in the licensed spectrum; thus, the exact locations need to be verified, as well as intercell synchronization for proper femtocell deployments [9]. Also, location information is essential to provide tracking in emergency calls. However, global positioning system (GPS), which is used in macro/micro-BSs for synchronization and location, cannot be used in FBSs due to the lack of the coverage of GPS indoors, since the typical deployed FBSs are indoors [23].

■ *Integration of FBS into the CN:* Traffic between FBS and the CN sends/receives through broadband networks, so it is necessary to determine how FBSs integrate with CN, with or without gateways, and what interface they want to connect FBS with CN, or there might a need to upgrade the CN (software/hardware) to be connected to femto-GW. Many possible configurations are available [21].

■ *Self-organization network (SON) and autoconfigurations:* FBSs as a consumer premises equipment (CPE) are deployed as plug-and-play devices, so they shall integrates themselves into the mobile network without user intervention [5, 24]. Hence, different SON and autoconfiguration algorithms and techniques are needed.

■ *Mobility and handoff management:* Considering that FBSs will be deployed densely and by the millions [24], and may not be accessible to all users, mobility management in femtocell networks (such as searching for FBS, handoff from/to macro/micro-BS, access control) becomes one of the most challenging issues [23]. (More details can be found in Section 9.3.6.)

9.3 Mobility Management in Femtocell Networks

In Section 9.3.1, we present principles of mobility management in wireless networks. An overview of mobility management in femtocell networks is provided in Section 9.3.2.

9.3.1 Mobility Management in WBNs

Mobility management is a set of tasks for controlling and supervising mobile user terminals or equipments (UEs) in a wireless network to find them for delivery services, as well as to maintain their connections while they are moving [6]. Mobility management is concerned with various aspects, e.g., quality of service (QoS), power management, location management, handoff management, and admission control. It is one of the most essential features in wireless communications due to the direct impact on user experience, network performance, and power consumption [5]. The two core components of mobility management are location management and handoff management [7]. In the following subsections, different mobility management procedures and aspects are presented.

9.3.1.1 Location Management

Location management includes two componenets, namely, registration and paging. Registration is the task of knowing where the UE is located to handle incoming and outgoing calls. The registration process is performed via a database called the Home Locations Register (HLR). The HLR contains information about the UE and its capabilities. It also describes the home area of the user [8]. Another database is the Visitor Location Register (VLR); it is attached to the HLR to maintain the current location of the user. The VLR is updated whenever the user moves from one area to another area [5]. Paging is a process that allows the network to page the UE when it is setting up a call. The paging process is a message to be sent to the serving BS in order to get a response from the UE before sending a call or message [8].

9.3.1.2 Mobile Modes

1. *Idle mode:* This mode applies when the UE has no ongoing service (data, voice). The UE is in this mode most of the time after turning on and registering its location; it monitors for page messages from the networks [5]. When the UE is moving with idle mode, it performs a reselection of BS on the way. Cell reselection helps to transfer registration (VLR) and aims to keep camped on the best available cell during the idle mode. UE periodically searches for a better cell according to the reselection criteria [5]. Another mobility mechanism is selection. Cell selection is the mechanism of selecting an appropriate cell to camp on when a UE is powered on or after having lost coverage.

2. *Connected mode:* This mode is when the UE has ongoing service (data, voice). After the UE releases its active session, it will switch to the idle mode. When the UE is moving with connected mode, handoff occurs from one BS to the next. More details are in Section 2.2.3 of [5].

9.3.1.3 Handoff Management

Handoff (HO) management is the main function by which wireless networks support mobility and maintain quality of service. HO enables the network to keep the UE's connection while it moves from the coverage area of one cell/sector to another [8]. HO is the process of transferring an ongoing voice call or data session from one cell connected to the CN to another. HO is divided into two broad categories: hard and soft HOs [9]. In hard HO, current resources are released before new resources are used. However, in soft HO, both existing and new resources are used during the HO process.

HO initiation: HO initiation is the process of deciding when to request a HO. The four basic HO initiation techniques are [10]:

1. *Relative signal strength (RSS):* In this technique, the received signal strength indicator (RSSI) is measured over time, and the UE chooses the BS with the strongest signal to handoff.

2. *Relative signal strength with threshold:* This technique introduces a threshold value. If the current signal is weak (less than a threshold) and the other signal is stronger than the current signal and the threshold, then the UE will hand off.

3. *Relative signal strength with hysteresis:* In this technique, the UE will be allowed to hand off if the new BS is sufficiently stronger than the current BS (via hysteresis).

4. **Relative signal strength with hysteresis and threshold:** This technique allows a UE to hand off to a new BS if and only if the current signal level is below a certain threshold and the signal of the target BS is stronger than the current BS by a given hysteresis margin.

HO decisions: There are several methods for performing HO. The primary HO decision protocols include [10, 11]:

1. *Network-controlled HO:* The network is responsible for the overall HO decision and handles the necessary RSS measurements.

2. *Mobile-controlled HO:* The UE totally controls the HO process. The UE and BS make the necessary measurements and the BS sends them to the UE. The UE then decides when to hand off.

3. *Mobile-assisted HO:* The UE makes RSS measurements and then sends them to the network or BS to determine when to hand off. In this section, we describe the mobility management procedures and issues in femtocell networks.

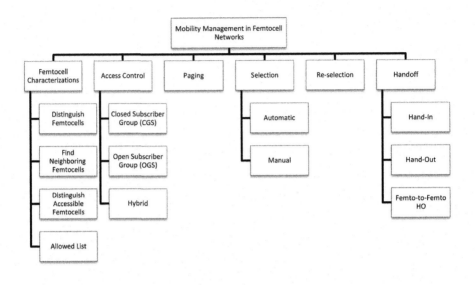

Figure 9.3: Overview of mobility management functionalities in femtocell networks.

9.3.2 Femtocell Mobility Elements

Nowadays, there are more than 100 million users of femtocells on more than 30 million BSs [7]. Hence, with a large number of femtocells randomly deployed, it is difficult to treat them as a normal macro/microcell. In addition, the network cannot afford broadcasting the femtocell information, as this will affect the overall performance of the network [26]. However, some identifiers and techniques for femtocells are needed to reduce the impact and improve the mobility management of femtocell networks.

9.3.2.1 Femtocells vs. Macro/Microcells

According to identifiers and techniques that can distinguish femtocells from macro/microcells, the network can be divided into two tiers [1], as shown in Figure 9.4. Such division minimizes the signaling overhead across tiers and shortens the neighbor cell list (NCL) that the UE scans when performing a HO. The methods proposed for such classification are listed below:

- *Hierarchical cell structure* (HCS): HCS can be used to distinguish between the different network cells. Each tier can be assigned individual access priority (i.e., HCS_0, HCS_1).

- *Separate femtocell PLMN ID:* This technique uses a different public land mobile network identity (PLMN ID) for femtocells [27]. PLMN ID is an identifier for the operator's networks, and each operator has its own PLMN [1]. Hence, femtocells are assigned a different PLMN ID from macro/microcells to ensure femtocell selection and minimize the impact on macro/microcell

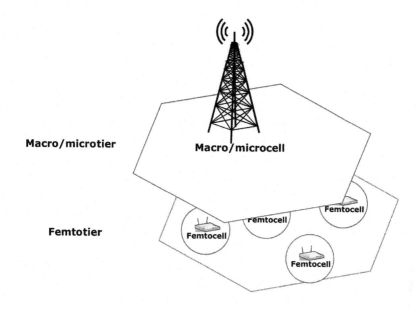

Figure 9.4: Two-tier network.

users (i.e., PLMN ID_0, PLMN ID-1) [1]. Hence, more PLMN IDs are required for operators.

- *CSG PSC/PCI:* A set of primary scrambling codes (PSCs) in UMTS or physical cell identifier (PCI)[3] in LTE is reserved for identifying CSG cells of a specific frequency [27].

- *Dedicated CSG frequency list:* Specifies the frequencies dedicated for UMTS CSG cells only [27].

- *CSG indicator:* Another approach is to use a CSG indicator to determine whether a femtocell is a CSG or not [26]; this is used in LTE networks.

- *CSG ID* [1]: One or more CSG cells are identified by a unique numeric identifier called CSG identity (CSG ID). When the UE is not authorized to access the target femtocell, a new reject message is used. The UE will then block the corresponding CSG ID, for a configurable duration, instead of the whole frequency (in LTE system).

- *Femtocell name* [27]: A text-based name for the femtocell sent only by CSG and hybrid cell. UE may display the femtocell name.

---

[3]Physical cell identifier (PCI) is an ID used to identify a cell for radio purposes.

9.3.2.2 Finding Neighboring Femtocells

UEs can distinguish femtocells from other cellular tiers, but it is not easy to decide on joining a neighboring femtocell due to the large numbers of FBSs. The following are techniques that have been proposed to find femtocells:

- *NCL:* The NCL can be created by the FBS through self-configuration algorithms implemented by the vendor, and it might be updated when the FBS senses any changes of the neighboring cells or when the FBS is turned on [1].

- *UE autonomous search:* The objective of the autonomous search is to determine when the UE should start searching for a femtocell to which it can have access and whether the CSG cell is valid or not [26]. Due the limited area coverage of femtocells, the UE only starts searching for a femtocell when the UE is in its vicinity [26]. The autonomous search function is not specified and is left to UE manufacturers.

- *Manual CSG search:* This method is specified in LTE and UMTS, where a user can find a CSG cell. On request of the user (e.g., via UE application), the UE is expected to search for available CSG IDs by scanning all frequency bands for CSG cells, and then the UE reports the CSG ID of the strongest or higher priority to higher tiers [26].

9.3.2.3 Distinguishing Accessible (CSG and Hybrid) Femtocells

By knowing whether a femtocell is accessible or not, unnecessary signaling overhead can be avoided. The following are CSG-related identification parameters used to identify accessible femtocells:

- *LAI/TAI* [1]: For femtocells, the location area identity (LAI)/tracking area identity (TAI) of neighbors needs to be different for the purpose of user access control. The LAI of unauthorized femtocells will be put in the UE's Universal Subscriber Identity Model (USIM) after it receives location area update (LAU) rejections from these femtocells.

- CSG indicator, CSG ID, and femtocell name are also used to identify accessible femtocells, whereas only CSG or hybrid cells broadcast their CSG ID or femtocell name.

9.3.2.4 Handling Allowed List

The allowed list is necessary in order to check whether or not the UE is allowed to access the target femtocell. The allowed list could be in:

- *FBS or femto-GW:* This is a list stored locally in the FBS or femto-GW that contains the UEs that are allowed access [1]. The operator or owner can manage the allowed list. This technique is used in UMTS.

- *CN:* In LTE, a UE's allowed CSG list (ACL) [26] is provided. ACL is the list of CSG IDs (FBSs) that the UE belongs to [26]. The ACL is stored with the user's subscriber information in the CN, and it may keep a copy in UEs.

9.3.3   Access Control

Access control mechanisms play a vital role in HO management, as well as when a user tries to camp on a femtocell to prevent unauthorized use of that femtocell and in mitigating cross-tier interference. Users of femtocells in two-tier networks are classified into [20]:

- *Subscribers* of a femtocell are those that are registered to have the right to use it. Femtocell subscribers may be any UEs, for instance, cell phone, laptop, etc.

- *Nonsubscribers* are users that are not registered in a femtocell.

There are many possible locations of access control in the femtocell network. Locations of access control may be [1, 26]:

- *Access control in FBS or femto-GW:* Access control is performed in FBS or femto-GH for femtocells and in CN for macro/microcell. For example, in LTE, femto-GW shall perform access control, and FBS may optionally perform access control.

- *Access control in UE:* In LTE, UE can perform the basic access control in the registration procedure to enhance the mobility procedure.

- *Access control in CN:* In WiMAX and LTE, one of the CN entities (such as admission server) checks the access of the UE to the target femtocell by the ACL after it receives information from the femtocell.

Femtocells support flexible access control mechanisms.

9.3.4   Paging

For OSG and hybrid femtocells, the CN pages the UE in all cells that the UE is registered in [1]. For CSG femtocells, there may be many CSG femtocells that the UE is registered in but the UE is not allowed to camp on. Perhaps, the paging procedure needs to be optimized, in terms of minimizing the amount of paging messages used to page a UE in femtocells [27]. CN and femto-GW can perform the paging optimization [1]. Management of paging by the Femto-GW is left to vendor implementation [27].

9.3.5   Idle Mode, Cell Selection, and Cell Reselection

It is desirable for the UE to switch to its femtocell when the received signal from the macro/microcell is strong enough to support service, even when the macro/microcell can still provide reliable service [28]. Hence, cell selection and reselection in femtocell environments are more complicated than in macro/microcell networks [1]. There are few alternatives to enabling cell selection and reselection in femtocells depending on the technology used.

> *Cell reselection:* A UE in idle mode changes (reselects) the cell it is camped at as it moves across cells. Reselection requires parameters broadcast by the cell sites [28]. To extend battery life of the UE (with idle mode), the UE scans other radio channels only when the signal-to-noise ratio (SNR or S/N) of the current cell is lower than a certain threshold. In UMTS and LTE networks, this threshold is defined by parameters called *S intersearch* and *S intrasearch* [28]. Cell reselection may also be as an autonomous search function that is intended to find CSG/hybrid femtocells to the UE to camp on it [27]. The autonomous search function is not specified and is left to vendor implementations [27].

> *Cell selection:* There are two modes for network selection: manual and automatic. Automatic cell selection is similar to that for macro/microcells. An extra CSG ID check is performed when the target cell is CSG or hybrid, to check whether the CSG/hybrid cell is suitable for the UE or not [28]. This technique is used in WiMAX and LTE. On the other hand, in manual cell selection, the UE is allowed to choose its serving CSG manually. Manual selection is not allowed in connected mode, though [1, 26,]. This technique is used in UMTS and LTE.

9.3.6   Connected Mode and Handoff

There are three scenarios for HO in femtocell networks: hand-in, hand-out, and femto-to-femto HO, as shown in Figure 9.5.

1. *Hand-in* takes place when a UE moves from a macro/microcell to a femtocell. Hand-in is a complex scenario since it requires the UE to select an appropriate FBS, while considering neighbor cells. As well, there is no straightforward mechanism for the macro/micro-BS to determine the identity of the target FBS from the measurement reports sent by the UE.

2. *Hand-out* takes place when a UE moves out of a femtocell to a macro/microcell. The hand-out process is supported in femtocells almost without any changes to the existing macro/microcell network or to the UE. The handling neighbor cell is easier than the hand-in case since the target cell is always one.

3. *Femto-to-femto HO* takes place when a UE moves out of a femtocell to another femtocell, and requires handling long NCLs.

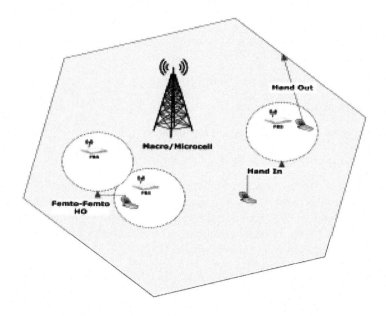

Figure 9.5: HO scenarios in femtocell networks.

Access control in femtocell networks makes the HO procedures more complex than in conventional networks, especially in hand-in and femto-to-femto HO [1]. In addition, in macro/microcell networks, HOs are triggered when users enter the coverage area of other cells. However, given the coverage size of open/hybrid access femtocells, this occurs more often than in the macro/microcell case. Hence, different HO management procedures are needed to allow nonsubscribers to camp for longer periods near femtocells. In 3G femtocell networks, there is no support for soft HO [1]. In LTE Release 8, there is no support for hand-in and femto-to-femto HO. However, 3GPP Release 9 supports hand-in. 3GPP Release 10 enables femto-to-femto HO [1, 28].

9.3.7 Mobility Management Issues

Mobility management in femtocells should offer a seamless experience for users as they move in and out of femtocell coverage. Since existing cellular networks and mobile devices have been designed without awareness of femtocells, these requirements must be met without requiring changes to existing infrastructure or to mobile devices. Dense deployments will cause serious issues on mobility management between the macro/microcells and femtocells. The importance of mobility management in femtocell networks is due to the following reasons:

■ Large number of FBSs that are usually overlaid with macro/micro-BS coverage

- High density of FBSs

- Dynamic neighbor cell lists

- Variant access control mechanisms

- Different user preferences

- Different operator policies and requirements

The above characteristics pose the following issues:

1. *Neighbor advertisement lists and messages:* Since large numbers of FBSs may be within the range of a single macro/microcell, a long list of neighboring FBSs would be broadcast via neighbor advertisement messages. This leads to a waste of the wireless resources and makes the process of scanning all neighboring femtocells time-consuming [29].

2. *HOs:* Current macro/microcells share the radio frequency with potentially large numbers of femtocells. Hence, a UE may face continual HOs, especially when it moves around the home or enters areas where the received signal from the macro/microcell is greater than that from the femtocell [30]. In addition, leakage of coverage to the outside of a house may occur and can lead to a highly increased number of unnecessary HO of macro/microcell users, which may lead to higher call dropping probability. Femtocells also introduce complexities in hand-in and femto-to-femto HO [24, 29, 31, 32].

3. *HO decision parameters:* In femtocell environments, creative and flexible decision parameters will influence the HO other than existing parameters, such as serving cost, user's status and preferences, load balancing [42, 44], etc. In other words, the serving cell or UE should decide to hand off to a target cell based on multiple parameters. Hence, there is a need for algorithms to optimize and adapt these and other parameters.

4. *Searching for FBS in different access scenarios:* In order to manage the mobility procedures in both idle and connected modes, with the case of hybrid and CSG scenarios, there are two problems to be solved. The first problem is how the mobile devices will find out whether the target is the CSG or not. The second problem is how to identify target CSG cells as the mobile device's own accessible FBS among many FBSs [33, 26].

5. *Idle mode mobility procedures:* Additional energy consumption should be taken into account due to the dense deployment of femtocells and their continuous receiving and transmission signals [34]. The increase of the number of cells may result in a large increase in traffic and load of the CN for idle mode mobility procedures [34, 35].

9.4   Related Work

Several research efforts have been done to modify and adapt the existing mobility management procedures in cellular networks for femtocells. In this section, we present a survey of mobility management schemes in femtocell networks. In Section 9.4.1, we describe the mobility management techniques. In Section 9.4.2, we define a set of evaluation criteria. This is followed by a thorough review of proposed schemes in different categories in Section 9.4.3.

9.4.1   Mobility Management Techniques

Network-based mobility management: In this type of mobility management protocol, the network takes responsibility for all aspects of mobility management without requiring participation from the UE in any related mobility procedures and their signaling. This domain does not require the UE to be involved in the exchange of signaling messages between itself and the network.

Mobile-based mobility management: This type requires the participation of the UE in all aspects of the mobility management. However, the participation of the UE in the mobility management and associated resources and software has become a hurdle for standards and protocols.

Mobile-assisted mobility management: In this type, information and measurement from the UE are used by the network to take care of all aspects of the mobility management.

9.4.2   Evaluation Criteria

In order to evaluate the schemes proposed, we define a set of evaluation criteria to compare the different mobility schemes in the first and second categories. These are:

- *Access control type(s):* This criterion indicates the type(s) of access control supported (CSG, OSG, and hybrid).

- *HO scenario:* This criterion contains the HO scenario(s) supported (hand-in/out, femto-to-femto HO).

- *HO objective(s):* This criterion shows the target goal(s). The goal could be to reduce the number of HOs or unnecessary HOs, decrease signaling overhead, etc.

- *Additional HO parameter(s):* This criterion shows the additional HO parameter(s) used, such as QoS, user's state, etc.

- *HO latency:* This criterion relates to the HO latency in proposed mechanisms. The shorter, the better.

- *Signaling traffic overhead:* This criterion relates to control data load. The amount of signaling packets should be within an acceptable range.

■ *QoS support:* A HO algorithm should be reliable and the call should have good quality after handoff, and it should be fast so that the user does not experience service degradation or interruption. This criterion shows the QoS types used, such as real-time support, packet lost, HO delay, etc.

■ *Special support required:* This criterion discusses change required to the infrastructure or the UE. The fewer the required changes, the better.

9.4.3 *Proposed Mobility Management Schemes*

Existing mobility management schemes can be categorized based on the target problem for each scheme, such as HO schemes, reselection, and scanning.

9.4.3.1 *Handoff Schemes*

Schemes belonging to this category are not complete mobility management schemes. They provide new HO protocols and algorithms in femtocells, in addition to schemes that deal with different aspects that are related to the HO processes, such as HO optimization process, new HO decision, context transfer stages, reading system information in hand-in, etc. In this section, many protocols and algorithms are discussed. Table 9.2 shows a comparison of the schemes surveyed below based on the evaluation criteria in Section 9.4.2.

Zhang et al. [29] propose a modified signaling flow of HO in the LTE network with CSG scenario for hand-in and hand-out. This method is applied in the femto-GW. The HO algorithm is based on the user's speed and QoS. The proposed scheme integrates the measurement value, maximal capacity, and current load of the cell as the input of HO judgment. This algorithm does not allow high-speed user HOs (>30 Kph) from macrocell to femtocell, while low-speed users are allowed. The algorithm distinguishes between real-time users and non-real-time users with moderate speed (>15 Kph); it allows the real-time moderate speed users to hand off. However, it does not allow moderate, speed users HOs without real time. The proposed algorithm performs better in reducing the unnecessary HOs and the numbers of HOs compared to the traditional HO algorithms, especially in medium- and high-speed users with a small penalty of signaling overhead.

Wang et al. [24] propose two mobility management schemes applied in the femto-GW at the radio network layer (RNL) for LTE femto-to-femto HO. In method I, the authors propose the femto-GW to serve as a mobility anchor, and let it make the HO decisions. When the femto-GW receives a HO request from the source FBS, the Femto-GW checks the target ID. If the target cell is under its control, it will handle the HO. In method II, Femto-GW operates as a transparent node and simply forwards the HO messages between the FBS and MME. S-GW has to be notified with the change of endpoint after HO. Moreover, method I is more suitable for enterprise use, because it reduces the signaling traffic with the CN. On the other hand, method II is more suitable for home use, because more signaling messages are exchanged.

Table 9.2   Comparison of HO Schemes

	Access Control Type(s)	HO Scenario(s)	HO Objective(s)	Additional HO Parameter(s)	HO Latency	Signaling Overhead	QoS Support	Special Support Required
Zhang et al. [29]	CSG	In, out	Reduce number of unnecessary HO	User's velocity, QoS load balancing	High	High	Real-time	Femto-GW
Wang et al. [24] Method I	All	F-F	Enhance HO	N/A	Medium	Low	No	Femto-GW
Wang et al. [24] Method II	All	F-F	Enhance HO	N/A	High	High	No	Femto-GW
Chowdhury et al. [36]	All	In, out	Reduce number of unnecessary HO	CAC, interference level, time duration[a]	High	Medium	No	Femto-GW
Chowdhury and Jang [37]	All	In, out	Reduce number of unnecessary HO	CAC, interference level, time duration	Low	Medium	Call/HO blocking probability	Small deployments: New server Medium deployments: Femto-GW
Kim and Lee [38]	Hybrid	In, out	Reduce number of unnecessary HO	Hybrid access CAC and other	High	Medium	No	Femto-GW
Ulvan et al. [31, 39]	OSG	All	Prevent frequent and unnecessary HOs	Movement prediction	Medium in RHO, low in PHO	Low in RHO	Minimized PL in PHO	Femto-GW
Ellouze et al. [40]	OSG	In, out	Achieve load balancing and enhance QoS	Load balancing, QoS	Low	Low	Yes	No
De Lima et al. [41]	OSG	In	Enhance the HO	CINR	High	Medium	Real-time	No

(continued)

Table 9.2 Comparison of HO Schemes (continued)

	Access Control Type(s)	HO Scenario(s)	HO Objective(s)	Additional HO Parameter(s)	HO Latency	Signaling Overhead	QoS Support	Special Support Required
Xu et al. [42]	OSG	In	Solve the asymmetry transmit power in hand-in	User's state, SINR	Medium	High	Minimized PL	No
Shaohong et al. [43]	OSG	In	Cut down the number of unnecessary HO and increase throughput	User's velocity	High	High in UHO	No	No
Wu [44]	Hybrid	In, out	Reduce number of unnecessary and failure HOs	Velocity, bandwidth, QoS, interference	High	High	Real-time	Femto-GW
Becvar and Mach [45]	OSG, hybrid	All	Reduce redundant HO	CINR	High	N/A	N/A	No
Becvar and Mach [50]	OSG, hybrid	In	Eliminate redundant HOs	FBS backbone, time spent in a FBS	Medium	High	Yes	Modifications in backbone management
Moon and Cho [46, 47]	OSG	In	Enhance assignment probability	Combine RSS from serving and target BSs	High HO failure	High	N/A	No
Jeong et al. [48]	OSG	In	Reduce number of unnecessary HO	Mobility pattern prediction	N/A	Medium	Yes	New server
Fan et al. [32]	CSG, hybrid	All	Interference mitigation and reduce number of HOs	Interference threshold	N/A	Low	No	Femto-GW
Li et al. [49]	CSG, hybrid	All	Avoid collision interference	Interference threshold	N/A	Low	No	No

aDuration of a UE maintains the minimum required signal level. F-F-femto-to-femto, in-hand-in, out-hand-out.

Chowdhury et al. [36] propose a signaling flow for hand-in and hand-out in UMTS networks with call admission control (CAC). In the proposed signaling flow, there are two phases: HO preparation phase (information gathering about HO candidates and authentications, HO decision to determine the best HO candidate) and HO execution phase. The proposed method considers the interference level as an additional HO decision parameter for hand-in procedure, and uses the proposed CAC to reduce unnecessary HOs. Three parameters are considered for the proposed CAC: received signal, duration of a UE maintains the minimum required signal level, and signal-to-interference level. The results show that the number of unnecessary HOs is minimized due to the proposed CAC.

As an extension work of [36], Chowdhury and Jang [37] propose a modified signaling flow for hand-in and hand-out for small- and medium-scale femtocell network deployments. The authors present the detailed HO call flow for these two femtocell deployments and the proposed CAC scheme to reduce the unnecessary HO. The proposed queuing scheme optimizes the new call blocking probability, HO call blocking probability, and bandwidth utilization. Simulation results show that the number of unnecessary HOs is minimized due to the proposed CAC. As well, the proposed scheme is able to provide a seamless and reliable HO for both small- and medium-scale deployments.

Kim and Lee [38] propose a signaling flow for hand-in and hand-out in UMTS networks with CAC in the hybrid access mode. The proposed hand-in procedure makes a decision based on the new CAC. The new CAC takes into consideration the residence time in a cell, user types, RSS level, the duration a UE maintains the signal level above the threshold level, the signal-to-interference level, and the capacity that one FBS can support. If the received signal level from the femtocell is higher than the threshold, the femto-GW checks whether the UE is preregistered or not. If the UE is preregistered, the next handover procedure is performed. If the UE is not preregistered, UE must remain in the femtocell area for the threshold time interval during which a signal level is higher than the threshold signal level before continuing to the next handover procedure. Results show that the number of unnecessary HOs is reduced via the hybrid access CAC.

Ulvan et al. [31, 39], propose an adapted signaling flow for the three types of HO-based LTE networks. The proposed scheme considers the movement prediction mechanism as an additional parameter for HO decision. This HO is a clientbased HO. Reactive and proactive HO (PHO and RHO) procedures are proposed to initiate the HO, since the HO procedure may be initiated by FBS, macro/microcell, and UE. In RHO, the HO is trigged when the UE almost loses its serving cell signal or the most probable position of UE is predicted. RHO aims to delay the HO as long as possible to prevent the frequent and unnecessary HO, and reduce the generated overhead of HO. However, in PHO, the HO may occur any time before the level of RSSI for the current BS reaches the HO threshold via estimate of a specific position before the UE reaches that position. After the UE discovers the new target cell RSSI, the UE calculates the time remaining before the normal HO is triggered, and then the HO triggers before the HO threshold. RHO is expected to minimize packet loss (PL) and latency in HO.

Ellouze et al. propose a modified HO procedure between WiMAX macro-BSs and FBSs. The proposed HO scheme takes QoS and load balancing into account in two ways, first, by limiting the break time connection caused by the scanning interval, and second, the HO selection procedure decides to connect the user to either FBS or other BSs according to the user's QoS profile. The proposed solution reduces the HO delay, and balances the load over the network.

De Lima et al. [41] propose a stochastic association mechanism for hand-in with OSG. The proposed system introduces a new distributed HO procedure, which does not depend on any centralized coordination. The proposed solution uses a modified multistage Dutch auction to autonomously coordinate and prioritize bidding FBSs. As well, the solution uses a stochastic process to separate candidate FBSs to reduce the probability of collision. A macrocell user plays an "auctioneer", while target FBSs play "bidders". A stochastic election process is incorporated in to the selection process to reduce the chances of multiple bidders. Femtocell users face some QoS degradation.

Xu et al. [42] propose a user's state and signal-to-interference ratio (SINR)-based hand-in algorithm for 3G networks, to overcome the drawbacks of the large asymmetry transmit power between FBSs and macro-BSs in two-tier networks. The proposed algorithm uses SINR to avoid the asymmetry transmit power, and user's state to reduce the unnecessary HOs. The authors add a user state such as velocity, QoS, with SINR, as a HO decision parameter. On the one hand, the results show that the new algorithm cuts down the number of unnecessary HOs due to taking the user's state into account. On the other hand, the total number of HOs is increased.

Shaohong et al. [43] propose two HO algorithms for 3G networks using the mathematical concept of set. In the velocity and signal HO algorithm (VSHO), they consider the velocity and RSS of the UE; hence, the frequent HOs of high-speed UEs are avoided. In the other improved algorithm, called unequal HO algorithm (UHO), the scheme considers the difference between macro-BSs and FBSs and the issue that the UE receives a higher signal strength from a FBS in a house than from the macrocell outside. In other words, UHO sets a higher signal level limit for femtocells than macrocells to serve as a chief BS. The results of the two proposed algorithms show that HO probabilities decreased for high-speed users, and the total throughput of macrocell networks increased. The above algorithms do not represent the power asymmetry by adding a constant value to the received signal from FBSs. In addition, the user state is not considered a factor for HO. Hence, the hand-in process may initiate with no guarantee of QoS from the femtocells, and then it may lead to HO failure.

Wu [44] proposes hand-in and hand-out procedures for LTE femtocells. The author considers a group of parameters for the HO decision: interference level, RSS, user's velocity, available bandwidth, and QoS level. The hand-in has two kinds of procedures. The first is for CSG users where the UE shall choose the most appropriate target FBS. The second is for non-CSG users; if a non-CSG UE causes too much interference, it can hand off to FBS to minimize interference. This HO is different from the normal situation, because the HO is triggered by FBS. The proposed solution does not consider the co-tier interference.

Becvar and Mach [45] propose an adaptive hysteresis margin for HO for LTE networks. The proposed solution utilizes the reported metrics (RSSI or CINR) for the dynamic adaptation of an actual value of hysteresis margin according to the position of the user in a cell. The hysteresis margin decreases with UE's moving closer to the cell border. This proposed solution shows reduction of redundant HO by mainly focusing on avoiding ping-pong effects. However, this is not an appropriate way to prevent unnecessary HOs caused by femtocell visitors.

In another paper, Becvar and Mach [50] propose an enhanced HO decision for a hand-in procedure with OSG and hybrid scenarios. The proposed scheme takes into account the delay and capacity of FBS's backbone as an additional decision parameter to achieve an acceptable level of QoS for femtocell users. Also, the authors consider the time duration spent by users in femtocell coverage as a decision parameter to reduce the number of unnecessary and frequent HOs. Results show that signaling overhead is increased, and the HO latency remains within the acceptable range.

Moon and Cho [46, 47] propose a modified HO decision algorithm for the hand-in procedure in LTE networks based on RSS. They combine the value RSS from the serving macro-BS and a target FBS to derive a reasonable HO criterion using the concept of combination factor, and take into consideration that the UE has an ability to detect neighboring femtocells. Results show that there is an enhancement of the assignment probability to the femtocell while keeping the same level of the number of HOs. In the problem of the asymmetry transmit power for the previous two proposed algorithms, the user's state is not considered as a factor for HOs; hence, the HOs may initiate with no guarantee of QoS from the femtocells, and then it may lead to HO failure.

Jeong et al. [48] propose a smart hand-in procedure to reduce the number of unnecessary HOs of temporary visitors that stay a short time in the femtocell. They keep UEs connected to a macrocell rather than conducting a hand-in when staying a short time in a femtocell, based on next movement pattern analysis. The proposed scheme applies a location prediction algorithm to identify temporary femtocell visitors while the users move along random movement patterns. Results show that the proposed scheme reduces the number of unnecessary HOs.

In order to mitigate the interference problem that may take place when a non-CSG user comes to a femtocell, **Fan** and Sun [32] propose a method for access and HO management for OFDAM femtocell networks. In the CSG scenario, when a UE comes near a FBS, its serving BS will check the UE's ID; if within the allowed list for the target FBS, the BS informs the FBS to start the HO procedure. Otherwise, the BS should notify the FBS to start the proposed proactive interference management procedure. As well, the authors propose a hybrid access for the same situation. After a non-CSG UE enters a femtocell, a FBS measures the UE's signal strength and decides whether the potential interference caused by the UE is above the interference threshold or not. If so, the FBS will request a HO procedure from the serving BS for the UE, and informs that this is an avoid interference HO. The CSG scenario reduces the unnecessary HOs and signaling load. However in the hybrid scenario, the number of HOs is increased.

To solve the same problem presented in [32], **Li** et al. [49] propose a pseudo-HO based on the scheduling information exchange method, subchannel, and power adaptation to avoid collision interference in LTE-A networks. The pseudo-HO is executed in the Radio Access Network (RAN), not referring to MME, which significantly reduced signaling overhead. When the UE tries to camp on a CSG FBS, if the UE belongs to this femtocell, the regular HO is triggered; or else, the pseudo-HO is triggered. The FBS will set up and maintain a table that contains the ID of non-CSG called pseudo-HO users.

9.4.3.1.1 Discussion

Although there are several proposed solutions for HO in femtocells, most of the solutions have targeted only one or two parts of the HO procedure, such as HO preparation, HO decision parameter, HO signaling, and hand-in algorithm. A few solutions have proposed a comprehensive HO procedure. Table 9.2 provides a detailed comparison between the schemes. Zhang et al. [29], Wang et al. [24], Chowdhury et al. [36], Chowdhury and Jang [37], Kim and Lee [38], Ulvan et al. [31, 39], and Wu [44] provide schemes for the signaling flow of the HO process with different additional parameters to reduce the number of unnecessary HOs. For example, Wang et al. [24] support CSG and OSG scenarios in the femto-to-femto HO. The scheme uses the user speed, QoS, and load balancing as additional parameters for the HO decision. The HO latency and signaling overhead are increased due to additional gateway (femto-GW) that is installed.

Ellouze et al. [40], de Lima et al. [41], Xu et al. [42], Shaohong et al. [43], Becvar and Mach [45], and Moon and Cho [46, 47] provide HO algorithms and decision parameters with more details to enhance the HO process and reduce the number of unnecessary HOs. In [52], the scheme just targets the OSG scenario to enhance the hand-in process. The HO process triggers by using the CINR as an additional parameter for the HO. The proposed scheme enhances the selection process of a target FBS to hand-in, where there is an obvious degradation of the QoS. Reference [51] provides a hand-in/out algorithm under the OSG scenario. This scheme is a complete solution due to inclusion of scanning and selecting procedures that are the two main processes before the HO process, in addition to the proposed HO procedures.

9.4.3.1.2 Summary

A complete HO solution should take into account a number of factors in order to be comprehensive and reliable. For instance, an additional HO parameter must be considered for the HO decision and triggering process in the femtocells, viz. the user state (e.g., velocity, preference) in Shaohong et al. [43] and Zhang et al. [29]. Also, the operator's preference, bandwidth, and load balancing should take place in any HO process as in Ellouze et al. [40] and Wu [44]. The HO latency and signaling overhead must be minimized as in Ellouze et al. [40].

9.4.3.2   Scanning and Selection Schemes

As aforementioned, scanning is the process used to find a cell for the UE to camp on. Schemes in this category propose mechanisms and algorithms to improve the scanning and selection processes.

Nam et al. [51] propose a network-assisted FBS management scheme in Mobile WiMAX networks using a triangulation mechanism and FBS monitoring scheme to reduce the number of scanning operations, as well as the size of neighbor advertisement messages. The proposed scheme uses a FBS monitoring mechanism to provide UE with the nearest FBS information under the OSG scenario. The authors assume that every FBS has two interfaces. One is used to communicate with the attached UE, while the other is used to monitor signals of candidate UE. Results show that the proposed scheme improves scanning performance and reduces wasting air resources.

Han et al. [52] propose an automatic generation scheme of neighboring BS lists for femtocell networks under the OSG scenario. This scheme is utilized by a UE to specify a target FBS to hand off. A neighbor list automatically generates by jointly utilizing the measurement of multiple neighbor BSs to include all the neighboring BSs for a proper HO. The proposed scheme operates in three steps. First, a BS measures the RSS of the neighboring BSs locally and reports the measurement results. Then the BS requests the other BSs' measurements. Second, a BS reconstructs the topology of the identified neighbors, and then the topology is used to find hidden neighboring FBSs. Third, a BS discovers hidden neighboring BSs with the support of other identified neighbors. These three steps can repeat periodically or be triggered whenever the network topology changes. The scheme shows acceptable results regardless of the shadowing effects.

Jung et al. [53] propose a scanning scheme to reduce unnecessary scanning procedures for an accessible FBS (CSG) to reduce the power consumption. The scheme uses an adaptive threshold with a margin for RSSI. Using thresholds within a serving macrocell, the target region is separated into smaller regions; hence, the UE only scans for the FBS within a small region satisfying triggering conditions. Results show that the proposed scheme can reduce scanning time and power consumption.

Chowdhury et al. [54] propose an optimized NCL for femto-to-femto HO and hand-in in dense femtocell networks. The proposed algorithm considers the received signal level from FBSs, open or closed scenarios, detected frequencies from the serving FBS and the neighbor FBSs, and location information (using SON capabilities of the FBSs) for the optimal neighbour FBS. The authors try to reduce power consumption for scanning many FBSs and the media access control (MAC) overhead. Femtocells are categorized in two categories. The first category contains the FBSs from which the received signals are greater than or equal to a threshold level. The FBSs in the second category are the received signals that are less than a threshold level or the serving FBS and the neighbor FBSs use the same frequency. The results show that the proposed scheme is able to maintain a minimum number (not optimal) of neighbor FBS list for the femto-to-femto HO.

Kwon and Cho [57] propose a load-based cell selection algorithm for faulted HO. The proposed cell selection algorithm allows each UE to select a target femtocell

Table 9.3   Comparison of Scanning Schemes

	Access Control	Scheme Objective(s)
Nam et al. [51]	OSG	Reduce scanning overhead
Han et al. [52]	OSG	Generate automatic neighboring list
Jung et al. [53]	CSG	Minimize unnecessary scanning
Chowdhury et al. [54]	OSG, CSG	Optimize the NCL
Kwon and Cho [57]	OSG	Balance load among cells; minimize HO blocking probability

based on the information of other UEs that have previously made their selection. The proposed scheme minimizes HO blocking probability and achieves load balancing between neighboring cells when the FBS generates a fault. In contrast, the scheme increases signaling overhead.

9.4.3.3   Reselection Schemes

Reselection schemes are used to allow the UE to change its serving cell to another, without having an active session.

In [55], the authors target the issue of when a UE searches for a femtocell. The UE needs to scan the entire femtocell spectrum in order to switch from macrocell to femtocell. The authors propose a cache scheme for femtocell reselection. The proposed scheme considers the UE's movement history by storing the cell information of the recently visited FBSs. The aim is to obtain the most recently visited order of FBSs that have been stored in the cache. The scheme seems useful in the OSG scenario with a large number of FBSs; otherwise, it is inefficient.

The authors in [34] propose two dynamic idle mode procedures for femtocells. The first procedure activates the FBS only to serve active calls from its registered users. This is achieved by a low-power "sniffer" capability in the FBS that allows the detection of an active call from a UE to the underlying macro/microcell based on a measured rise in noise that activates the FBS upon request. The second procedure reduces the pilot power and adjusts the cell's reselection thresholds when it is not serving an active call.

The authors in [56] address the issue of discovering a 3G femtocell in multiple frequencies, and the impact of this issue on the UE battery life, as well as the effect of different cell reselection parameters on capacity offloads. For instance, with good macrocell coverage, a UE may never initiate searches and would remain on the macrocell even when it is in the vicinity of its own FBS. The authors improve the idle mobility procedure to enable UEs to discover and camp on FBSs via three potential techniques: search threshold optimization, beacon-based approach for enhanced reselection procedure, and UE enhancement for enhanced idle mode to the UE.

9.5 Conclusion and Open Issues

This chapter presents a comprehensive study of mobility management in femtocell networks. Issues in handoff and other mobility management procedures in femtocells are identified. Several research efforts are presented and classified.

Building efficient mobility mechanisms will play a vital role for successful deployment of femtocells and for providing seamless services. We highlight some open problems and issues that can be derived from our study of the research efforts discussed in this chapter, as follows:

■ The HO decision parameters must be adaptive and flexible based on the situation. For example, the HO decision parameter for the hand-in in existing schemes is typically based on the user preferences, CINR or RSSI. There are cases where other factors must be considered as well. For instance, a UE may receive a stronger RSSI from a macrocell while it is under the coverage of a femtocell; hence, the UE will not hand off to the femtocell. This is due to the large asymmetry in transmitting power. Therefore, techniques to optimize and adapt multiple HO parameters depending on the situations need to be addressed.

■ The service interruption time caused by reading system information of a target CSG FBS during hand-in and femto-to-femto HO should be minimized. This is because UEs cannot receive data while reading system information. Existing schemes do not address such issues in a satisfactory manner.

■ Mobile-based or mobile-assisted HOs will play a key role in femtocell networks to meet user needs. Some research work has been proposed on proactive HO in femtocell networks. These solutions focus on satisfying user needs only and ignore other networking requirements leading to unexpected results (e.g., whether they are also interested in service quality or cost, etc.).

■ Mobile femtocells on transit systems (e.g., buses and trains) will become more prominent in the future. These will typically be used to aggregate user traffic and relay it to the macrocell networks or to other access networks. Means of offloading this aggregated traffic of these fast-moving femtocells are needed. Hence, modified protocols for HO and location management for mobile femtocells need to be developed.

■ Methods and techniques that assist in managing and updating the network topology are essential for effective mobility procedures between macro/microcells and femtocells and among femtocells. All macro/micro-BSs and FBSs have to be aware if a FBS enters or leaves their coverage, hence changing the mobility conditions. For UEs to perform handoff and cell searching in a more efficient way, the UE and FBS should acquire network topology.

References

1. Z. Jie and D. Guillaume, Femtocells: technologies and deployment, Wiley, 2010.

2. C. Patel, M. Yavuz, and S. Nanda, Femtocells (industry perspectives), *Wireless Communications, IEEE*, vol. 17, pp. 6–7, 2010.

3. D. Knisely and F. Favichia, Standardization of femtocells in 3GPP2, *Communications Magazine, IEEE*, vol. 47, pp. 76–82, 2009.

4. I. Guvenc, S. Saunders, O. Oyman, H. Claussen, and A. Gatherer, Femtocell networks, *EURASIP Journal on Wireless Communications and Networking*, 2010.

5. S. Yeh, S. Talwar, S. Lee, and H. Kim, WiMAX femtocells: a perspective on network architecture, capacity, and coverage, *Communications Magazine, IEEE*, vol. 46, pp. 58–65, 2008.

6. S. Ortiz, The wireless industry begins to embrace femtocells, *Computer*, vol. 41, pp. 14–17, 2008.

7. Smallcell Forum, http://www.smallcellforum.org/.

8. F. Chiussi, D. Logothetis, I. Widjaja, and D. Kataria, Femtocells (guest editorial), *Communications Magazine, IEEE*, vol. 48, pp. 24–25, 2010.

9. M. Neruda, J. Vrana, and R. Bestak, Femtocells in 3G mobile networks, in *Systems, Signals and Image Processing (IWSSIP), 16th International Conference on*, 2010.

10. J. Boccuzzi and M. Ruggiero, *Femtocells: design and application*, McGraw-Hill, 2011.

11. G. Korinthios, E. Theodoropoulou, N. Marouda, I. Mesogiti, E. Nikolitsa, and G. Lyberopoulos, Early experiences and lessons learned from femtocells, *Communications Magazine, IEEE*, vol. 47, pp. 124–130, 2009.

12. D. Knisely, T. Yoshizawa, and F. Favichia, Standardization of femtocells in 3GPP, *Communications Magazine, IEEE*, vol. 47, pp. 68–75, 2009.

13. K. Elleithy and V. Ra, Femto cells: current status and future directions, *International Journal of Next-Generation Networks (IJNGN)*, vol. 3, 2011.

14. WiMAX femtocells, white paper, ARICENT, 2010.

15. E. Seidel and E. Saad, *LTE home node Bs and its enhancement in release 9*, white paper, Nomor Research, May 2010.

16. S. Choi, T. Lee, M. Chung, and H. Choo, Adaptive coverage adjustment for femtocell management in a residential scenario, in *Management enabling the future internet for changing business and new computing services*, vol. 5787 of Lecture Notes in Computer Science, Springer, Berlin, 2009, pp. 221–230.

17. L. Mohjazi, M. Al-Qutayri, H. Barada, K. Poon, and R. Shubair, Deployment challenges of femtocells in future indoor wireless networks, in *GCC Conference and Exhibition (GCC), IEEE*, February 2011, pp. 405–408.

18. F. Hasan, H. Siddique, and S. Chakraborty, Femtocell versus WiFi—a survey and comparison of architecture and performance, in *Wireless Communication, Vehicular Technology, Information Theory and Aerospace Electronic Systems Technology (Wireless VITAE), 1st International Conference on*, May 2009, pp. 916–920.

19. H. Claussen, L. Ho, and L. Samuel, An overview of the femtocell concept, *Bell Labs Technical Journal*, vol. 13, pp. 221–245, 2008.

20. G. de la Roche, A. Valcarce, D. Lopez-Perez, and J. Zhang, Access control mechanisms for femtocells, *Communications Magazine, IEEE*, vol. 48, pp. 33–39, 2010.

21. V. Chandrasekhar, J. Andrews, and A. Gatherer, Femtocell networks: a survey, *Communications Magazine, IEEE*, vol. 46, pp. 59–67, 2008.

22. R. Ellouze, M. Gueroui, and A. Alimi, Femtocells QoS management with user priority in Mobile WiMAX, in *Wireless Communications and Networking Conference (WCNC), IEEE*, March 2011, pp. 84–89.

23. R. Kim, J. K., and K. Etemad, WiMAX femtocell: requirements, challenges, and solutions, *Communications Magazine, IEEE*, vol. 47, pp. 84–91, 2009.

24. L. Wang, Y. Zhang, and Z. Wei, Mobility management schemes at radio network layer for LTE femtocells, in *Vehicular Technology Conference (VTC Spring), IEEE 69th*, April 2009, pp. 1–5.

25. *Femtocell access point fixed-mobile convergence for residential, SMB, and enterprise markets*, research report, ABI Research, 2010.

26. A. Golaup, M. Mustapha, and B. Patanapongpibul, Femtocell access control strategy in UMTS and LTE, *Communications Magazine, IEEE*, vol. 47, pp. 117–123, 2009.

27. G. Horn, *3GPP femtocells: architecture and protocols*, report, QUALCOMM, September 2010.

28. P. Humblet and A. Richardson, Femtocell radio technology, white paper, Airvana, May 2010.

29. H. Zhang, X. Wen, B. Wang, W. Zheng, and Y. Sun, A novel handover mechanism between femtocell and macrocell for LTE based networks, in *Communication Software and Networks (ICCSN), Second International Conference on*, February 2010, pp. 228–231.

30. Femtocells—the gateway to the home, white paper, Motorola, May 2008.

31. A. Ulvan, R. Bestak, and M. Ulvan, The study of handover procedure in LTE-based femtocell network, in *Wireless and Mobile Networking Conference (WMNC), Third Joint IFIP*.

32. Z. Fan and Y. Sun, Access and handover management for femtocell systems, in *Vehicular Technology Conference (VTC Spring), IEEE 71st*, May 2010, pp. 1–5.

33. H. Kwak, P. Lee, Y. Kim, N. Saxena, and J. Shin, Mobility management survey for home-eNB based 3GPP LTE systems, *Journal of Information Processing Systems*, vol. 4, pp. 145–152, 2008.

34. H. Claussen, I. Ashraf, and L. Ho, Dynamic idle mode procedures for femtocells, *Bell Labs Technical Journal*, vol. 15, no. 2, pp. 95–116, 2010.

35. Y. Li, G. Su, P. Hui, D. Jin, L. Su, and L. Zeng, Multiple mobile data offloading through delay tolerant networks, in *Proceedings of the 6th ACM Workshop on Challenged Networks*, CHANTS, New York, 2011, pp. 43–48.

36. Z. Chowdhury, W. Ryu, E. Rhee, and Y. Jang, Handover between macrocell and femtocell for UMTS based networks, in *Advanced Communication Technology (ICACT), 11th International Conference on*, 2009.

37. M. Chowdhury and Y. Jang, Handover control for WCDMA femtocell networks, *Journal of Korea Information and Communication Society*, vol. abs/1009.3779, 2010.

38. J. Kim and T. Lee, Handover in UMTS networks with hybrid access femtocells, in *Advanced Communication Technology (ICACT0, the 12th International Conference on*.

39. A. Ulvan, R. Bestak, and M. Ulvan, Handover scenario and procedure in LTE-based femtocell networks, in *Fourth International Conference on Mobile Ubiquitous Computing, Systems, Services and Technologies (UBICOMM)*, 2010.

40. R. Ellouze, M. Gueroui, and A. Alimi, Macro-femto cell handover with enhanced QoS in Mobile WiMAX, in *Wireless Telecommunications Symposium (WTS)*, April 2011, pp. 1–6.

41. C. de Lima, K. Ghaboosi, M. Bennis, A. MacKenzie, and M. Latvaaho, A stochastic association mechanism for macro-to-femtocell handover, in *Signals, Systems and Computers (ASILOMAR), Conference Record of the Forty-Fourth Asilomar Conference on*, November 2010, pp. 1570–1574.

42. P. Xu, X. Fang, J. Yang, and Y. Cui, A user's state and SINR-based handoff algorithm in hierarchical cell networks, in *Wireless Communications Networking and Mobile Computing (WiCOM), 6th International Conference on*, September 2010, pp. 1–4.

43. W. Shaohong, Z. Xin, Z. Ruiming, Y. Zhiwei, F. Yinglong, and Y. Dacheng, Handover study concerning mobility in the two-hierarchy network, in *Vehicular Technology Conference (VTC Spring), IEEE 69th*, April 2009, pp. 1–5.

44. S. Wu, A new handover strategy between femtocell and macrocell for LTE-based network, in *Ubi-Media Computing (U-Media), 4th International Conference on*, July 2011, pp. 203–208.

45. Z. Becvar and P. Mach, Adaptive hysteresis margin for handover in femtocell networks, in *Wireless and Mobile Communications (ICWMC), 2010 6th International Conference on*, September 2010, pp. 256–261.

46. J. Moon and D. Cho, Novel handoff decision algorithm in hierarchical macro/femto-cell networks, in *Wireless Communications and Networking Conference (WCNC), IEEE*, April 2010, pp. 1–6.

47. J. Moon and D. Cho, Efficient handoff algorithm for inbound mobility in hierarchical macro/femto cell networks, *Communications Letters, IEEE*, vol. 13, pp. 755–757, 2009.

48. B. Jeong, S. Shin, I. Jang, N. Sung, and H.Yoon, A smart handover decision algorithm using location prediction for hierarchical macro/femto-cell networks, in *Vehicular Technology Conference (VTC Fall), IEEE*, September 2011, pp. 1–5.

49. H. Li, X. Xu, D. Hu, X. Chen, X. Tao, and P. Zhang, Pseudo-handover based power and subchannel adaptation for two-tier femtocell networks, in *Wireless Communications and Networking Conference (WCNC), IEEE*, March 2011, pp. 980–985.

50. Z. Becvar and P. Mach, On enhancement of handover decision in femtocells, in *Wireless Days (WD), IFIP*, October 2011, pp. 1–3.

51. J. Nam, W. Seo, D. Kum, J. Choi, and Y. Cho, A network-assisted femto base station management scheme in IEEE 802.16e system, in *Consumer Communications and Networking Conference (CCNC), 7th IEEE*, January 2010, pp. 1–2.

52. K. Han, S. Woo, D. Kang, and S. Choi, Automatic neighboring BS list generation scheme for femtocell network, in *Ubiquitous and Future Networks (ICUFN), Second International Conference on*, June 2010, pp. 251–255.

53. B. Jung, J. Moon, and D. Cho, Scanning time reduction based on adaptive threshold in hierarchical cellular networks, in *Personal, Indoor and Mobile Radio Communications, IEEE 20th International Symposium on*, September 2009, pp. 2861–2865.

54. M. Chowdhury, M. Bui, and Y. Jang, Neighbor cell list optimization for femtocell-to-femtocell handover in dense femtocellular networks, in *Ubiquitous and Future Networks (ICUFN), Third International Conference on*, June 2011, pp. 241–245.

55. K. Lee, J. Lee, Y. Yi, I. Rhee, and S. Chong, Mobile data offloading: how much can WiFi deliver? in *Proceedings of the 6th International Conference (Co-NEXT)*, November 2010.

56. F. Meshkati, Y. Jiang, L. Grokop, S. Nagaraja, M. Yavuz, and S. Nanda, Mobility and capacity offload for 3G UMTS femtocells, in *Global Telecommunications Conference (GLOBECOM), IEEE*, December 2009, pp. 1–7.

57. Kwon, Yong Jin, and Dong-Ho Cho. Load based cell selection algorithm for faulted handover in indoor femtocell network, in *Vehicular Technology Conference (VTC Spring)*, 2011 IEEE 73rd, pp. 1–5. IEEE, 2011.

SERVICES AND APPLICATION IN FUTURE WIRELESS NETWORKS

Chapter 10

Multimedia Streaming over Mobile Networks: Exploring the Quality Gain Potentials of P2P Coordinative Multisource Selection

Lin Xing

Yunan Normal University, Kunming, China

Wei Wang

San Diego State University, San Diego, CA, USA

Sunho Lim

Texas Tech University, Lubbock, TX, USA

Onyeka Ezenwoye

Georgia Regents University, Augusta, GA, USA

Kun Hua

Lawrence Technological University, Southfield, MI, USA

CONTENTS

The availability of multiple data sources provides extra potential to improve wireless multimedia service quality. However, the exploration of such multisource selection has largely been ignored in literature. In this chapter we present a new solution to optimize the receiver-side multimedia quality by coordinating the transmission of multiple media sources, using a lossy wireless peer-to-peer (P2P) network an an example scenario, while assuring the latency constraint. The major advantages of the coordinative multisource selection solution are twofold. First, it distributes workload to each media source by optimal grouping of multimedia frames on each peer host. Second, optimal channel coding rates are allocated to multimedia frames transmitted on each path. To reduce the computing complexity, the global optimal solution for all the multimedia frames is divided into multiple local optimal solutions. Specifically, (1) we divide all transmission paths into two groups in light of related bit error rates and path-pass probability, and allocate the same channel coding rate to each group, and we (2) find the optimal data source for each multimedia frame according to the value of effective transmission capacity (ETC). The simulation results show that the simplified strategy works as good as the global optimal solution, and it significantly improves the end-received multimedia quality under different latency constraints. This work casts new insights to future wireless multimedia streaming solution provision, by exploring the potential of coordinative multisource selection.

10.1 Introduction

With recent technological advancements of broadband radio frequency access, multimedia information delivery over modern wireless communication networks (such as P2P networks [1, 2]) demonstrates considerable potential for a wide variety of applications, such as IPTV [3], video on demand (VoD) [4], file and media content sharing [5], etc. In such systems, media flows transmitted from multiple sources can be jointly decoded at the receiver side, providing considerable source diversity gains. Compared with traditional data P2P networks, wireless multimedia P2P networks have two major unique characteristics: stringent multimedia service quality requirement and real-time streaming latency requirement. Thus, how to provide quality multimedia service with bounded latency performance becomes the key research challenge.

P2P networks have two major advantages: (1) The first one is multiple media sources can work in a coordinated manner to provide extra data delivery reliability. For example, the same video frames stored on different video servers can be transmitted simultaneously via different paths. So even if some video frames are lost in the wireless network or are delayed over the decoding deadline, the video can still be reconstructed when other video frames arrive at the receiver on time without transmission errors. The disadvantage, however, its that more bandwidth is consumed. (2) Another advantage is different sources can send different multimedia frames to the user simultaneously, so during the same period, the user can receive more multimedia frames. In this chapter we explore such multisource diversity to improve multimedia service quality while satisfying latency requirements. The quality-driven multisource selection problem studied in this chapter is illustrated in Figure 10.1. This figure shows some user wanting to obtain a multimedia file that contains some multimedia frames with different quality contributions (i.e., distortion reduction), where the multimedia file is wholly or partly stored on some multimedia servers. After searching for the desired frames in the wireless multimedia P2P network, available frames are transmitted simultaneously from servers to the user through different paths. The paths have different error rates and transmission speeds. Due to the real-time characteristic of multimedia applications, a latency constraint is specified for the multimedia file transmission. Within the latency requirement, the more video frames with higher-quality contribution received correctly at the destination, the better the multimedia service quality that will be achieved.

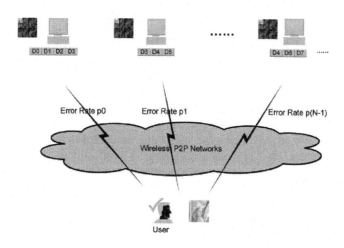

Figure 10.1: Illustration of the quality-driven multisource selection problem.

The same multimedia frame may be stored on more than one data server. In order to maximize the multimedia service quality, multiple available frames on these servers are transmitted to the end user simultaneously in spite of the fact that the same multimedia frame may be transmitted multiple times. In addition, channel coding redundancy is added to the multimedia frames to reduce the end-to-end path-loss probability. The end-to-end path-loss probability is the probability of losing a data frame from the source node to the destination through a single communication path. The more channel coding redundancy added to the multimedia frame, the lower the path-pass probability will be, and consequently, the better the service quality that will be achieved. However, due to the delay constraint of real-time multimedia applications and the limited bandwidth of the wireless multimedia P2P networks, it is unrealistic to transmit all multimedia frames stored on each server and to apply the maximal channel coding rate on each path and link. Thus, we address two key issues for improving the service quality of P2P-based live-streaming multimedia applications while satisfying the latency requirement: (1) improving the coordination of participating peers in the P2P networks by applying the optimal transmission scheduling for each server, and (2) allocating the optimal channel coding rate to multimedia frames passing through different paths.

Recent research in the area of multi-source-based coordinative selection for multimedia applications in wireless multimedia P2P networks focuses on the protocols and network structures, and seldom considered the optimal workload distribution and the optimal channel coding rate allocation. In [6], the authors investigated the impact of different wireless channels, traffic densities, and road layouts to establish feasible multisource video streaming over a wireless vehicular ad hoc network (VANET). A spatial partitioning of video streaming was introduced in that paper to achieve quality video streaming.

Si et al. [7] proposed a distributed algorithm for maximizing the receiving data rate and minimizing the power consumption. They formulated the problem as a multiarmed bandit system and selected the optimal sender in light of the Gittins indices of the senders. Another scheme based on restless bandit algorithms is proposed to solve the sender selection problem in [8]. In order to tackle the challenge raised by information dependency, an information plane called MANIP (Manet information plane) was proposed in [9] to support the autonomous reconfiguration of the P2P network. The authors in [10] proposed a P2P live streaming system called LayerP2P to improve current live P2P streaming systems in which a majority of users lacked incentives to contribute resources. The proposed LayerP2P system combined layered video, mesh P2P distribution, and a tit-for-tat-like algorithm to achieve better video quality by giving incentives to users to contribute more upload bandwidth to receive more video streaming layers. In [12], a new solution was proposed to solve the same problem. The authors designed an optimal content uploading policy as incentive for participating peers to contribute bandwidth by imposing an upload download fraction constraint that is enforced by either a central authority or a decentralized coalition of peers. Under the constraint, users were allowed to download new content only when their content contribution exceeded a threshold. In [11], the authors investigated the impact of user mobility with various practical issues related to addressing and forwarding strategies for P2P video implementations in wireless mesh networks (WMNs). The performance of the four most popular P2P video streaming applications was evaluated when the users crossed the coverage of the WMN testbed. In [13], an approach using multilayered/multiviewed content coding techniques was proposed for delivering seamless content services across P2P networks. The system was tested in the framework of a European project named seamless content delivery. In research [14], delay-sensitive multimedia transmission over wireless multihop enterprise mesh networks was studied. The authors proposed a distributed framework for P2P resource exchange, that considered QoS requirements, channel conditions, and network topologies for enabling participating peers to collaboratively distribute available wireless resources. In addition, a real-time cross-layer optimization algorithm was designed to enable resource exchange by efficiently adapting wireless resource usage to varying channel conditions. The problem of multiuser resource management was investigated in [15] for improving delay-sensitive applications. The authors proposed a distributed resource management strategy to enable participating nodes to exchange information; the delay and cost of exchanging information were also studied. Other related methods were proposed in [17–21] to improve the performance of wireless P2P networks.

From the above literal review, we can see that previous studies do not solve the optimal transmission scheduling problem to improve multiple media source coordinative selection. Fundamentally different from the above research, in this chapter, we present a novel algorithm to maximize multimedia service quality while assuring the latency constraint by improving the coordinative selection of participating peers in lossy wireless multimedia P2P networks. The contribution of the proposed algorithm is to achieve the optimal transmission scheduling for each server and the optimal channel coding allocation for each path by jointly considering end-to-end multimedia quality, delay constraint, channel condition, available bandwidth, and channel coding allocation. Specifically, the proposed optimal scheduling algorithm defines which video frames on each multimedia server will be transmitted at the next time slot, and then allocates the optimal channel coding redundancy to each video frame in the process of data transmission. Time slot is a scheduling period. The proposed strategy is performed in the beginning of each time slot at the user side. In one period, some multimedia frames go through multiple hops to be received by the user. The necessary information is collected in a feedback fashion.

The rest of the chapter is organized as follows. In Section 10.2 we present the scenario of wireless multimedia P2P networks, and then formulate the quality-driven multisource coordinative selection problem with latency constraint. In Section 10.3 we analyze the receiver-end distortion after multimedia frames pass through different paths and are reconstructed at the user side. In Sections 10.4 and 10.5, we present the proposed optimal multi-source-based data transmission scheduling. Several cross-layer parameters are estimated for specifying the optimal multimedia frame group for each server. The optimal channel coding redundancy will be allocated on each path. Simulation results are shown in Section 10.6. In Section 10.7 we conclude the chapter. Table 10.1 summarizes the symbols used in this chapter.

Table 10.1   Symbols and Equations

Symbol	Notation
$U[D]$	Expected receiver-end service quality
S	Optimal frame group to be transmitted
C	Optimal channel coding allocation
T_{max}	Latency constraint
T_S	Transmission latency on each path
d	Multimedia frame
D_i	Quality contribution of frame i
q	Receiving probability
M	Total multimedia frames desired by user
N	Total number of servers
P	Path-loss probability
$PATH$	Total path
$\{S_i, C_i\}$	Optimal transmission scheduling
L	Size of original multimedia frame
R	Transmission rate
G	Channel coding rate
t_o	Protocol overhead
p	Link-loss probability
H	Total number of hops in each path
α	Channel coding redundancy
b	Bit-loss probability

10.2   Problem Statement

Wireless P2P networks have been widely applied to various applications. In this chapter, we focus on improving the service quality of real-time multimedia applications in bandwidth-limited lossy wireless P2P networks. Assume M multimedia frames $\{d_0, d_1, ..., d_{M-1}\}$ with various quality contributions $\{D_0, D_1, ..., D_{M-1}\}$ are requested by the user for the next time slot. The lookup functionality provided by the network layer routing protocol finds that M servers $\{server_0, server_1, ..., server_{M-1}\}$ in the P2P networks store at least one complete union of these frames. Existing ad hoc routing protocols will be used to select the transmission path $PATH = \{path_0, path_1, ..., path_{M-1}\}$ for each server. The server and path are marked with the same subscript. Since we focus on coordinative source selection and communication resource allocation, routing selection is not considered in this chapter. Due to the limited network bandwidth and the latency constraint of real-time multimedia applications, not all multimedia frames can be sent to the user on time during any time slot. In addition, the induction of channel coding redundancy increases the traffic load of wireless P2P networks. Thus, we propose a new strategy to optimize transmission scheduling and coordinate the peers in the network. The address of potential servers is obtained by the user by running the P2P lookup protocol. An overlay feedback network is used to collect necessary information such as channel condition [22, 23].

The feedback information is used by the proposed strategy on the user side for estimating the best result. The user broadcasts dictate to all potential servers which multimedia frames will be transmitted on each server and the channel coding rate on each path in order maximize the multimedia quality. To simplify the problem, we consider a node-disjoint wireless mesh network architecture in which the transmitted data packets merely pass along a single path without crossing another, similar to [24, 25]. The multimedia service quality maximization can be formulated as

$$\left\{ S_{i|i\in[0,N-1]}, C_{i|i\in[0,N-1]} \right\} = \arg\max \left\{ U[D] \right\} \tag{10.1}$$

subject to the latency constraint

$$\max \left\{ \exp(T_{S_i}) \right\} \leq T_{\max} \tag{10.2}$$

where $U[D]$ denotes the expected service quality perceived on the user side; S_i denotes the optimal frame selection for server i, that is, which multimedia frames will be transmitted on each server during the next time slot; C_i denotes the optimal channel coding allocation on path i; and $\left\{ S_{i|i\in[0,N-1]}, C_{i|i\in[0,N-1]} \right\}$ is the optimal transmission scheduling strategy that leads to the maximal service quality obtained by the user. In the constraint equation (10.2), $T_{\max}$ denotes the latency constraint of a specific time slot, and $\exp(T_{S_i})$ denotes the expected maximum transmission time of multimedia frames on server i. Equation 10.2 evaluates the transmission time of each path and guarantees the longest transmission time is less than the latency requirement. The proposed optimal transmission scheduling in this chapter integrates both Equations 10.1 and 10.2 to maximize the end-received multimedia service quality while satisfying the latency constraint. That is achieved by selecting the optimal frames transmitted from each server and allocating the optimal channel rates for communication paths.

10.3   Coordinative Multisource Distortion Analysis

Traditional approaches try to send all desirable multimedia frames to the users while seldomly considering the coordinative selection of participating peers in the P2P network. We propose a new transmission scheduling approach to maximize service quality with latency constraint by improving multisource coordinative selection. The expected end-received multimedia service quality and the maximum transmission time on each path are evaluated in order to specify the optimal multimedia frames sent by each server while allocating the optimal channel coding rate on each path.

First we analyze the expected end-received multimedia quality on the user side by considering three factors: (1) the quality contribution of each multimedia frame, (2) what redundancy a certain frame has among different servers, and (3) the related path-pass probability for packet transmission (i.e., the error rate). The expected end-received multimedia quality can be estimated as

$$U[D] = \sum_{i=0}^{M-1} D_i \times q_i \tag{10.3}$$

where D_i denotes the expected quality contribution of frame i after the frame passed across the P2P networks hop by hop and is decoded by the user, and q_i denotes the receiving probability, which is the probability that the multimedia frame d_i is received successfully by the user. Since the same frame may be transmitted by more than one server, the frame can be decoded by the user as long as one of them passes through the network successfully. So the probability d_x of a frame being received on the user side can be expressed as

$$q_x = 1 - \prod_{i \in PATH} P_{xi} \tag{10.4}$$

where P_{xi} denotes the path-loss probability of frame d_x on $path_i$. We define $P_{xi} = 1$ if $server_i$ does not transmit frame d_x. $\prod_{i \in PATH} P_{xi}$ presents the probability that all frames d_x transmitted on different paths do not pass across the P2P networks, which is expressed as a consecutive multiplication of path-loss probability of frame d_x on each path. The evaluation of path-loss probability is discussed in Section 10.5.

10.4 Quality-Driven Coordinative Multisource Selection

Due to the characteristics of multisource opportunity, limited bandwidth, and latency constraint of real-time multimedia applications in P2P networks, the coordinative selection of participating peers in the P2P networks should be considered in improving service quality. The total workload should be optimally distributed to each peer to avoid sending identical frames excessively. In addition, due to differing packet lengths and quality contributions of multimedia data frames, bandwidth should be smartly utilized by the multimedia frames with higher-quality contribution. In order to achieve this, the proposed approach improves the coordinative multisource selection in the P2P networks in three steps: (1) finds out all possible multimedia frame combinations on each server, (2) evaluates the expected end-received multimedia quality for all possible arrangements between frame combinations with different channel coding allocations, and (3) obtains the optimal frame allocation strategy for each server and the optimal channel coding rate for each path.

The proposed optimal transmission scheduling integrates the optimal frame distribution strategy and the optimal channel coding allocation for the purpose of achieving the maximum end-received multimedia quality and satisfying the latency requirement. We can express the optimal scheduling as follows:

$$\left\{ S_{i|i \in [0,N-1]}, C_{i|i \in [0,N-1]} \right\} = \left\{ \begin{bmatrix} S_0 \\ C_0 \end{bmatrix} \begin{bmatrix} S_1 \\ C_1 \end{bmatrix} \quad \cdots \cdots \quad \begin{bmatrix} S_{N-1} \\ C_{N-1} \end{bmatrix} \right\} \tag{10.5}$$

The optimal channel coding strategy C is applied to the optimal frame group S on each server to obtain the optimal result.

In the process of finding out the optimal transmission scheduling, the transmission time of each frame in combination with different channel coding rates should be estimated to satisfy the latency requirement. The transmission time on each path is expressed as

$$
\exp(T_{S_i}) =
\begin{cases}
T_0 = \sum_{i \in S_0} \left(\frac{\sum L_{i,j}}{R_0} \times G_0 + T_o \right) \\
\ldots \ldots \\
T_{N-1} = \sum_{i \in S_{N-1}} \left(\frac{\sum L_{i,j}}{R_{N-1}} \times G_{N-1} + T_o \right)
\end{cases}
\tag{10.6}
$$

where T_{N-1} denotes the transmission time along $path_{N-1}$, . $L_{i,j}$ is the packet size of the $j-th$ frame in group S_i, R_{N-1} is the end-to-end transmission rate along $path_{N-1}$, and G_{N-1} is the channel coding rate allocated to the frame group S_{N-1}. T_o denotes the network protocol overhead. If the transmission time of a frame group overruns the latency constraint, the transmission schedule will not be selected as a potential transmission solution. In this chapter, we do not consider the collision in the last hop that links to the destination. The collision model also can be applied by the proposed approach. We can set $T_{collision} = N_{path} \times T_{overhead}$, where N_{path} is the number of paths, and then add this value into the T_o in the Equation 10.6.

We focus on high-level resource allocation in this chapter, and the channel BER is the input of our proposal framework. Simply speaking, any channel model can be integrated into the framework, where the BER output of channel models becomes the input of the proposed framework. In a realistic environment, BER can be affected by many factors, such as multipath fading, propagation, noise, distortion, interference, etc. To accommodate such a realistic environment, in our work we target a more generic framework, so typical channel models can be seamlessly integrated into this framework.

10.5 Channel Coding Resource Allocation

Due to the error correction capability of forward error coding (FEC), channel coding redundancy is often added into original multimedia frames to reduce packet loss in transmission. In general, under the same network channel condition, more channel coding redundancies lead to higher data transmission reliability with higher transmission latency. In order to achieve the maximal end-received multimedia service quality while satisfying the latency constraint, the optimal channel coding allocation strategy should be achieved for each path. Assume a multimedia frame transmitted in a certain path has H hops from server to the destination; the path-loss probability for the frame can be expressed as

$$
P = 1 - \prod_{i=0}^{H} (1 - p_i)
\tag{10.7}
$$

where p_i denotes the frame link-loss probability, which is the frame loss probability from one node to the next node, and $\prod_{i=0}^{H} (1 - p_i)$ denotes the probability of a frame passing through all hops from server to the destination successfully. The frame link-loss probability can be presented as follows according to [16]:

$$p = \sum_{i=L+1}^{L+a} \binom{L+a}{i} b^i (1-b)^{L+a-i} \tag{10.8}$$

In this equation L denotes the length of the original multimedia frame, a means how many bits of channel coding redundancy are added to the original multimedia frame, and b denotes the bit-loss probability during data transmission. The equation evaluates the link-loss probability by summing up all possibilities of different bit-loss quantities, weighted with related bit-success probability and bit-loss probability. From this equation we can see that adding channel coding redundancy to the original multimedia frames can reduce the link-loss probability, and consequently improve the received multimedia quality. Under different channel coding allocation rates, the proposed optimal transmission scheduling evaluates the expected end-received multimedia service quality and related transmission latency, and then obtains the optimal channel coding allocation strategy for each path. The channel coding rate is the ratio of total multimedia frame length to the payload; this is shown as

$$G = \frac{L+\alpha}{L} \tag{10.9}$$

10.6 Revised Algorithm with Reduced Complexity

The design guideline of the proposed quality-driven coordinative multisource selection transmission scheduling is described as three parts: (1) find out all possible transmission arrangements with various channel coding allocations, (2) evaluate the expected receiver-end multimedia quality and the related transmission latency for each transmission arrangement, and (3) find the optimal transmission arrangement that leads to the maximum multimedia quality while satisfying the latency constraint. The above approach provides the complete solution for achieving the optimal transmission scheduling during the next time slot. However, it is unpractical to compute the complete optimal solution since the computational overhead is too high. For example, assume there are ε servers in the P2P networks, and each of them contains δ multimedia frames, and the number of available channel coding rate choices is four; then the complexity of achieving the complete optimal solution is $\left(2^\delta \times 4\right)^\varepsilon$. Thus, we have to simplify the proposed optimal transmission scheduling to reduce the complexity, yet approximating to the best solution. We simplify the complete optimal solution with two steps:

1. Divide all paths in the P2P network into two groups according to the related bit-pass probability. We use the average bit-pass probability as a metric to group paths. All paths in the same group will be allocated with the same channel coding rate.

2. Find out the optimal source server for each multimedia frame. The global optimal solution for all multimedia frames is divided into the local optimal solution for each frame. In other words, we find the best server for each multimedia frame. In this chapter, we define effect transmission capacity (EFC) as an indicator of the transmission capacity of each path. The optimal server for a certain multimedia frame is the server storing this frame and whose transmission path has the highest EFC.

The effect transmission capacity J is presented as follows:

$$J = R \times T_{remain} \times (1 - P) \tag{10.10}$$

where T_{remain} denotes the time remaining on a certain path. In the beginning, all remaining times T_{remain} on each path are initialized as $T_{remain} = T_{max}$. If remaining time becomes negative after transmitting a multimedia frame, it means that the path has no capacity to send this frame, so the remaining time should be evaluated in advance. From the equation, we can see that ETC expresses how many data can be transmitted to the destination correctly during the remaining time. The remaining time can be expressed as

$$T_{remain} = T_{remain} - \frac{L \times G}{R} \tag{10.11}$$

where L is the length of a multimedia frame specified to be transmitted on this path. The remaining time guarantees that the transmission latency on each path satisfies the latency constraint, so it is not necessary to check whether the maximal transmission time on each path is greater than the latency requirement.

Furthermore, due to the limited bandwidth and latency constraint, not all multimedia frames can be transmitted to the user. In order to maximize the end-received service quality, the multimedia frame with the highest quality contribution should be selected and transmitted first. Thus, the simplified optimal transmission scheduling is described as three steps: (1) Divide paths into two groups according to packet-pass probability. Frames transmitted along the paths in the same group will be applied with the same channel coding allocation strategy. (2) Find the optimal server for each multimedia frame in decreasing order of quality contribution, and then estimate the expected multimedia service quality and the transmission latency under different channel coding allocation schemes. (3) Find the optimal transmission scheduling leading to the maximum multimedia quality while satisfying the latency requirement.

The proposed simplified strategy is illustrated in Algorithm 10.1. The simulation results in the next section show that the function of simplified strategy works as well as the complete global optimal solution.

10.7 Simulation Study, Numerical Results, and Discussion

In this section, we present simulation studies to evaluate the performance gain of the proposed quality-driven coordinative multisource selection strategy in terms of

---

Algorithm 10.4 : Quality-driven coordinative multisource selection strategy

---

Step 1. Initialization. Define system configuration parameters: latency constraint T_{max}, transmission data rate R, the protocol overhead t_o, the total server number N, the total multimedia frames M, the length of original multimedia frame L, the quality contribution D of each multimedia frame, the number of hops H on each transmission path, the link-level bit error rate b , and the preselected channel coding rate G.

Step 2. Divide all available paths into two groups according to whether the related bit error rate is greater than the average or not.

Step 3. Initialize the remaining time on each path as $T_{remain} = T_{max}$.

Step 4. Under different channel coding allocations, find the optimal server to transmit multimedia frames and estimate the expected receiver end multimedia quality.

  1. Select the optimal server for each multimedia frame in contribution-descending order.

      a. Find potential servers for the multimedia frame.

      b. Estimate the frame link error probability p on each hop according to Equation (10.8).

      c. Estimate the frame path-pass probability q on each path according to Equations (10.4) and (10.7).

      d. Determine the remaining time according to Equation (10.10), and select the optimal server for the frame, which has the path with the highest transmission capacity.

      e. Update remaining time T_{remain} for the path that sends the current frame according to Equation (10.11).

      f. Go to step 4.1.a until done.

  2. Estimate the final receiver-end multimedia quality $U[D]$ contributed by all multimedia frames transmitted during the next time slot according to Equation (10.3).

Step 5. Search the optimal transmission scheduling leading to the maximum receiver-end multimedia quality according to Equation (10.1).

---

multimedia quality with latency constraint. Two traditional approaches are evaluated for comparison: in traditional approach 1, each server sends multimedia frames in the increasing order of frame number; in traditional approach 2, the multimedia frame with the highest multimedia contribution is sent with higher priority. By default, either the maximum or minimum channel coding rates are applied to each frame in traditional approaches. If not pointed out particularly, the maximal channel coding strategy is used.

The simulation parameters are stated as follows; The multimedia file is a JPEG 2000 picture, which includes five different frame types in terms of frame length and quality contribution. The distortion reduction of each multimedia packet can be either measured or estimated. In terms of measurement, it is done by calculating the decoded image quality improvement in a way similar to existing research [26]. In terms of estimation, it is done according to the wavelet coefficient square error units [27]. We follow the estimation approach, where the distortion reduction of each packet can be directly acquired from the JPEG 2000 codec. The length of five frame types and the related distortion reduction are illustrated in Table 10.2. As the figure shows, frames with shorter length have higher-quality contribution. The frame header of each multimedia frame is 6 bytes. There are four preselected channel coding rates: 9/9, 10/9, 13/9, and 15/9. There are three hops from server to the user on each path, and each transmission path is independent. The bit error rate on each hop is randomly allocated from 5/10,000 to 1/100,000. The server number, stored frames

Table 10.2   Frame Distortion Reduction and the Related Frame Length of Five Frame Types

Frame Type	0	1	2	3	4
Length	911	3649	4679	8835	9196
Distortion Reduction	290,515,906	43,366,978	25,885,684	11,863,336	3,642,474

(Length unit is byte, distortion reduction unit is MSE)

on each server, and transmission rates vary as defined under each result figure. The frame type is frame number MOD 5.

Figure 10.2 illustrates the receiver-end multimedia quality achieved by using global optimal scheduling and the simplified scheduling version with different latency requirements. In the rest of the chapter, we call the simplified version the proposed approach. From this figure we can see under all latency constraints, the results of the optimal strategy and the proposed approach are the same. Although the global optimal scheduling strategy tests all possible frame arrangements with different channel coding rates, the proposed approach still can achieve almost the same multimedia quality by specifying the optimal server for each frame. Under more complex situations, such as more servers in the P2P networks and more multimedia frames on

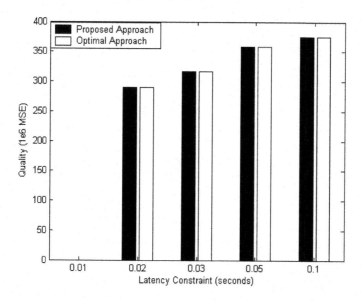

Figure 10.2: Multimedia quality with different latency constraints.
(Server 0: $R_0 = 0.5$ Mbps, $M_0 = D_0D_1D_2$; Server 1: $R_1 = 1.0$ Mbps, $M_1 = D_1D_2$; Server 2: $R_2 = 1.5$ Mbps, $M_2 = D_2D_3D_4$; R_0 is transmission rate in path 1; M_0 means stored frames on server 1.)

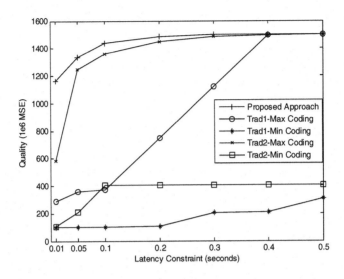

Figure 10.3: Multimedia quality with different latency constraints.
(Server 0: $R_0 = 0.5$ Mbps, $M_0 = D_0\text{-}D_{19}$; Server 1: $R_1 = 1.0$ Mbps, $M_1 = M_0$; Server 2: $R_2 = 1.0$ Mbps, $M_2 = M_0$; Server 3: $R_3 = 2.0$ Mbps, $M_3 = M_0$.)

each server, maybe the proposed approach cannot gain the same service quality as the global optimal strategy. However, in view of the complexity of the global optimal scheduling, the proposed approach is more practical, making the simplified approach a better option for solving the complex transmission scheduling in P2P networks.

Figure 10.3 shows the service quality received on the user side by different transmission strategies. As the figure shows, performance of the proposed approach under all latency requirements is higher than that of other strategies. Traditional approach 1 sends frames without transmission scheduling, so the performance totally depends on the random organization of the frames of each server. Traditional approach 2 obtains a better multimedia quality because it selects the frame with the highest multimedia contribution to transmit, so the limited latency resource is allocated to the most important frame. In addition, due to the characteristic of multimedia data, multimedia frames with higher multimedia contribution typically have shorter packet lengths. Thus, the strategy that sends important data first can achieve higher service quality and satisfy a more strict latency requirement. However, if a frame is stored on several servers, then the frame will be sent to the user a few times from different servers, which is a waste of limited resources. To solve these issues, the proposed strategy specifies which frames on each server will be sent to the user during the next time slot; in addition, the optimal channel coding allocation is applied to the packet transmitted along each path to achieve the maximal service quality. When the latency constraint is 0.01 s, the quality achieved by the proposed approach is almost twice that of the Trad2-Max coding. When the latency constraint is greater than 0.05 s, the performance of Trad2-Max coding is close to the proposed approach. However,

0.05, 0.1, 0.2, 0.3, 0.4, and 0.5 s are 5, 10, 20, 30, 40, and 50 times 0.01 s, respectively. Compared with 0.01 s, 0.05 to 0.5 s are too long. So the proposed approach is significantly greater than the Trad2-Max coding approach.

Figures 10.4 and 10.5 illustrate the multimedia quality at the user side with different latency constraints and different server numbers. To evaluate the exact impact of sender side server number, we set the same parameters on each path, such as stored frames on each server, transmission rate, and bit-loss probability. The number of participating peers in the P2P networks impacts the service quality. In Figure 10.4, the performance of the proposed transmission strategy with three senders is higher than that of traditional approaches, especially when the latency constraint is strict. As this figure shows, when the latency requirement is 0.01 s, the proposed approach with three servers achieves much higher quality than others. Compared with traditional approaches, the proposed approach with more senders achieves better service quality. This is because the proposed approach distributes transmission workload to proper servers by finding the optimal server for each multimedia frame. Traditional approaches did not achieve increased multimedia quality when given more senders. Traditional approach 1 strategy ignores the coordination of participating peers in the P2P networks and just sends frames according to the existing order of frames on the servers. So traditional approach 1 cannot adapt to changing conditions, and the multimedia quality depends on the frame order on each server. Traditional approach 2 does not consider the servers' coordination either and just selects the most important frame to transmit. The quality of the proposed approach and that of traditional ap-

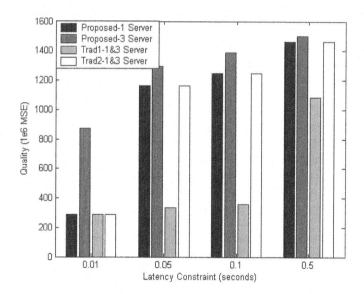

Figure 10.4: Multimedia quality with different latency constraints and server numbers.

(R = 1.0 Mbps on each path, M = D_0–D_{19} on each server.)

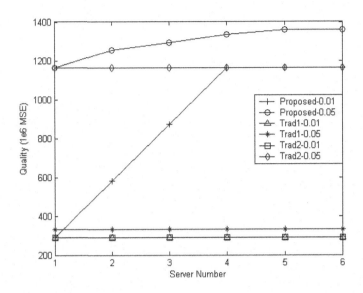

Figure 10.5: Multimedia quality with different server number and latency constraints.

(R = 1.0 Mbps on each path, M = D_0–D_{19} on each server, 0.01 and 0.05 are latency constraint.)

proach 2 with one server are the same, because the proposed approach has no chance to distribute transmission workload to different servers, and these two cases obtain the same transmission scheduling.

In Figure 10.5 we can see, under a strict latency constraint, performance of the proposed approach increases rapidly with more senders, but the two traditional approaches do not have obvious changes. In traditional approaches, if all servers have the same content, multimedia frames on different servers will be sent to the user in the same order, and thus the performance will not change in spite of increasing number of servers. However, the performance of the proposed approach does not always improve by increasing the number of senders. When the latency requirement is 0.01 s, the multimedia quality of the proposed approach does not change a lot if the server number is greater than four. That is because there are no more available multimedia frames to be assigned to more servers.

Figure 10.6 shows the service quality of different transmission strategies with different frame numbers on each server. The number of frames on each server has no significant influence on each strategy. Because the proposed approach assigns the best data source for each transmitted frame by jointly considering multisource coordination, channel condition, and multimedia contribution, in spite of more or less data resources on each server, the strategy can achieve higher multimedia service quality. In addition, more data resources on each server provide more opportunities to select the optimal source for each frame to achieve better results. The performance of traditional approach 1 does not change with increasing frame numbers. This is because

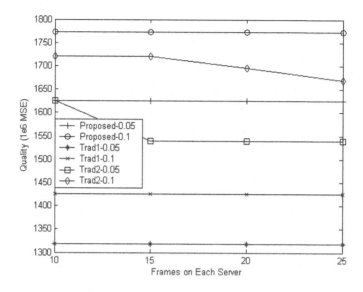

Figure 10.6: Multimedia quality with different frame numbers on each server.
($R_0 = 0.5$ Mbps, $R_1 = 1.0$ Mbps, $R_2 = 1.0$ Mbps, $R_4 = 2$ Mbps, Repeated frame $= 5$)

it has no more data resources to use due to limited data transmission capacity and latency constraint. The quality of traditional approach 2 is reduced with increasing frame amount stored on each server. This is because it repeats transmitting duplicated multimedia frames.

10.8 Summary

In this chapter, we have presented a quality-driven multisource transmission coordination/selection strategy for streaming services over wireless multimedia networks. A P2P network is utilized as an example scenario for discussion. The multisource coordination strategy maximizes the multimedia quality while assuring latency constraint by improving the coordinative selection of participating peers and allocating the optimal channel coding redundancy along each transmission path. To reduce the computational overhead of the global optimal scheduling, a simplified version is investigated. The simulation results show the simplified strategy achieves the same multimedia quality as the global optimal solution, while significantly improving multimedia transmission quality via coordinative selection among multiple peers. We expect future wireless multimedia streaming services may widely explore various coordinative multisource selection strategies, which have significant quality gain potential.

References

1. L. Caviglione and F. Davoli, Using P2P overlays to provide QoS in service-oriented wireless networks, *IEEE Wireless Commun.*, vol. 16, no. 4, pp. 32–38, 2009.

2. L. Garcia, L. Arnaiz, F. Alvarez, J. M. Menendez, and K. Gruneberg, Protected seamless content delivery in P2P wireless and wired networks, *IEEE Wireless Commun.*, vol. 16, no. 5, pp. 50–57, 2009.

3. D. Kim, E. Kim, and C. Lee, Efficient peer-to-peer overlay networks for mobile IPTV services, *IEEE Trans. Consumer Electronics*, vol. 56, no. 4, pp. 2303–2309, 2010.

4. Y. He, I. Lee, and L. Guan, Distributed throughput maximization in P2P VoD applications, *IEEE Trans. Multimedia*, vol. 11, no. 3, pp. 509–522, 2009.

5. E. Kim, Y. Lee, and W. Song, JXTA based content-sharing application, in *Proceedings of IEEE 35th Conference on Local Computer Networks*, October 2010, pp. 993–996.

6. N. N. Qadri, M. Fleury, M. Altaf, and M. Ghanbari, Multi-source video streaming in a wireless vehicular ad hoc network, *IET Commun. Video Commun. Wireless Networks*, vol. 4, pp. 1300–1311, 2010.

7. P. Si, F. R. Yu, H. Ji, and V. C. M. Leung, Distributed sender scheduling for multimedia transmission in wireless mobile peer-to-peer networks, *IEEE Trans. Wireless Commun.* vol. 8, no. 9, pp. 4594–4603, 2009.

8. P. Si, F. R. Yu, H. Ji, and V. C. M. Leung, Distributed multisource transmission in wireless mobile peer-to-peer networks: a restless-bandit approach, *IEEE Trans. Vehic. Technol.*, vol. 59, no. 1, pp. 420–433, 2010.

9. D. F. Macedo, A. L. Santos, J. M. S. Nogueira, and G. Pujolle, A distributed information repository for autonomic context-aware MANETs, *IEEE Trans. Netw. Service Management*, vol. 6, no. 1, pp. 45–55, 2009.

10. Z. Liu, Y. Shen, K. W. Ross, S. S. Panwar, and Y. Wang, LayerP2P: using layered video chunks in P2P live streaming, *IEEE Trans. Multimedia*, vol. 11, no. 7, 2009.

11. I. M. Moraes, M. E. M. Campista, J. L. Duarte, D. G. Passos, L. H. M. K. Costa, M. G. Rubinstein, C. V. N. D. Albuquerque, and O. C. M. B. Duarte, On the impact of user mobility on peer-to-peer video streaming, *IEEE Wireless Commun.*, pp. 54–62, 2008.

12. M. v. d. Schaar, D. S. Turaga, and R. Sood, Stochastic optimization for content sharing in P2P systems, *IEEE Trans. Multimedia*, vol. 10, no. 1, 2008.

13. L. Garccia, L. Arnaiz, F. Alvarez, and J. M. Menendez, Protected seamless content delivery in P2P wireless and wired networks, *IEEE Wireless Commun.*, vol. 16, no. 5, pp. 50–57, 2009.

14. N. Mastronarde, D. S. Turaga, and M. V. D. Schaar, Collaborative resource exchanges for peer-to-peer video streaming over wireless mesh networks, *IEEE J. Select. Areas Commun.*, vol. 25, no. 1, 2007.

15. H. Shiang, and M. V. D. Schaar, Distributed resource management in multihop cognitive radio networks for delay-sensitive transmission, *IEEE Trans. Vehic. Technol.*, vol. 58, no. 2, pp. 941–953, 2009.

16. B. Sklar, Digital communications, 2nd ed., Pearson Prentice Hall, Upper Saddle River, New Jersey, USA

17. J. Zhao, P. Zhang, G. Cao, and C. R. Das, Cooperative caching in wireless P2P networks: design, implementation, and evaluation, *IEEE Trans. Parallel Distrib. Syst.*, vol. 21, no. 2, pp. 229–241, 2010.

18. D. Li, Y. Xu, and J. Liu, Distributed cooperative diversity methods for wireless ad hoc peer-to-peer file sharing, *IET Commun.*, vol. 4, no. 3, pp. 343–352, 2010.

19. C. Huang, T. Hsu, and M. Hsu, Network-aware P2P file sharing over the wireless mobile networks, *IEEE Select. Areas Commun.*, vol. 25, no. 1, 2007.

20. C. Canali, M. E. Renda, P. Santi, and S. Burresi, Enabling efficient peer-to-peer resource sharing in wireless mesh networks, *IEEE Trans. Mobile Computing*, vol. 9, no. 3, pp. 333–347, 2010.

21. H. Chou, S. Wang, S. Kuo, I. Chen and S. Yuan, Randomised and distributed methods for reliable peer-to-peer data communication in wireless ad hoc networks, *IET Commun.*, vol. 1, no. 5, pp. 915–923, 2007.

22. Y. Andreopoulos, N. Mastronarde, and M. v. d. Schaar, Cross-layer optimized video streaming over wireless multihop mesh networks, *IEEE J. Select. Areas Commun.*, vol. 24, no. 11, pp. 2104–2115, 2006.

23. H. P. Shiang and M. v. d. Schaar, Informationally decentralized video streaming over multi-hop wireless networks, *IEEE Trans. Multimedia*, vol. 9, no. 6, pp. 1299–1313, 2007.

24. Y. Yang and J. Wang, Routing permutations on baseline networks with node-disjoint paths, *IEEE Trans. Parallel Distributed Syst.*, vol. 16, no. 8, pp. 737–746, 2005.

25. X. Huang and Y. Fang, Performance study of node-disjoint multipath routing in vehicular ad hoc networks, *IEEE Trans. Vehic. Technol.*, vol. 58, no. 4, pp. 1942–1950, 2009.

26. M. v. d. Schaar and D. Turaga, Cross-layer packetization and retransmission strategies for delay-sensitive wireless multimedia transmission, *IEEE Trans. Multimedia*, vol. 9, no. 1, pp. 185–197, 2007.

27. D. Taubman, High performance scalable image compression with EBCOT, *IEEE Trans. Image Process.*, vol. 9, no. 7, pp. 1158–1170, 2000.

Chapter 11

SNS-Based Mobile Traffic Offloading by Opportunistic Device-to-Device Sharing

Xiaofei Wang and Victor C. M. Leung

Department of Electrical and Computer Engineering, University of British Columbia

CONTENTS

Abstract

Rapidly increasing mobile traffic has become a serious concern of mobile network operators. In order to alleviate this traffic explosion problem, there have been some efforts to research offloading traffic from cellular links to local short-range communications among mobile users in proximity of each other. In this chapter, we first survey related studies on opportunistic user-to-user (device-to-device) content sharing, and then present a new proposal to carry out social-aware mobile *traffic offloading* assisted by *social network services via opportunistic sharing in mobile networks, called TOSS*. TOSS initially selects a subset of mobile users as seeds, depending on their content spreading impact in online social network services (SNSs) and their mobility patterns in mobile social networks (MSNs). Then users share the content with each other via opportunistic local connectivity (e.g., Bluetooth, Wi-Fi Direct) according to their social relationships. Due to the distinct access patterns of individual SNS users, TOSS further exploits the user-dependent access delay between the content generation time and each user's access time for opportunistic content sharing. We model and analyze the process of traffic offloading and content spreading by taking into account various options in linking SNS and MSN data sets. We present trace-driven evaluations to show that TOSS can reduce 63.8 to 86.5% of the cellular traffic while satisfying the access delay requirements of all users. Thus, SNS-based traffic offloading by opportunistic sharing can provide an effective and efficient content delivery service that is promising for the future wireless and mobile networks.

11.1 Introduction

Recent advances in mobile communication technologies have made ubiquitous Internet access available to a large number of mobile devices. Increasingly, users are downloading contents, e.g., articles, images, and videos, to their smart phones and tablets for consumption. The ever-increasing traffic load caused by content downloading is becoming a serious concern of mobile network operators (MNOs) [1], but studies [2–4] have revealed that much of the traffic load is due to duplicated downloads of the same contents; e.g., the top 10% of videos in YouTube account for nearly

80% of all the views [4]. Therefore, how to effectively reduce the downloads duplicated via cellular links by *offloading* the traffic via other networks has become a hot topic.

Recently, there have been many studies to exploit local communications, i.e., user-to-user or device-to-device (D2D)[1] opportunistic sharing, during intermittent meetings of mobile users for Traffic offloading in mobile social networks (MSNs). This is a special form of delay-tolerant network (DTN) [5–11] that further considers the social relationship of network users, where the content dissemination procedure is tolerant to a certain delay so that exploitation of opportunistic networking [12] is possible. We consider MSNs in which users are able to discover their neighbors [13] and set up temporary local network connections via technologies such as Bluetooth, Wi-Fi Direct, near-field Communication (NFC) [14], or D2D in Long Term Evolution (LTE)-Advanced [15, 16], for sharing delay-tolerant contents with each other. Particularly, the D2D technique is under detailed design in the third Generation Partnership Project (3GPP) as an underlay to LTE-Advanced networks [16], whereby users can use operator-authorized spectrum for direct communications between their devices.

Regarding D2D sharing, it is shown that by selecting an appropriate initial set of *seeds*, which have a large potential to deliver a content object downloaded via the cellular link to others by D2D sharing later, the peak traffic load can be reduced by 20 to 50% [9]. The study in [10] also proves that content dissemination with a small number of initial seeds can meet the delay requirements of all users while substantially reducing the amount of cellular traffic. However, there are still several important issues in related areas that call for further research, such as:

■ *How to determine or how to predict the access delay of each user for each content?* Recent studies [10, 11, 17, 18] assume the same dissemination deadline of the same content for all users; however, users indeed have various delay requirements [19].

■ *How to design the seeding strategy to minimize the cellular traffic while satisfying the delay requirements of all users.* Strategies of selecting initial seeds are discussed in prior work [8, 9, 20], but most of them focus on user mobility while ignoring the practical social relationships among users.

■ *How to efficiently make mobile users share contents with others.* Studies in [8, 10, 11] assume people will always exchange content gratuitously. However, in reality, people mostly share information due to the word-of-mouth" propagation [21, 22], and the real social relationships among users.

Regarding the above issues, we seek to exploit the relationship between the online social network services (SNSs) and the offline MSNs.[2] Recently, there have been

---

[1]For simplicity, in this chapter we use the term *D2D sharing* synonymously for user-to-user sharing or people-to-people sharing methods.

[2]"Online SNSs" here indicate the virtual social networks in the Internet formed by the accounts of people, while "offline MSNs" here mean the real MSNs formed by people in a physical environment.

dramatic increases in the numbers of mobile users who participate in various online SNSs, e.g., Facebook [23], Twitter [24], Sina Weibo [25], and so on, where larger and larger amounts of new contents are recommended by users and spread rapidly and widely [22, 26]. By investigating related studies in measurement and modeling of MSNs and SNSs, we discovered the following key points, which can be exploited for efficient content dissemination:

■ In online SNSs, the access pattern of each user can be measured, statistically modeled, and thus predicted; i.e., we can analyze the *access delay* between the content generation time and the user access time [27], which is user-dependent mainly due to people's different lifestyles [19, 22, 28]. We can disseminate the contents of interest to users considering their different delay sensitivities (requirements).

■ In online SNSs, a user's influence, or *spreading impact*, to other users, can be modeled based on the analysis of historical social behaviors, e.g., the forwarding probability.

■ In offline MSNs, the mobility patterns of users can be measured and modeled [10, 17, 29–31], and hence a different offline mobility impact of each user to disseminate some contents to others can be derived.

■ User relationships and interests in online SNSs have significant *homophily* and locality properties (to be detailed in Section 11.2), which are similar to those of offline MSNs [21,32,33]. Users are mostly clustered by geographical regions and interests, which can be exploited for content sharing.

In this chapter, we mainly discuss the proposal of a framework for Mobile *T*raffic *O*ffloading by *S*NS-Based opportunistic *S*haring in MSNs (**TOSS**) [34]. TOSS pushes the content object to a properly selected group of seed users, who will opportunistically meet and share the content with others, depending on their spreading impact in the SNS and their mobility impact in the MSN. TOSS further exploits the user-dependent access delay between the content generation time and each user's access time for traffic offloading purposes. From trace-driven evaluation and model-based analysis, it is shown that TOSS reduces the related cellular traffic by 63.8 to 86.5% while still satisfying the delay requirements of all users. To the best of our knowledge, this is the first study that seeks to combine online SNSs with offline MSNs for traffic offloading while considering user access patterns.

The advantages of offloading the cellular traffic by opportunistic D2D sharing have been also discussed in prior studies [8, 18, 35, 36]. We compare the pushing and sharing approach with other strategies of content dissemination as follows:

■ *Pull-based unicast:* In the traditional pull-based delivery, the file of interest may be downloaded via cellular links as many times as the number of subscribers [3, 4]. Meanwhile, the social-aware sharing-based offloading can leverage the social encounters of users to offload the redundant downloads from the cellular links to opportunistic local links.

- *Broadcast/multicast:* When multiple users (in the same cell) wish to receive the same content, broadcasting (or multicasting) might be efficient. To improve the reliability of broadcasting, the lowest bit rate is normally used to cover all the mobile users in its cell, which reduces the efficiency substantially. However, the reliability of content delivery may still be less than desired. There is also a security issue since nonsubscriber users can also receive the content.

The rest of the chapter is organized as follows. After reviewing the related work in Section 11.2, we discuss the TOSS framework in Section 11.3. System optimization is discussed in Section 11.4. Measurement and evaluation results are presented in Section 11.5 and Section 11.6, respectively. We discuss the practical deployment in Section 11.7, followed by concluding remarks in Section 11.8.

11.2 Related Work

11.2.1 Opportunistic Sharing in DTNs/MSNs

Figure 11.1 illustrates some scenarios of DTNs/MSNs in the real world, where the user may go around to meet friends, families, and colleagues, and also to take public transportation, where many other passengers are in proximity. Therefore, it is expected that the user can have a large potential to carry and exchange interesting contents with other nearby users while moving.

Content exchanges among encountered users are considered "epidemic content delivery" in DTNs/MSNs, which have been extensively studied recently. Zhang et al. [17] have developed a differentiation-based model to study the delay of epidemic content delivery. Another related study done by Li et al. [11] has designed an energy-efficient opportunistic content delivery framework in DTNs by exploiting users' contacts. This work models the content dissemination by a continuous-time

Figure 11.1: Illustration of opportunistic meetings in DTNs/MSNs.

Markov framework, and then formulates the optimization problem of opportunistic forwarding, with the constraint of energy consumed by the message delivery.

Regarding the scalability and optimality of content dissemination by exploiting the epidemic user-to-user sharing, the study in [8] has proposed a social welfare maximization problem for the best pushing and sharing strategies offering the best offloading performance. Similarly, the study in [18] has solved the maximization of traffic offloading utility in DTNs as a knapsack problem. Regarding the slow start and long completion time of the epidemic delivery, strategic pushing is studied to expedite the dissemination in [9]. Whitebeck et al. have also demonstrated the effectiveness of opportunistic sharing and offloading strategies based on practical mobility traces [7]. While the above studies were limited to single-cell environments, Wang et al. extend the pushing and sharing model into multicell cellular network environments in [10].

While researching D2D sharing in epidemic content delivery, the acceleration of the content dissemination by leveraging users' social relationships has become a more popular topic recently. The study called BUBBLE Rap [36] seeks to utilize human mobility in terms of social structures, and to use these structures in the design of sharing and forwarding. Furthermore, authors discover that human interactions are heterogeneous in terms of both hubs (popular individuals) and groups or communities. Another related study in [20] investigates how to select the target set for information delivery based on social participation, such that the mobile data traffic over cellular networks can be minimized. Extensive trace-driven simulations show that up to 73.66% mobile traffic can be offloaded through social participation in the MSN.

The study done by Gao and Cao in [37] has proposed to assign interest tags to the users and content objects to identify their preferences of contents, and then to carry out effective relay users selection based on a social centrality metric, which considers the social contact patterns and interests of mobile users simultaneously. The authors analytically investigate the trade-off between the effectiveness of relay selection and the overhead of maintaining network information. Similarly, ContentPlace [38] utilizes social central betweenness of mobile users to optimize mobile content sharing. The similarity concept is also exploited by [39], in which user encounter history is explored for getting the friendship similarity for delegation forwarding in the DTNs/MSNs. Recently, there has been one practical study, "DataSpotting" in [40], which shows real sharing-based offloading tests in Manhattan. The authors explore the possibility of serving user requests from other mobile devices located geographically close to the user. Note that in this chapter, security or privacy problems are not considered, but readers interested in these issues may refer to related studies, such as [41, 42].

All the sharing techniques in the studies reviewed above mostly rely on local communication techniques. Among existing sharing methods, e.g., Bluetooth, Wi-Fi Direct, and NFC [14], which are based on public short-range communication techniques, Wi-Fi Direct is very popular. For instance, Apple's Airdrop [43] provides a convenient user interface for a user to share a content with nearby users. Furthermore, the underlay of communications in 3GPP LTE-Advanced cellular networks using the operators D2D authorized spectrum is studied as an enabler of intelligent local services with limited interference impact on the primary cellular network [15]. This will further enhance the development of D2D sharing for traffic offloading in emerging mobile networks.

11.2.2   Mobile Traffic Offloading

There have been many studies focusing on mobile traffic offloading by deploying Wi-Fi access points (APs) in a large scale, such as [46, 47] in Korea, as well as [48, 49] in the United States. Depending on the density of AP deployment, the offloading based on Wi-Fi APs can have significantly different performances. For instance, realistic measurements from Korean Telecom (KT) [46] indicate that about 18 to 26% of the cellular traffic load is offloaded to KT's Wi-Fi APs. However, up to 65% of the traffic can be offloaded to Wi-Fi APs, as found in [47], in the downtown of Seoul, Korea.

Figure 11.2 illustrates offloading based on Wi-Fi APs, and it is obvious that the traffic load in the network backbone is actually not reduced at all. In contrast, offloading based on D2D opportunistic sharing is illustrated in Figure 11.2, where more than half of the transmissions are offloaded via local communications. Note that sharing-based offloading does not replace Wi-Fi APs, but the two methods can coexist and collaborate to solve the traffic explosion problem in the emerging future mobile wireless networks.

Therefore, regarding traffic offloading based on D2D opportunistic sharing, how to encourage people to share while they are moving becomes an important issue that does not depend only on technology. As sharing-based offloading will help the MNOs to reduce their traffic load significantly and accelerate content dissemination for the content providers (CPs), a popular research trend is to design incentive-based

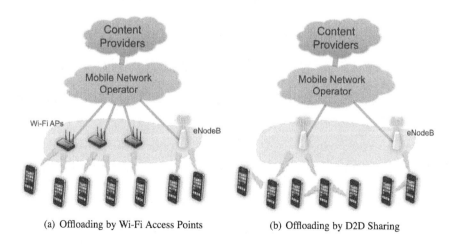

(a) Offloading by Wi-Fi Access Points   (b) Offloading by D2D Sharing

Figure 11.2: Illustration of mobile traffic offloading techniques.

business models for MNOs, CPs, and mobile users. In the following studies, it is advocated that financial benefits will motivate users to cache contents and share them with nearby users to reduce the costs of their cellular data plan, or even to earn them some money: the sharing-oriented pricing study, Win-Coupon [35]; the self-interest-driven incentive study for ad dissemination among mobile users, SID [50]; and the incentive-based study with conjunctive consideration of privacy, IPAD [51].

11.2.3 Information/Content Spreading in SNSs

In this chapter, we seek to exploit the social relationship of users for traffic offloading, so we briefly survey some studies on information and content spreading in SNSs. Figure 11.3 illustrates some social relationships that are collected from the SNS site of one of the authors, where different friends have different social relationships, and thus different impact strengths to the author. Undoubtedly, a tight relationship indicates a strong social influence, and thus a strong impact of spreading some information. In [52] the social influence of people has been researched and identified as a two-step flow of communications; i.e., most people form their opinions under the influence of "opinion leaders," who in turn are influenced by the media source. Also, the study in [53] found that a small number of opinion leaders, who have strong impact on spreading information, perform the key roles to broadcast information by socially connected networks.

In SNSs, due to the effect of word of mouth [21], users can significantly impact the information spreading to other users [54, 55]. Many studies have proposed to use probabilistic modeling to analyze the information/content commenting or re-sharing activities, and thus the information spreading impact among users [27, 28, 55–59]. Especially, the recommendation from famous people, who have potentially strong impact on others, may accelerate the topic spreading, as studied in [27, 60]. Also, [22]

Figure 11.3: Illustration of the social impacts in SNSs.

indicates that people's historical impact on information sharing can impact and thus enable the accurately forecasting of the future sharing activities. Furthermore, [22] points out that there are always some delays of resharing behaviors, while the spreading impact of each user can be accumulated hop by hop. This access delay between the content generation time and the user access time due to people's different lifestyles has been mentioned in many studies [19, 22, 28]. Researchers can obtain, analyze, and even predict the spreading impact and the access delays of SNS users based on measurement traces [26]. For example, the general retweeting delay in Twitter is about 100 to 1,000 s [27].

Recent studies in [21] and [33] report that user relationships and interests in SNSs have significant *homophily* and *locality* characteristics that are similar to those in MSNs. *Homophily* is the tendency of individuals to associate and bond with other similar individuals [61]. Here homophily means that online and offline users are both highly clustered by regions and interests, which also is studied as "birds of a feather" in [62]. User homophily significantly impacts the information diffusion in social media. People with similar interests like to share interesting information with each other. For instance, if one's good friends have watched a video, one will watch the same video with a very high probability. *Locality* is originally a phenomenon describing the same value, or related storage locations, being frequently accessed [63]. More specifically, in this chapter, locality means that people who are close graphically may have similar trends of accessing the content and sharing with each other [21]. So due to the homophily and locality characteristics of users, those who are near in proximity may have the same trends of accessing the content and sharing with each other. Even in online SNSs, users may significantly interact with and thus impact others in proximity, which also indicates the locality nature [32]. In other words, users within a short geographical distance have a higher probability of posting the same content than those users who are physically located farther apart. Thus, the locality characteristics of user interests can be utilized to facilitate the traffic load balancing [2] and content delivery [33].

11.3 Details of TOSS Framework

11.3.1 Preliminaries

The TOSS framework entails an online SNS and an offline MSN. Suppose there are totally N mobile users, u_i, $i = 1, ..., N$, who have corresponding SNS identities (i.e., virtual accounts). Because we focus on the content spreading in an online SNS, we use a directional graph to model the SNS,[3] e.g., Twitter [24] and Sina Weibo [25]. The online SNS can thus be represented by $G(V, E)$, where V is the set of users in the online SNS, and E is the set of directional edges. If u_j follows u_i, u_j is one *follower* of u_i and u_i is one *followee* of u_j. As we focus on content spreading, the directional

---

[3]TOSS can also work with any SNS based on the bidirectional graph model (e.g., Facebook [23]) since it is a subset of the directional graph model.

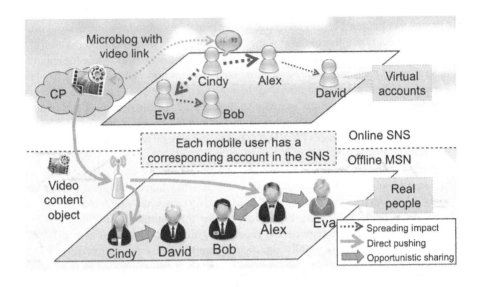

Figure 11.4: Example of the TOSS framework.

edge (represented by an arrow in Figure 11.4) is from u_i to u_j, denoted by v_{ij}. That is, u_i has a direct impact on u_j for content spreading. There can also be a bidirectional relationship where two users follow each other.

We define the home site, where a user creates and shares content in the SNS platform, as the *microblog*, and we define a short message posted by a user containing the content (or link to the content) as a *micropost*, which can be, e.g., a tweet in Twitter or a post in Facebook, and the content file is called a *content object*. Furthermore, we define the *timeline* of a user in online SNS as the series of all microposts published by the user in her microblog, sorted by time.

At any time, a user may find or create a new interesting article, image, or video, and share it in the SNS as an *initiator*. All her followers will then be able to access the content, and some of them will further reshare in their timelines. Making comments will not induce any information spread; thus, we only consider the resharing activities. Afterwards, what TOSS seeks to achieve is that while the micropost with the content is being spread to other users in the online SNS, the content object will be accessed and delivered among user devices in the offline MSN. Note that the TOSS framework is not confined strictly to the dissemination of one popular content to all the users, but applies to general deliveries of any content to a group of potential recipients with any size.

TOSS defines four factors for user u_i: two for the online SNS, including (1) the outgoing spreading impact, $I_i^{S \rightarrow}$, and (2) the incoming spreading impact, $I_i^{S \leftarrow}$, which indicate how important the user is for propagating the micropost (to others or from others), and two for the offline MSN, including (3) the outgoing mobility impact, $I_i^{M \rightarrow}$, and (4) the incoming mobility impact, $I_i^{M \leftarrow}$, which indicate how important the user is for sharing the content object (to others or from others) via physical encounters. We shall discuss how to calculate these factors in Sections 11.3.3 and 11.3.2.

Considering the above factors, TOSS seeks to select a proper subset of users as seeds for pushing the content object directly via cellular links, and to exploit the D2D sharing in the offline MSN, while satisfying different access delay requirements of different users. We define a vector $\overrightarrow{p}$ to indicate whether to push the content object to a user via cellular links or not; i.e., $p_i = 1$ means pushing the content object directly to user u_i.

In the scenario of TOSS illustrated in Figure 11.4, in the online SNS, Cindy shares a video (or a link to the video) with Eva and Alex, who may in turn share with Bob and David, respectively. Meanwhile, the video content is first downloaded via a cellular link and stored in Cindy's phone. However, in the offline MSN, Cindy is geographically distant from other people, but David is in proximity. Although David may not know Cindy, TOSS detects that the $I^{S\to}$ impact of Cindy on David via Alex is also very strong, and thus lets Cindy share the video with David via a local Wi-Fi connectivity. Furthermore, TOSS evaluates the $I^{M\to}$ impact of Alex, and pushes another copy to him via a cellular link, because Alex is likely to meet Bob and Eva in the offline MSN frequently, and Bob and Eva often access content with some delays. Then the content object will be propagated by local connectivity from Alex to Bob and to Eva at a later time. In this example, TOSS reduces the cellular traffic by 3/5.

11.3.2 Mobility Impact in the Offline MSN

It has been shown that mobile users in the offline MSNs (or DTNs) have different mobility patterns [10, 17, 29–31], and hence different potentials for sharing content. Thus, the mobility impact, I^M, is defined to quantify the capability of a mobile user to share a content object with other users via opportunistic meetings, or *contacts*, while roaming in the MSN. Temporary connectivity with nearby users mostly relies on active discovery mechanisms; thus, we assume all mobile users are synchronized with a low duty cycle for probing as proposed by eDiscovery [13].

Referring to [8, 10, 11, 17, 29–31, 37], we assume that the inter-contact intervals of any two mobile users follow the exponential distribution. We use λ_{ij} to denote the opportunistic contact rate of user u_i with user u_j. Note that there are many practical methods to measure λ_{ij} values, e.g., centralized measurement by the location management entity in the MNO [68] or by distributed user-to-user exchanges [38]. Note that the contact duration may be ignored in TOSS, because we assume the content delivery is always finished successfully during the contact due to the high bandwidth of local communications (e.g., Wi-Fi) [8, 10, 17, 37].

We adopt the epidemic modeling from [11, 17] to model the opportunistic sharing in TOSS with the continuous-time Markov chain. Let $S_i(t)$ be the probability that user u_i has the content until t, $0 \leq S_i(t) \leq 1$, while $1 - S_i(t)$ is the probability that user u_i has not received the content until t. $S_i(t)$ will be increasing over t while roaming and meeting users in the offline MSN. The increment of $S_i(t)$ within a period Δt, that is, $S_i(t + \Delta t) - S_i(t)$, will be calculated in the following procedure.

The probability of user u_i meeting user u_j during Δt is $1 - e^{-\lambda_{ij}\Delta t}$ due to the exponential decay of intercontact intervals. The probability that user u_i can get the content from another user u_j via opportunistically meeting, denoted by ε_{ij}, can be calculated by

$$\varepsilon_{ij} = \left(1 - e^{-\lambda_{ij}\Delta t}\right) \cdot \gamma_{ji}^* \cdot S_j(t), \tag{11.1}$$

where the $I^{S\rightarrow}$ impact factor from u_j to u_i, γ_{ji}^*, is considered as both (1) the spreading probability that u_j will reshare microposts from u_i and (2) the sharing probability that u_i can obtain the content object from u_j.

Considering the ε_{ij} of u_i from all users, the probably that u_i can get the content from others within Δt is

$$1 - \prod_{j=1, j\neq i}^{N} (1 - \varepsilon_{ij}) \tag{11.2}$$

Hence, based on the probability that u_i has not received the content,

$$S_i(t + \Delta t) - S_i(t) = (1 - S_i(t)) \cdot \left(1 - \prod_{j=1, j\neq i}^{N} (1 - \varepsilon_{ij})\right) \tag{11.3}$$

Letting $\Delta t \rightarrow 0$, the derivative of $S_i(t)$ is

$$\dot{S_i}(t) = \lim_{\Delta t \rightarrow \infty} \frac{S_i(t+\Delta t) - S_i(t)}{\Delta t} = (1 - S_i(t)) \cdot \sum_{j=1, j\neq i}^{N} \lambda_{ij} \cdot \gamma_{ji}^* \cdot S_j(t) \tag{11.4}$$

where initially $S_i(0) = p_i$ from $\overrightarrow{p}$.

Solving the above matrix of the ordinary differential equation system is complicated. However, we can find a numerical solution easily by approximation with power series [69, 70]. We skip the details of the procedure for getting numerical solutions, since this is straightforward.

Given a pushing vector $\overrightarrow{p}$, we can calculate how long it will take for any user u_i to obtain the content by the inverse function of $S_i(t)$ with $S_i(t) = 1$, defined as the *content obtaining delay* of u_i, denoted by t_i^*:

$$t_i^* = S_i^{-1}\left(\{\gamma_{ji}^*\}, \{\lambda_{ij}\}, \overrightarrow{p}\right), \quad j = 1, ..., N, j \neq i \tag{11.5}$$

where $\left\{\gamma_{ji}^*\right\}$ is the series of $I^{S\leftarrow}$ factors from all other users to u_i in the SNS, and $\{\lambda_{ij}\}$ is the series of meeting rates of user u_i to all other users in the MSN. Note that TOSS mainly seeks the optimal $\overrightarrow{p}$ to match the content obtaining delays of all users with their access delay probability distribution functions (PDFs).

$I_i^{M\rightarrow}$ is actually the same as $I_i^{M\leftarrow}$ since $\lambda_{ij} = \lambda_{ji}$ for any u_i and u_j due to the symmetric nature of contacts. Hereby, we define the I^M factor for u_i as

$$I_i^{M\rightarrow} = I_i^{M\leftarrow} = \lambda_i^* = \sum_{j=1}^{N} \lambda_{ij} \tag{11.6}$$

And then we will only use I^M to denote the mobility impact. We can use approximation methods, e.g., the Newton method, to get the numerical result of the inverse function of $S_i(t)$.

Note that the above content obtaining delay is the expected delay that a user can get a content object based on opportunistic sharing while moving, which is an objective factor depending on the mobility traces given an initial pushing vector. The aforementioned content access delay is instead a subjective factor depending on user behaviors (lifestyles); TOSS fits the access delays of users by Weibull function, which converts the subjective access delays into objective PDFs, and then uses it for indicating a user's delay QoS requirement. So TOSS is just seeking a match between content access delay and content obtaining delay, so that the user can mostly get expected content objects when she needs them the most.

11.3.3 *Spreading Impact in the Online SNS*

We extend the previous probabilistic models [55–58, 60] to quantify the content spreading impact in the SNS. Hereby we define, the $I^{S\rightarrow}$ factor of user u_i to user u_j, denoted by γ_{ij}, $0 \leq \gamma_{ij} \leq 1$, as the ratio of the number of microposts of u_i that u_j accesses and re-shares to the number of all microposts of u_i in u_j's timeline. Thus for a given object of u_i in the future, γ_{ij} is the probability that u_j will reshare the micropost from u_i [26].

Based on the SNS graph G, we define U_i^h as the set of h-hop upstream neighbors (followees) of user u_i through all possible shortest h-hop paths without a loop, and likewise, D_i^h as that of h-hop downstream neighbors (followers). And we use γ_{ij}^h to denote the $I^{S\rightarrow}$ factor from user u_i to u_j by any h-hop path (inversely γ_{ji}^h as the $I^{S\leftarrow}$ factor from user u_j to u_i). From u_j's point of view over a certain period, we need to consider (1) the number of microposts that u_j has created by herself, c_j, (2) the number of reshared microposts by u_j from u_i, r_{ij}, and (3) the number of reshared microposts from all h-hop followees, to calculate $I_i^{S\rightarrow}$ as follows:

$$\gamma_{ij}^1 = \frac{r_{ij}}{c_j + \sum\limits_{u_k \in U_j^1} r_{kj}} \tag{11.7}$$

$$\gamma_{ij}^2 = 1 - \prod_{k \in D_i^1 \cap U_j^1} \left(1 - \gamma_{ik}^1 * \gamma_{kj}^1\right), \quad \gamma_{ij}^3 = 1 - \prod_{k \in D_i^2 \cap U_j^1} \left(1 - \gamma_{ik}^2 * \gamma_{kj}^1\right), \dots \tag{11.8}$$

$$\gamma_{ij}^h = 1 - \prod_{k \in D_i^{h-1} \cap U_j^1} \left(1 - \gamma_{ik}^{h-1} * \gamma_{kj}^1\right) \tag{11.9}$$

We use γ_{ij}^* to denote the impact from user u_i to user u_j via all possible paths with hop less than or equal to H computed by

$$\gamma_{ij}^* = 1 - \prod_{n=1}^{H} \left(1 - \gamma_{ij}^n\right) \tag{11.10}$$

where H is less than or equal to the maximal diameter of the SNS graph G. Then $I_i^{S \rightarrow}$ and $I_i^{S \leftarrow}$ of u_i to and from the whole user base can be respectively calculated by

$$I_i^{S \rightarrow} = \sum_{j=1}^{N} \gamma_{ij}^*, \quad I_i^{S \leftarrow} = \sum_{j=1}^{N} \gamma_{ji}^* \qquad (11.11)$$

Note that it is reported in [22, 62] that the average path length in SNS graphs is about 4.12, and the spreading impact after three hops becomes negligible.

11.3.4   Access Delays of Users

Different users have different patterns of accessing contents via the online SNS. Some may access the SNS frequently, while others access the SNS at relatively longer intervals. Thus, the access delay between the content generation time and user's access time becomes different for each user [19, 22, 28].

As illustrated in Figure 11.5, user A creates a micropost for an interesting video in the SNS at t_0. One of A's followers, B, happens to see A's micropost after a certain delay at t_1 due to B's personal business. Once B clicks to play it, a buffering delay is needed until t_2; B will reshare the video at t_3 after watching it. According to the definition of access delay given above, the access delay for B should hence be $t_2 - t_0$. However, in practice, only the SNS application provider can obtain t_1 and t_2 data. Since for texts, images, and most videos, t_1, t_2, and t_3 may be close, in our study, we consider B's access delay as $t_3 - t_0$, which can be captured from SNS measurement traces by checking B's resharing time from SNS measurements.

To investigate access delays, we collected the SNS trace data of approximately 2.2 million users from the biggest online SNS in China, Sina Weibo (measurement

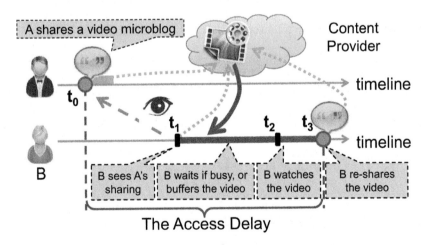

Figure 11.5: Illustration of the content access delay.

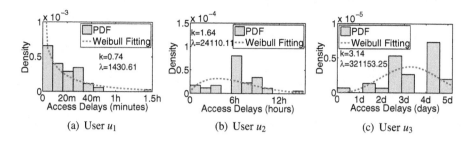

Figure 11.6: Access delay distributions of three real users with Weibull fitting.

details will be explained in Section 11.5). The access delay of a follower to access the content that is shared by the followee is gathered as the time difference between the generation time of the original micropost and the time of resharing by the follower. We then pick up three real users from the online SNS traces as examples, and plot their access delay PDFs as shown in Figure 11.6. User u_1 is likely to access the content frequently with short delays. But users u_2 and u_3 have significantly larger delays, on the order of hours and days, respectively. Generally, TOSS tends to initially push content to users with large social impact and small access delays similar to u_1, but seeks to utilize the opportunistic sharing in the MSN to disseminate the object to users with small social impact and large access delays similar to u_2 and u_3.

We use a PDF to model the access delays of each user, say u_i, in terms of the probability to access the content at t, denoted as $A_i(t)$. Similar to [8], $A_i(t)$ can be considered the access utility function. If the content object is already obtained locally in the user's device when she has the highest probability to access the content, she will be mostly satisfied. In order to model the various distributions of access delays with different shapes of PDF curves, we choose to use the Weibull distribution for fitting, which is commonly used for profiling user behaviors in SNSs [64]:

$$A_i(t, \beta_i, k_i) = \frac{k_i}{\beta_i}\left(\frac{t}{\beta_i}\right)^{k_i-1} e^{-\left(\frac{t}{\beta_i}\right)^{k_i}}, \quad t \geq 0 \tag{11.12}$$

where the fitting parameters β_i and k_i can identify the access pattern of user u_i (note that k_i controls the shape of the curve). Note that it is possible to fit many different functions to the access delay statistics, but we choose the Weibull function because a large number of studies have used the Weibull function for fitting the statistics of data generated based on human behaviors, such as [64] for user behaviors in online SNSs, [65] for user web browsing activities, and [66] for traffic flows in online games.

11.4   System Optimization

TOSS aims to choose a proper set of seeds for initial pushing, $\overrightarrow{p}$, by evaluating I^S (both incoming and outgoing) and I^M values of all users, to get the content obtaining delay t^* for each user in order to maximize the sum of the access utilities (access probabilities) for all users:

$$
\begin{aligned}
\text{Maximize}_{\vec{p}} : \quad & \sum_{i=1}^{N} A_i \left(t_i^*, \beta_i, k_i \right) \\
= \quad & \sum_{i=1}^{N} A_i \left(S_i^{-1} \left(\left\{ \gamma_{ji}^* \right\}, \left\{ \lambda_{ij} \right\}, \vec{p} \right), \beta_i, k_i \right) \\
& (j = 1, ..., N, j \neq i) \\
\text{Subject to} : \quad & |\vec{p}| \leq C
\end{aligned}
\tag{11.13}
$$

where the number of initial pushing seeds, C, is a constraint controlled by the MNO, and we call $\sum A_i(t)$ the total access utility function of the whole user base.

This problem is similar to the social welfare maximization problem discussed in [8] as well as some other studies in the literature. It is hard to analytically solve the above optimization problem, since all related equations are not in closed form. With power series approximations [69, 70], we can find the maximum values by general numerical methods. Also, we can even tune and find the needed C given a target total access utility value. One of the key remaining future works will be the reduction of the complexity of the equations, and thus the optimization problem.

From the maximization objective function, we can see that as the number of initial seeds, C, increases, $\sum A_i(t)$ will increase and converges to the maximum. Due to different characteristics of the mobility traces and the SNS traces, different initial pushing vectors will have different converging speed, and thus will lead to different maximal values of $\sum A_i(t)$.

For getting the best pushing vector, we design a heuristic algorithm to find the near-optimal solution $\vec{p}$ for maximizing $\sum A_i(t)$ numerically, based on the well-known hill-climbing method [71]. Due to the space limit as well as the popularity of the hill-climbing method, the detailed algorithm is skipped in this chapter. In the algorithm, initially we select the top C users from all users sorted by I^M in descending order ($I^{S \rightarrow}$ or $I^{S \leftarrow}$ works similarly) and iteratively exchange the p_i and p_j values of any two users u_i and u_j if a larger $\sum A_i(t)$ can be obtained, until the increment of $\sum A_i(t)$ is smaller than a specified threshold. Note that MATLAB [67] is used to numerically evaluate the above model and the heuristic algorithm.

11.5   Trace-Driven Measurements

To evaluate the effectiveness of the TOSS framework, we need SNS trace data to quantify the spreading impact factors and access delays, as well as MSN trace data to analyze the mobility impact. However, no trace data are available in the public domain, that measure both the SNS and MSN activities. Thus, we choose to take separate measurements, and combine them by some mapping strategies, which will be explained in Section 11.6.1.

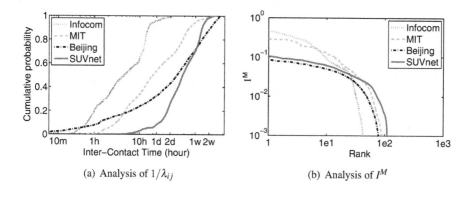

(a) Analysis of $1/\lambda_{ij}$ (b) Analysis of I^M

Figure 11.7: Measurement results of λ_{ij} and I^M.

11.5.1 Measurement of Mobility Impact, I^M

We choose four mobility traces, MIT [80], Infocom [81], Beijing [82], and SUVnet [83], in order to evaluate the performance of TOSS. These traces record either direct contacts among users carrying mobile devices or GPS coordinates of each user's mobile route. The four traces differ in their scales, durations, and mobility patterns. The MIT and the Infocom traces are collected by people, but the Bejing and SUVnet traces are collected by vehicles. The Beijing and SUVnet traces have no record of contacts, but only GPS coordinates with time. We assume two users have a contact once they are separated by a sufficiently small distance (20 m) during a short interval (20 s).

Recall that the λ_{ij} is the intercontact rate of two users, which indicates the mobility impact between them. And the I_i^M is the overall mobility impact factor of a user to the whole user base in the MSN base on (11.6). We analyze the traces and obtain the intercontact intervals $(1/\lambda_{ij})$ of all user pairs, as shown in Figure 11.7(a). The Infocom trace has the highest contact rate because users are at a conference spot, and thus have high contact rates. The MIT trace also has high contact rate since users are friends within the campus. The Beijing and the SUVnet traces have large intercontact intervals because they have relatively low frequency of GPS records and a large user base, which is considered as sparse user density. I^M values of all users of the traces (values smaller than 0.001 are ignored) are plotted in Figure 11.7(b), which indicates the similar trends of the traces as discussed above. Users in the Infocom trace have the highest potential to obtain the content by sharing, but users in the Beijing trace have the weakest potential.

11.5.2 Measurement of Spreading Impact, I^S

We kept track of 2,223,294 users in the most popular online SNS in China, Sina Weibo, for 4 weeks during July 2012, and finally we collected 37,267,512 microposts generated (and partially reshared) by the users, and further obtained the list of all the resharing activities for each micropost. We implemented the data collection software, which starts from 15 famous users distributing popular video clips, and expands the user base from their followers. Capturing the next hop followers is carried out iteratively. The captured data include details of owner's account profile, all microposts with timestamps of the owner, all comments and reposts with timestamps, as well as the profile of the users that make comments and reposts to the owner. Note that there are some robots in Sina Weibo, which always reshare some microposts of famous people with extremely short delays, and thus we exclude users with no followers, no followees, or no self-created microposts. How to precisely exclude all the robots in the SNS trace is out of the scope of this chapter, and there are many related studies for reference, such as [72]. In all, we believe that the 2.2 million user base can reflect the ground truth of the social impact factor and the access delay statistics.

I^S is the overall spreading impact of the user to all users in the SNS, calculated by (11.11). However, calculating I^S for the whole user base takes a substantially long time. Thus, we analyze the subgraphs of a corresponding number of users from the whole social graph by random walking according to the scale of four mobility traces. One important issue here is whether the randomly selected subgraph of the SNS can still reflect the characteristics of the whole SNS user base. There have been some related measurement studies pointing out that the SNS is a *scale-free* network [22, 73–77]. A scale-free network is a complex network whose degree distribution follows the power-law, at least asymptotically, which means that in such network, a small number of nodes make a dominant impact on the network, while many nodes make little contribution, if we consider the node degree or the spreading impact (resharing ratio) as the impact of a node to the network [73, 74, 78].

As researched in [74, 75, 78], due to the natural characteristics of scale-free complex networks, no matter any subgraph we choose from the whole network graph (with not too small size) by random walking, similar characteristics (power-law distribution of node strength) can still be obtained. We then check the subgraphs that we abstract from the online SNS graph with the sizes corresponding to the mobility traces, and for each trace we abstract subgraphs five times, and then make an average value. As shown in Figure 11.8, a smaller number of people have significant outgoing impact ($I^{S\rightarrow}$) to the whole SNS, while many users have very small impact. Also, we see that many users are more likely to be impacted rather than impacting others ($I^{S\rightarrow} < I^{S\leftarrow}$). All of the figures are able to reflect the asymptotical power-law trend [79]. So conclusively, all of the subgraphs with different sizes can still represent the SNS characteristics, and it will be an acceptable methodology to map the SNS subgraphs to the mobility traces.

11.5.3 Measurement of Access Delay, $A_i(t)$

Measurement results of the access delays on the whole user base, $A_i(t)$ from (11.12), are shown in Figure 11.9. From the cumulative distribution function (CDF) of the average of all the access delays of each user in Figure 11.9(a), half of the users have an average access delay larger than 23,880 s, which is about 6 h 38 min. Taking a closer look, we find (1) 3.67% of users have an average access delay less than 10 min,

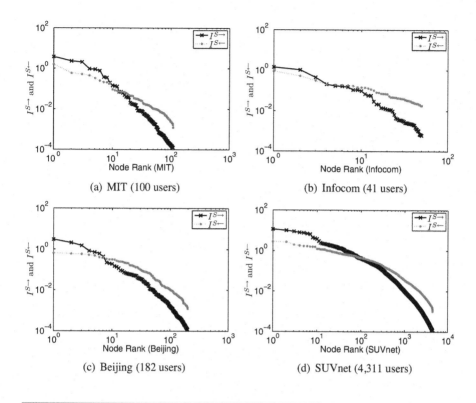

(a) MIT (100 users)

(b) Infocom (41 users)

(c) Beijing (182 users)

(d) SUVnet (4,311 users)

Figure 11.8: Measurement results of I^S for subgraphs sampled from the SNS graph with different sizes corresponding to the mobility traces.

(2) 20.38% of users have a delay smaller than 1 h, and (3) 26.79% of users access the SNS with average delay larger than 1 day. We verify that a substantial number of users access the SNSs with sufficiently large delays, which TOSS can utilize to disseminate the content object by offline opportunistic sharing.

11.6 Performance Evaluation

To evaluate the TOSS framework, we now consider how the spreading and mobility impact factors (I^S and I^M) affect the total access utility function ($\sum A_i(t)$).

11.6.1 Mapping Schemes of Online SNSs and Offline MSNs

Due to the lack of traces that contain the activities of the same users in both online SNSs and offline MSNs, we consider three choices for mapping SNS users to MSN users in each of the four mobility traces: (1) *random*, SNS users are randomly

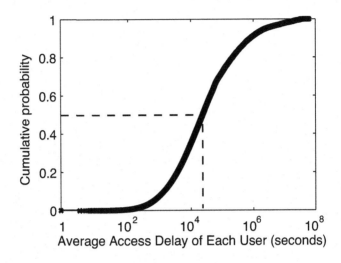

Figure 11.9: Measurement results of access delays.

mapped to MSN users; (2) *h-h*, both SNS and MSN users are sorted in descending order of $I^{S\rightarrow}$ and I^M, respectively, and then are mapped correspondingly; and (3) *h-l*, both users are sorted as similar to *h-h*, but an SNS user with high $I^{S\rightarrow}$ is mapped to an MSN user with low I^M. In the following parts, we will average the evaluation results across the three mapping schemes to reflect the general cases in the SNS and MSN.

Since the number of SNS users is much larger than that of MSN users in each trace, we pick accounts from the SNS trace by random walk sampling to match the number of MSN users in each trace. Regarding the methodology of mapping a sub-graph of online SNS by random-walk sampling to the offline MSN trace, we carry out the following discussion. It was already studied that when we consider the mobility impact (meeting rate) of two users as their vector strength, and the overall mobility impact of one user (sum of all mobility impacts to all other users) as the node strength, the MSN can also be classified as a scale-free network [81, 84–88]. That is, in the MSN, a small number of users are always moving quickly and meet many other users, while many of the users are relatively stable to meet a limited number of other users. So regarding each mobility trace with different numbers of mobile users, as we discussed in Section 11.5.2, we take random walk-based sampling to obtain the subgraphs from the SNS trace with a corresponding number of user accounts, and then map one SNS account to one mobile user by the above mapping choices. Conclusively, it is a reasonable methodology to map between online SNS traces and offline MSN traces when a trace with information for both is not available. To seek or carry out such a measurement study to track both the online SNS activities and offline MSN activities for a group of people is one important future work.

11.6.2 *Initial Pushing Strategies*

In order to select initial seeds, $\vec{p}$, constrained by the allowed total number of seeds, C, we consider the following five pushing strategies based on the impact factors:

- p-λ: We sort users by I^M ($\sum \lambda_i^*$) in descending order and choose the top C ones (similar to [8]).

- p-$\gamma^{\rightarrow}$: We sort users by $I^{S\rightarrow}$ ($\sum \gamma_{ij}^*$) in descending order and choose the top C ones (similar to [36, 38]).

- p-$\gamma^{\leftarrow}$: We sort users by $I^{S\leftarrow}$ ($\sum \gamma_{ji}^*$) in descending order and choose the top C ones.

- p-$\lambda * \gamma^{\rightarrow}$: We sort users by $I^M * I^{S\rightarrow}$ conjunctively in descending order and choose the top C ones.

- p-$\lambda * \gamma^{\leftarrow}$: We sort users by $I^M * I^{S\leftarrow}$ conjunctively in descending order and choose the top C ones.

There are many *viral marketing* methods to evaluate a SNS user's strength regarding information spreading; e.g., we can easily qualify by node degree, including outgoing degree (number of followees) and incoming degree (number of followers). Note that here the arrow direction is the following/followed relationship, reverse to the spreading direction. Furthermore, Google's PageRank algorithm [89, 90] is applied to the selected SNS subgraphs for getting the PageRank scores of all nodes. We also consider a random pushing and the heuristic algorithm, and hence we have five more initial pushing strategies based on the graphs:

- p-R: We randomly choose C users.

- p-$D^{\rightarrow}$: We sort users by outgoing node degree in descending order and choose C users.

- **p-$D^{\leftarrow}$**: We sort users by incoming node degree in descending order and choose C users.

- p-Pr: We sort users by PageRank score in descending order and choose the top C users.

- p-H: We run the hill-climbing heuristic algorithm to obtain the near-optimal pushing vector.

Note that we call the aforementioned nine pushing strategies, except p-H, simple pushing strategies.

11.6.3 Satisfying 100, 90, and 80% of Users

Recall that the access utility function of u_i is $A_i(t)$. A user is *satisfied* if she can obtain the content when her access probability $(A_i(t))$ approaches its maximum in the fitted Weibull probability density function (PDF). If we aim to make 100% of users obtain the content by initial pushing and sharing, substantially large delays may take place for certain users (e.g., a user with low γ and λ values [10]). Therefore, we investigate what percentage of users (initial pushing ratio) should be initial seeds to satisfy the access delay requirements of 100, 90, and 80% of users depending on different pushing strategies.

From Section 11.4, $\sum A_i(t)$ is an increasing function of C ($|\vec{p}|$), and the number of satisfied users is also an increasing function of C. The C value that makes $\sum A_i(t)$ approach its maximum will be the standard number of initial pushing seeds for satisfying 100% of user. We examine how C can be reduced (for higher offloading gains) if we target the satisfaction of 90 and 80% of users.

Figure 11.10 shows that to satisfy 100% of all users, p-H always finds the best initial pushing vector (i.e., the least number of seeds), and p-R performs the poorest, while p-$D^{\rightarrow}$ and p-$D^{\leftarrow}$ also perform poorly, so simply pushing by node degree is not preferred. In most cases, p-$\lambda * \gamma^{\rightarrow}$ and p-$\lambda * \gamma^{\leftarrow}$ perform the second best, which implies that we can conjunctively consider the I^S and I^M factor by simple multiplication to achieve near-optimal performance. p-Pr achieves not so good performance compared with strategies by impact factors, as it focuses on the connections of the network graph but ignores the historical spreading impact, while our proposed factors (γ) make better sense. In MIT and Infocom traces, λ-based strategies perform better than γ-bases ones, which means the mobility factor decides more on the sharing process when nodes are with high mobility. In Beijing and SUVnet traces, γ-base ones perform better, which means the social factor controls more when nodes are with low mobility. Note that the Infocom trace always has the best performance; only a 13.5% initial pushing ratio can satisfy all users by the p-H.

When we target to satisfy 90% of all users, the required initial pushing ratio is reduced significantly. With simple pushing strategies, for the MIT and the Infocom traces, only 15.4 and 10.5% of users need to be the initial seeds on average. The number of initial seeds is further dramatically reduced, when satisfying 80% of users. An approximately 10% initial pushing ratio is needed for all traces except the Beijing trace, which requires about a 17% initial pushing ratio. The Beijing and SUVnet traces always need a relatively larger number of initial seeds due to their low contact rates and large user bases.

For some worst case users, opportunistic sharing is ineffective, but it may be better to push the content to them in the beginning if they have tight access delay requirements, or it will be better to let them carry out on-demand fetching when they approach the peaks of their access delay PDFs.

Generally, p-H is about 15 to 24% better than p-R, and 12 to 16% better than p-λ and p-γ, and the multiplication of p-λ and p-γ will be quite a good solution in practice. It is a balance between performance and complexity. The implication is that if we focus on the best performance, we can run the heuristic algorithm; if we want a

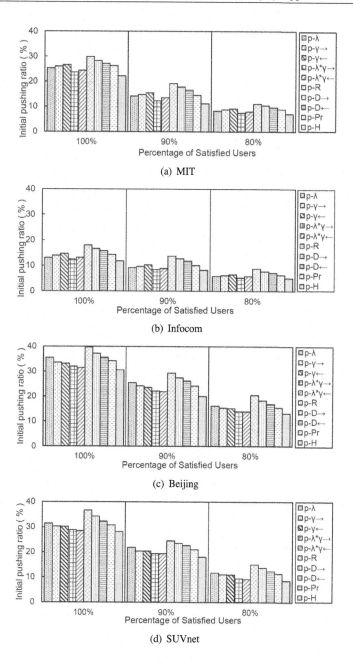

Figure 11.10: Initial pushing ratios to satisfy 100, 90, and 80% of all users.

balance between complexity and performance, we can evaluate user online spreading impact and offline mobility impact, and choose a proper strategy for offloading. p-R can still offload a certain amount of traffic, which indicates that sharing-based offloading can work very well in practice, because this is mainly due to the potential of the user access delays as discussed in Section 11.3.4.

11.6.4 On-Demand Delivery

If a user does not obtain the content (by initial pushing or sharing) until she actually accesses it, we have to deliver it over a cellular link, which is called on demand delivery. Then the traffic of the content delivered on-demand is not offloaded. We now compare the three target percentages of satisfied users (investigated above) in terms of total offloaded traffic. For example, in the case of 90% of satisfied users, 10% of remaining users (i.e., those who have not received the content) will access the content via cellular links. Table 11.1 shows how much traffic is offloaded from cellular links for the three cases, where the offloaded traffic ratios of the nine pushing strategies are averaged, which are juxtaposed with that of p-H. Note that the boldfaced numbers represent the highest amount of traffic reduction for each trace across the three target satisfaction cases (i.e., 100, 90 and 80%). When lowering the percentage of satisfied users from 100% to 90% and to 80%, although the initial pushing ratios become reduced, in some cases, the on-demand delivery for the abandoned 10 and 20% of users may increase the total cellular traffic instead. In the MIT, Beijing, and SUVnet traces, initial pushing for 90% of the users plus on-demand delivery for 10% of the users actually reduces the cellular traffic the most. Overall, TOSS can reduce 63.8 to 86.5% of the cellular traffic load while satisfying the access delay requirements of all users.

We notice a balance between the traffic reduction due to the initial pushing and the traffic increment by the on-demand delivery, as the satisfaction percentage of users changes. The balance is about how to deal with those worst case users (with both low online mobility impact and low offline mobility impact). For some of them who have urgent requirement of access delays, TOSS can just push in the beginning, but those who have large access delays will be a burden on selecting the optimal

**Table 11.1   Percentage (%) of the Traffic
Reduction with On-Demand Delivery**

Trace	100%	90%	80%
MIT [80]	73.6/76.3	**74.6/76.9**	70.9/72.2
Infocom [81]	**85.3/86.5**	79.5/80.4	73.4/74.1
Beijing [82]	65.3/68.4	**65.0/68.9**	63.8/65.2
SUVnet [83]	68.5/70.3	**68.7/71.0**	68.3/70.7

Reduced traffic ratios by the nine pushing strategies are averaged shown before /, while the result by heuristic algorithm is shown after it.

initial pushing seeds by TOSS, as they are hard to reach even by many hops. Instead, it will be better for TOSS to exclude them for a better solution to satisfy a part of other users at the beginning, and then they will carry out on-demand delivery. Note that the Infocom trace can achieve the highest traffic reduction, with the target percentage being 100% due to its high contact rates and small user base, and there is very little worst case users who will not impact the system at all.

11.7   Discussion of Practical Deployment

An easy beginning for deploying the sharing-based offloading framework can be a mobile content-sharing application embedded with SNSs. The application can temporarily cache content objects that the user has accessed from the SNSs, and also it has functionality to establish D2D sharing links with other users by any possible D2D communication techniques. While MNOs and CPs can track users' sharing activities via the application in order to count for the incentives under certain appropriate business models, mobile users can discover nearby SNS friends, friends of friends, and strangers, for exploring and transmitting files with them by active request-for-sharing or proactive background-sharing modes.

- In *request-for-sharing* mode, users can actively explore and list the available content objects that are (1) accessed and shared by friends in the SNSs, (2) published by interesting or famous publishers, and (3) shared by other users in proximity. Then users can send requests to the current content owners for fetching the content objects by establishing D2D links as soon as the owners are within the transmission range.

- Users can also mark an interesting content as "accessing it later," and hence can rely on the proactive *background-sharing* mode. Depending on the user's activity pattern and lifestyle, he or she can specify a deadline for obtaining the content by opportunistic sharing. While the user moves around and meets people, the pending content objects will be collected opportunistically. When the user wants to directly access the object, which is not prefetched yet, on-demand delivery will be carried out. It is also possible to set up a threshold of the social impact factor, so that if there are some content objects shared from the user's good friends who have large impact factors (above the threshold) to the user, the application may automatically fetch and cache those content objects without the user's requests, which can be considered social-aware prefetching.

Furthermore, the D2D in 3GPP standards for LTE-Advanced networks [16] has been hotly discussed, by which users use operator-authorized spectrum for direct communication without the support of infrastructure. The transmission range of D2D communication can be much larger than for other local range communications. By optimal resource allocation and interference management, D2D can utilize the direct user-to-user communication, and thus can increase the total throughput (resource utilization) in the cell area as studied in [16, 44, 45]. Therefore, based on the discussions in this chapter, the D2D technique in LTE-Advanced will significantly facilitate sharing-based offloading for mobile networks in the very near future.

11.8 Conclusion

In this chapter, we have mainly focused on elaborating the SNS-based mobile traffic offloading by opportunistic device-to-device sharing. We have discussed the TOSS framework, which selects optimal seed users for initial content pushing, depending on their content spreading impact in online social network services and their mobility patterns in offline mobile social networks. Then users may share the content via opportunistic local connectivity with others in daily life. Also, TOSS exploits user-dependent access delay between the content generation time and each user's access time for opportunistic sharing purposes. Trace-driven evaluations have shown that the framework can reduce 63.8 to 86.5% of the cellular traffic while satisfying the access delay requirements of all users. In particular, from the evaluation, users with high mobility impact will play key roles for traffic offloading in scenarios of high user mobility or high user density. The social spreading impact will then control the content dissemination in scenarios of low user mobility or sparse user density. For those worst case users with both little spreading impact and little mobility impact, it may be better to send the content to them by on-demand delivery. Conclusively, social-aware traffic offloading by opportunistic D2D sharing has been proven to be effective and efficient for offloading mobile traffic, and therefore can be considered a promising methodology for mobile networks in the future.

References

1. CISCO, *Cisco Visual Networking Index: Global Mobile Data Traffic Forecast Update, 2010–2015*, Technical Report, CISCO, 2011.

2. S. Scellato, C. Mascolo, M. Musolesi, and J. Crowcroft, Track Globally, Deliver Locally: Improving Content Delivery Networks by Tracking Geographic Social Cascades, in *Proceedings of the 20th International World Wide Web Conference (WWW)*, Bangalore, 2011.

3. Y. Chen, L. Qiu, W. Chen, L. Nguyen and R. Katz, Clustering Web Content for Efficient Replication, in *Proceedings of the 10th IEEE International Conference on Network Protocols (ICNP)*, Paris, 2002.

4. M. Cha, H. Kwak, P. Rodriguez, Y. Y. Ahn, and S. Moon, I Tube, You Tube, Everybody Tubes: Analyzing the World's Largest User Generated Content Video System, in *Proceedings of the 7th ACM Internet Measurement Conference (IMC)*, San Diego, 2007.

5. K. Fall, A Delay-Tolerant Network Architecture for Challenged Internets, in *Proc. of ACM Special Interest Group on Data Communication (SIGCOMM)*, Karlsruhe, Germany, 2003.

6. A. Voyiatzis, A Survey of Delay- and Disruption-Tolerant Networking Applications, *Journal of Internet Engineering*, vol. 5, no. 1, pp. 331–344, 2012.

7. J. Whitbeck, M. D. Amorim, Y. Lopez, J. Leguay, and V. Conan, Relieving the Wireless Infrastructure: When Opportunistic Networks Meet Guaranteed Delays, in *Proceedings of the 12th IEEE International Symposium on World of Wireless, Mobile and Multimedia Networks (WoWMoM)*, Lucca, Italy, 2011.

8. S. Ioannidis, A. Chaintreau, and L. Massoulie, Optimal and Scalable Distribution of Content Updates over a Mobile Social Network, in *Proceedings of the 28th IEEE International Conference on Computer Communications (INFOCOM)*, Rio de Janeiro, 2009.

9. F. Malandrino, M. Kurant, A. Markopoulou, C. Westphal, and U. C. Kozat, Proactive Seeding for Information Cascades in Cellular Networks, in *Proceedings of the 31st IEEE International Conference on Computer Communications (INFOCOM)*, Orlando, 2012.

10. X. Wang, M. Chen, Z. Han, T. Kwon, and Y. Choi, Content Dissemination by Pushing and Sharing in Mobile Cellular Networks: An Analytical Study, in *Proceedings of the 9th IEEE International Conference on Mobile Ad Hoc and Sensor Systems (MASS)*, Las Vegas, 2012.

11. Y. Li, Y. Jiang, D. Jin, L. Su, L. Zeng, and D. Wu, Energy-Efficient Optimal Opportunistic Forwarding for Delay-Tolerant Networks, in *IEEE Transactions on Vehicular Technology (TVT)*, vol. 58, no. 9, pp. 4500–4512, 2010.

12. L. Pelusi, A. Passarella, and M. Conti, Opportunistic Networking: Data Forwarding in Disconnected Mobile Ad-Hoc Networks, in *IEEE Communications Magazine*, vol. 44, no. 11, pp. 134–141, 2006

13. B. Han and A. Srinivasan, eDiscovery: Energy Efficient Device Discovery for Mobile Opportunistic Communications, in *Proceedings of the 20th IEEE International Conference on Network Protocols (ICNP)*, Austin, 2012.

14. F. Michahelles, F. Thiesse, A. Schmidt and J. R. Williams, Pervasive RFID and Near Field Communication Technology, in *IEEE Pervasive Computing*, vol. 6, no. 3, pp. 94–96, 2007.

15. C. Yu, K. Doppler, C. Ribeiro and O. Tirkkonen, Resource Sharing Optimization for Device-to-Device Communication Underlaying Cellular Networks, in *IEEE Transactions on Mobile Computing (TMC)*, vol. 10, no. 8, pp. 2752–2763, 2011.

16. K. Doppler, M. Rinne, C. Wijting, C. Ribeiro and K. Hugl, Device-to-Device Communication as An Underlay to LTE-Advanced Networks, in *IEEE Communications Magazine*, vol. 47, no. 12, pp. 42–49, 2009.

17. X. Zhang, G. Neglia, J. Kurose, and D. Towsley, Performance Modeling of Epidemic Routing, in *Elsevier Computer Networks*, no. 10, no. 10, pp. 2867–2891, 2007.

18. Y. Li, G. Su, P. Hui, D. Jin, L. Su and L. Zeng, Multiple Mobile Data Offloading Through Delay Tolerant Networks, in *Proceedings ofThe 13th ACM International Conference on Mobile Computing and Networking (MobiCom), Workshop on Challenged Networks (CHANT)*, 2011.

19. J. Yang and J. Leskovec, Patterns of Temporal Variation in Online Media, in *Proceedings of the 4th ACM International Conference on Web Search and Data Mining*, Hong kong, 2011.

20. B. Han, P. Hui, V. S. A. Kumar, M. V. Marathe, J. Shao, and A. Srinivasan, 'Mobile Data Offloading through Opportunistic Communications and Social Participation, in *IEEE Transaction on Mobile Computing (TMC)*, vol. 11, no. 5, pp. 821–834, 2012.

21. T. Rodrigues, F. Benvenuto, M. Cha, K. Gummadi, and V. Almeida, On Word-of-Mouth Based Discovery of the Web, in *Proceedings of the 11th ACM Internet Measurement Conference (IMC)*, Berlin, 2011.

22. H. Kwak, C. Lee, H. Park, and S. Moon, What Is Twitter, a Social Network or a News Media? in *Proceedings of the 19th International World Wide Web Conference (WWW)*, Raleigh, NC, 2010.

23. Facebook, http://www.facebook.com.

24. Twitter, http://www.twitter.com.

25. Sina Weibo, http://weibo.com.

26. G. V. Steeg and A. Galstyan, Information Transfer in Social Media, in *Proceedings of the 21st International World Wide Web Conference (WWW)*, Lyon, France, 2012.

27. G. Comarela, M. Crovella, V. Almeida, and F. Benevenuto, Understanding Factors That Affect Response Rates in Twitter, in *Proceedings of the 23rd ACM Conference on Hypertext and Social Media (HT)*, Milwaukee, WI, 2012.

28. S. Myers, C. Zhu, and J. Leskovec, Information Diffusion and External Influence in Networks, in *Proceedings of the 18th ACM SIGKDD International Conference on Knowledge Discovery and Data Mining (KDD)*, Beijing, 2012.

29. T. Karagiannis, J. Y. Le Boudec, and M. Vojnovic, Power Law and Exponential Decay of Inter Contact Times between Mobile Devices, in *Proceedings of the 13th ACM International Conference on Mobile Computing and Networking (MobiCom)*, Montreal, 2007.

30. W. J. Hsu, T. Spyropoulos, K. Psounis, and A. Helmy, Modeling Time-Variant User Mobility in Wireless Mobile Networks, in *Proceedings of the 26th IEEE International Conference on Computer Communications (INFOCOM)*, Anchorage, AK, 2007.

31. W. Gao and G. Cao, On Exploiting Transient Contact Patterns for Data Forwarding in Delay Tolerant Networks, in *Proceedings of the 18th IEEE International Conference on Network Protocols (ICNP)*, Kyoto, Japan, 2010.

32. E. Jaho and I. Stavrakakis, Joint Interest- and Locality-Aware Content Dissemination in Social Networks, in *Proceedings of the 6th International Conference on Wireless On-Demand Network Systems and Services (WONS)*, Snowbird, UT, USA, 2009.

33. M. P. Wittie, V. Pejovic, L. Deek, K. C. Almeroth, and B. Y. Zhao, Exploiting Locality of Interest in Online Social Networks, in *Proceedings of the 6th International Conference on emerging Networking EXperiments and Technologies (CoNEXT)*, Philadelphia, 2010.

34. X. Wang, M. Chen, Z. Han, D. O. Wu, and T. T. Kwon, TOSS: Traffic Offloading by Social Network Service-Based Opportunistic Sharing in Mobile Social Networks, in *Proceedings of the 33st IEEE International Conference on Computer Communications (INFOCOM)*, Toronto, 2014.

35. X. Zhuo, W. Gao, G. Cao, and Y. Dai, Win-Coupon: An Incentive Framework for 3G Traffic Offloading, in *Proceedings of the 19th IEEE International Conference on Network Protocols (ICNP)*, Vancouver, 2011.

36. P. Hui, J. Crowcroft, and E. Yoneki, BUBBLE Rap: Social-Based Forwarding in Delay-Tolerant Networks, *IEEE Transaction ons Mobile Computing (TMC)*, vol. 10, pp. 1576–1589, 2011.

37. W. Gao and G. Cao, User-Centric Data Dissemination in Disruption Tolerant Networks, in *Proceedings of the 30th IEEE International Conference on Computer Communications (INFOCOM)*, Shanghai, 2011.

38. C. Boldrini, M. Conti, and A. Passarella, ContentPlace: Social-Aware Data Dissemination in Opportunistic Networks, in *Proceedings of the 11th ACM International Conference on Modeling, Analysis and Simulation of Wireless and Mobile Systems (MSWiM)*, Vancouver, 2008.

39. D. Rothfus, C. Dunning and X. Chen, Social-Similarity-Based Routing Algorithm in Delay Tolerant Networks, in *Proceedings of IEEE International Conference on Communications (ICC)*, Budapest, 2013.

40. X. Bao, Y. Lin, U. Lee, I. Rimac, and R. R. Choudhury, DataSpotting: Exploiting Naturally Clustered Mobile Devices to Offload Cellular Traffic, in *Proceedings of the 32nd IEEE International Conference on Computer Communications (INFOCOM)*, Turin, Italy, 2013.

41. A. Kate, G. M. Zaverucha, and U. Hengartner, Anonymity and Security in Delay Tolerant Networks, in *Proceedings of the 3rd International Conference on Security and Privacy in Communication Networks (SecureCom)*, Nice, France, 2007.

42. R. Lu, X. Lin, T. H. Luan, X. Liang, X. Li, L. Chen and X. Shen, PReFilter: An Efficient Privacy-Preserving Relay Filtering Scheme for Delay Tolerant Networks, in *Proceedings of the 31st IEEE International Conference on Computer Communications (INFOCOM)*, Orlando, FL, 2012.

43. AirDrop, Apple, Inc., `http://en.wikipedia.org/wiki/AirDrop`.

44. M. Meibergen, Device-to-Device Communications Underlaying a Cellular Network, *Master thesis*, Delft University of Technology, 2011.

45. M. Zulhasinine, C. Huang, and A. Srinivasan, Efficient Resource Allocation for Device-to-Device Communication Underlaying LTE Network, in *Proceedings of the 6th IEEE International Conference on Wireless and Mobile Computing, Networking and Communications (WiMob)*, Niagara Falls, Canada, 2010.

46. Y. Cho, H. Ji, J. Park, H. Kim, and J. Silvester, A 3W Network Strategy for Mobile Data Traffic Offloading, *IEEE Communications Magazine*, vol. 49, no. 10, pp. 118–123, 2011.

47. K. Lee, J. Lee, Y. Yi, I. Rhee, and S. Chong, Mobile Data Offloading: How Much Can WiFi Deliver? *IEEE/ACM Transactions on Networking (ToN)*, vol. 21, no. 2, pp. 536–550, 2013.

48. A. Balasubramanian, R. Mahajan, and A. Venkataramani, Augmenting Mobile 3G Using Wi-Fi, in *Proceedings of the 8th International Conference on Mobile Systems, Applications, and Services (MobiSys)*, San Francisco, 2010.

49. S. Dimatteo, P. Hui and B. Han, Cellular Traffic Offloading through Wi-Fi Networks, in *Proceedings of the 8th IEEE 8th International Conference on Mobile Adhoc and Sensor Systems (MASS)*, Valencia, Spain, 2011.

50. T. Ning, Z. Yang, H. Wu, and Z. Han, Self-Interest-Driven Incentives for Ad Dissemination in Autonomous Mobile Social Networks, in *Proceedings of the 32nd IEEE International Conference on Computer Communications (INFOCOM)*, Turin, Italy, 2013.

51. R. Lu, X. Lin, Z. Shi, B. Cao, and X. Shen, IPAD: An Incentive and Privacy-Aware Data Dissemination Scheme in Opportunistic Networks, in *Proceedings of the 32nd IEEE International Conference on Computer Communications (INFOCOM)*, Turin, Italy, 2013.

52. E. Katz and P. F. Lazarsfeld, *Personal Influence*, Free Press, New York, 1955.

53. D. J. Watts and P. S. Dodds, Influentials, Networks, and Public Opinion Formation, *Journal of Consumer Research*, vol. 34, no. 4, pp. 441–458, 2007.

54. W. Chen, Y. Wang, and S. Yang, Efficient Influence Maximization in Social Networks, in *Proceedings of the 15th ACM SIGKDD International Conference on Knowledge Discovery and Data Mining (KDD)*, Paris, 2009.

55. A. Goyal, F. Bonchi, and L. Lakshmanan, Learning Influence Probabilities in Social Networks, in *Proceedings of the 3rd ACM International Conference on Web Search and Data Mining*, New York, 2010.

56. S. Wu, J. M. Hofman, W. A. Mason, and D. J. Watts, Who Says What to Whom on Twitter, in *Proceedings of the 20th International World Wide Web Conference (WWW)*, Bangalore, 2011.

57. R. Xiang, J. Neville, and M. Rogati, Modeling Relationship Strength in Online Social Networks, in *Proceedings of the 19th International World Wide Web Conference (WWW)*, Raleigh, NC, 2010.

58. M. Gomez-Rodriguez, J. Leskovec, and A. Krause, Inferring Networks of Diffusion and Influence, in *Proceedings of the 16th ACM SIGKDD International Conference on Knowledge Discovery and Data Mining (KDD)*, Washington, DC, 2010.

59. M. D. Choudhurry, W. A. Mason, J. M. Hofman, and D. J. Watts, Inferring Relevant Social Networks from Interpersonal Communication, in *Proceedings of the 19th International World Wide Web Conference (WWW)*, Raleigh, NC, 2010.

60. V. Chaoji, S. Ranu, R. Rastogi, and R. Bhatt, Recommendations to Boost Content Spread in Social Networks, in *Proceedings of the 21st International World Wide Web Conference (WWW)*, Lyon, France, 2012.

61. Homophily, http://en.wikipedia.org/wiki/Homophily.

62. M. D. Choudhury, H. Sundaram, A. John, D. D. Seligmann, and A. Kelliher, Birds of a Feather: Does User Homophily Impact Information Diffusion in Social Media? *Computing Research Repository (CoRR)*, vol. 1006, 2010.

63. Locality of Reference (Principle of Locality), http://en.wikipedia.org/wiki/Locality_of_reference.

64. L. Gyarmati and T. A. Trinh, Measuring User Behavior in Online Social Networks, *IEEE Network Magazine*, vol. 24, pp .26–31, 2010.

65. C. Liu, R. White, and S. Dumais, Understanding Web Browsing Behaviors through Weibull Analysis of Dwell Time, in *Proceedings of the 33rd International ACM SIGIR Conference on Research and Development in Information Retrieval*, New York, 2010.

66. P. Svoboda, W. Karner and M. Rupp, Traffic Analysis and Modeling for World of Warcraft, in *Proceedings of IEEE International Conference on Communications (ICC)*, Glasgow, 2007.

67. MATLAB, Mathworks, Inc.

68. 3GPP, 3GPP TS 23.012, *Location Management Procedures*, 2011.

69. H. Aminikhah and J. Biazar, A New Analytical Method for System of ODEs, in *Numerical Methods for Partial Differential Equations*, vol. 26, no. 5, pp. 1115–1124, 2010.

70. N. Guzel and M. Bayram, Power Series Solution of Non-linear First Order Differential Equation Systems, *Trakaya University Journal of Science*, vol. 6, no. 1, pp. 107–111, 2005.

71. Hill-Climbing Algorithm, `https://en.wikipedia.org/wiki/Hill_climbing`.

72. W. Wei, F. Xu, C. Tan, and Q. Li, SybilDefender: Defend against Sybil Attacks in Large Social Networks, in *Proceedings of the 31st IEEE International Conference on Computer Communications (INFOCOM)*, Orlando, FL, 2012.

73. R. P. Satorras and A. Vespignani, Epidemic Spreading in Scale-Free Networks, *Physical Review Letter*, vol. 86, no. 14, pp. 3200–3203, 2001.

74. A. Hernando, D. Villuendas, C. Vesperinas, M. Abad, and A. Plastino, Unravelling the Size Distribution of Social Groups with Information Theory in Complex Networks, *European Physical Journal B*, vol. 76, no. 1, pp. 87–97, 2010.

75. A. Mislove, M. Marcon, K. Gummadi, P. Druschel, and B. Bhattacharjee, Measurement and Analysis of Online Social Networks, in *Proceedings of the 7th ACM Internet Measurement Conference (IMC)*, San Diego, 2007.

76. A. Barabasi, H. Jeong, Z. Neda, T. Vicsek, and A. Schuber, On the Topology of the Scientific Collaboration Networks, *Physica A*, vol. 311, pp. 590–612, 2002.

77. M. Girvan and M. E. Newman, Community Structure in Social and Biological Networks, *Proceedings of the National Academy of Science of USA (PNAS)*, vol. 99, no. 12, pp. 7821–7826, 2002.

78. D. Towsley, A Walk in the Dark: Random Walks and Network Discovery, in *The 12th IFIP International Conference on Networking, Keynote*, New York, 2013.

79. Power Law, `http://en.wikipedia.org/wiki/Power_law`.

80. N. Eagle and A. Pentland, Reality Mining: Sensing Complex Social Systems, *Springer Journal of Personal and Ubiquitous Computing*, vol. 10, pp. 255–268, 2006.

81. A. Chaintreau, P. Hui, J. Crowcroft, C. Diot, R. Gass, and J. Scott, Impact of Human Mobility on Opportunistic Forwarding Algorithms, *IEEE Transaction of Mobile Computing (TMC)*, vol. 6, no. 6, pp. 606–620, 2007.

82. Y. Zheng, L. Zhang, X. Xie, and W. Ma, Mining Interesting Locations and Travel Sequences from GPS Trajectories, in *Proceedings of the 18st International World Wide Web Conference (WWW)*, Madrid, Spain, 2009.

83. H. Huang, P. Luo, M. Li, D. Li, X. Li, W. Shu, and M. Wu, Performance Evaluation of SUVnet with Real-Time Traffic Data, *IEEE Transactions on Vehicular Technology (TVT)*, vol. 56, no. 6, pp. 3381–3396, 2007.

84. A. Pietilainen, Opportunistic Mobile Social Networks at Work, *PhD thesis*, Pierre-and-Marie-Curie University, 2010.

85. F. Bjurefors, Measurements in Opportunistic Networks, *PhD thesis*, Uppsala University, 2012.

86. T. Karagiannis, J. Boundec, and M. Vojnovic, Power Law and Exponential Decay of Inter Contact Times between Mobile Devices, in *Proceedings of the 13th ACM International Conference on Mobile Computing and Networking (MobiCom)*, Montreal, 2007.

87. O. R. Helgason, Opportunistic Content Distribution, *PhD thesis*, Royal Institute of Technology, 2009.

88. E. Cho, S. Myers, and J. Leskovec, Friendship and Mobility: User Movement in Location-Based Social Networks, in *Proceedings of the 17th ACM SIGKDD International Conference on Knowledge Discovery and Data Mining (KDD)*, San Diego, 2011.

89. L. Page, S. Brin, R. Motwani and T. Winograd, *The PageRank Citation Ranking: Bringing Order to the Web*, Technical Report, Stanford University, 1998.

90. S. Brin and L. Page, The Anatomy of a Large-Scale Hypertextual Web Search Engine, *Computer Networks and ISDN Systems*, vol. 30, no. 1–7, pp. 107–117, 1998.

Chapter 12

Wireless Security

Juan Chen

Information Science and Technology Department, Dalian Maritime University, Dalian, China

Xiaojiang Du

Department of Computer and Information Sciences, Temple University, Philadelphia, Pennsylvania

CONTENTS

As wireless technologies are developed so fast, they also need a generation of security and privacy control techniques. However, the broadcast nature of wireless communications, wireless device mobility, and resource limitations make security in wireless networks an especially challenging problem. This chapter will provide state-of-the-art research results and future trends related to security in wireless networks, including 4G network, ad-hoc networks, and sensor network.

12.1 Introduction

Wireless technologies have penetrated every aspect of our daily lives, including multimedia, information search and retrieval, entertainment, social, business, medical applications, and so on. At the same time, tremendous growth of wireless networks and applications attracts a lot of research interest. However, compared to traditional wired networks, the broadcast nature of wireless communications, wireless device mobility, and resource limitations have given rise to newer security challenges. An understanding of these challenges and potential solutions is necessary for designing secure wireless and mobile systems and applications and, at the very least, for increasing user awareness [8].

As the third generation (3G) communication is moving to the fourth generation (4G) communication, many societies are preparing themselves for 4G. However, security is no doubt a major factor in 4G [45]. The network and service providers must ensure their infrastructures and services to be adequately protected against all kinds of threats, as well as provide end users with secured access/services. This means they are required to secure their network infrastructure for successful commercialization of their kinds of services. Accordingly, the need for secure networks and services will continue to grow as security will soon become a key differentiator for them.

In the context of the heterogeneous and integrated 4G environment, ad-hoc networking is considered an important solution to extend the radio coverage of wireless systems and multimedia Internet services to wireless environments [8]. In the ad-hoc network, network topology may change rapidly and unpredictably. Due to the open and vulnerable communication medium, many types of attacks are possible in ad-hoc networks, such as packet dropping attack, resource consumption attack, fabrication attack, and so on. Thus, the key issue for ad-hoc networks is designing viable security mechanisms for the protection of confidentiality, integrity, and authentication to prevent malicious attacks.

Different from ad-hoc networks, nodes in wireless sensor networks (WSNs) are weaker, with limited capability in communication and computing. Besides, the deployment nature of sensor networks makes them more vulnerable to various attacks. Thus, existing network security methods, including those developed for mobile ad hoc networks, are not well suited for wireless sensor networks. As a result, security techniques in wireless sensor networks attract a lot of attention. Some of them have been deeply researched, and some of them, such as privacy preservation techniques [23], have recently been proposed and become a hot topic. In addition, advances in WSNs have extended their application from traditional WSNs to emerging sensor ones, including unattended wireless sensor networks (UWSNs) [37], underwater sensor networks (UWNs) [16], and wireless multimedia sensor networks (WMSNs). UWSNs are deployed in critical applications such as military or homeland security. And UWNs can be used to monitor the environment under the sea. WMSNs are able to collect, store, and process multimedia data. Because of their unique characteristics, these emerging networks are particularly vulnerable to new privacy and security threats.

After successful deployment of Wi-Fi and cellular networks in the past decade, wireless and mobile communication systems have become the fastest growing sector of the communication industry. Almost all businesses that rely on wireless and mobile networks expect the same level of security, privacy and trust that exists in wired networks to ensure the integrity and confidentiality of communications among terminals, networks, applications, and services. Security is of vital importance to ensure integrity of communications in wireless and mobile networks. This chapter will provide state-of-the-art research results, challenges, and future trends related to security in wireless networks, including the 4G system, ad hoc networks, and wireless sensor networks. Specifically, for the 4G system, we first provide an overview of security requirements and therefore summarize security issues in three main standards designed for 4G systems. After that, possible threats in 4G are discussed. For ad hoc networks, we review the network vulnerabilities, fundamental security requirements, possible security attacks, and defenses of ad hoc networks. For sensor networks, emerging privacy preservation techniques have been reviewed first and thereafter security issues in emerging sensor networks, including UWSNs, UWNs, and WMSNs, are discussed.

12.2 Security in 4G Wireless System

The 4G network is a convergence of multiple heterogeneous access networks such as WiMAX and 3G. Although a service subscriber uses any of these multiple access networks, 4G provides services from the same service unit, for example, IP Multimedia Subsystem (IMS). For security reasons, 4G uses a comprehensive, network-wide security architecture such as IMS architecture and Next Generation Network (NGN) architecture [45].

12.2.1 Security Objectives, Requirements, and Challenges

The key objectives in designing the security architecture can be summarized as follows [45]:

- Availability. Availability enforces networks and services not to be disrupted or interrupted by, for example, malicious attacks.

- Interoperability. Interoperability ensures the security solutions that can avoid interoperability problems, e.g., by using generic solutions applicable to most of the NGN applications and service scenarios.

- Usability. Usability makes it easy for the end users to use the security-enabled services.

- QoS. QoS guarantees that required security solutions, like cryptographic algorithms, to meet QoS constraints of voice and multimedia traffic.

- Cost-effectiveness. Cost-effectiveness minimizes the additional cost of security and makes it lower than the cost of risks.

From a broad perspective, the security architecture for 4G systems should meet the following security requirements: (1) increased robustness over 3G, (2) user identity confidentiality, (3) strong authentication of user and network, (4) data integrity, (5) confidentiality, and (6) interworking of security across other radio networks [46]. Key challenges faced by designers of 4G wireless security include the following:

1. Security issues from 4G mobile wireless devices and security issues of accessing the Internet from a fixed location with requirements for flexibility and mobility.

2. Every time additional cryptographic methods and security mechanisms are applied to IP networks, there is an impact on the performance and traffic handling capacity of the service provider's network. Balance the security performance against the costs of a particular security solution.

3. A new generation of 4G devices and applications is sure to emerge within the next decade. All these applications and devices will need to be protected from a growing range of security threats.

12.2.2 Security Systems

According to 4G standards, the current security systems include those of Wi-Fi, WiMAX, and 3GPP LTE, which we will review one by one.

12.2.2.1 Wi-Fi Security

Wireless local area network (WLAN) based on the Wi-Fi technique allows an electronic device to exchange data wirelessly (using radio waves) over a computer network, including high-speed Internet connections. WLAN is by far the most widely used wireless networking technique. However, it faces many security problems. In general, security threats in WLAN are classified as physical and logical attacks [28].

> Logical attacks and mitigation techniques: A logical attack always relates to the software, system, and sensitive data flowing in the network. In this type of attack the target of the intruder is to find the code and software or any drawback in the network that will help the intruder to access the network and alter the sensitive data easily. Logical attacks include denial of service (DOS) attack, man-in-the-middle attack, default access point configuration attack, and so on.

> Physical attack and mitigation techniques: A physical attack always relates to the hardware and design of the network. In this type of attack the target of the intruder is to interrupt or decrease the network performance rather than searching for sensitive data and then making some changes with the data. Existing physical attacks in WLAN are mainly physical placement of access points, illegal access points, spam attacks, and so on.

To provide security to WLANs, five security requirements should be achieved: data integrity, confidentiality, authentication, access control, and nonrepudiation. Three main security techniques for Wi-Fi are proposed [50]: Wired Equivalent Privacy (WEP), Wi-Fi Protected Access (WPA), and WPA2. WEP is the first security technique used in IEEE 802.11 standards. The main purpose of using WEP is to provide the security to WLANs like the security provided in the wired LAN. WEP helps to make the communication secure and provide the secret authentication scheme between the access point (AP) and the end user who is going to access the WLAN.

1. Architecture. Basically, WEP was implemented on initial Wi-Fi networks so that the user cannot access the network without the correct key. In the shared-key authentication process, the client sends an authentication request to the access point, which replies with a plain text test; the client then encrypts the test using a WEP key and sends it back. If the returned key matches, access is granted.

   The basic and standard way to make the integrity safe is to add some message authentication code to each part of data before transmitting them. WEP uses the 32-bit cyclic redundancy code (CRC-32) as an integrity algorithm that is generated at the transmitting side. It is generated for each frame of data that is to be transmitted by performing some polynomial calculation, and after that a checksum is added to each data frame. At the recipient side, similar polynomial calculations are performed on the data frames. If the checksums calculated at both sides are same, then it assumes that the data are safe; otherwise, it is assumed to be altered data.

   WEP uses the RC4 stream cipher, the same one used in secure socket layers (SSLs) to protect Internet traffic. Initially, 64-bit WEP used a 40-bit key (later 104 bits) that was concatenated with the 24-bit initialization vector (IV) to form the RC4 key. Unfortunately, the IV key was transmitted as plaintext and used repeatedly, making it fairly straightforward for an eavesdropper to recover the key.

2. Weakness. WEP is considered to be a weak security technique for WLAN nowadays. There are some major reasons for which WEP is unable to provide security to WLAN.

   - Use of master keys directly: From the cryptographic point of view, using master keys directly is unsecure. Master keys should only be used to generate other temporary keys.

   - Small key size: The key size for WEP is small.

   - Lack of key management: Without key management, keys will tend to be long-lived and of poor quality.

   - Use of RC4: RC4 has been considered to generate weak keys for high correlation between the key and the ciphertext.

   - Use the 24-bit-long IV, which is added as plaintext to each packet.

WPA, a new and better solution, was developed to address and fix the known flaws in WEP as well as improve its user authentication. Basically, WPA is an intermediate step between WEP and IEEE 802.11i specification and considered as a subset of IEEE 802.11i [28]. WPA vastly improves WEP's encrypting process and adds a concrete user authentication mechanism. WPA technology mainly includes three improvements over WEP. WPA improves data encryption through the Temporal Key Integrity Protocol (TKIP), which generates a new 128-bit key for each packet. Furthermore, WPA relies on IEEE 802.1X, which defines an authentication mechanism for 802.11 networks. For enterprise users, WPA uses the Extensible Authentication Protocol (EAP), specifically EAP-TLS, which provides transport layer security; for residential and consumer users, WPA uses a preshared key (PSK) system. To verify the integrity of packets, WPA uses much stronger message authentication codes than the CRC used by WEP.

While WPA is far more secure than WEP from passive attacks, its PSK implementation can be fairly easily cracked by a brute-force attack if you have a weak password. WPA is far more robust than WEP but not nearly as strong as WPA2.

WPA2 was later released to further improve the authentication and encryption. For encryption, WPA2 utilizes the Counter Mode with Cipher Block Chaining Message Authentication Code Protocol (CCMP), which does Advanced Encryption Standard (AES) encryption using a 128-bit key and a 128-bit block size. CCMP replaced TKIP, which had proved vulnerable to a variety of attacks. WPA2 uses a large number of IEEE 802.11 MAC layer protocols, which helps to provide further key management and authentication to the networks. Specifically, WPA2 provides more excellent security than WEP and WPA by techniques as follows: using AES, using stronger key management, protecting against the man-in-the-middle attacks by using a two-way authentication process, and providing improved message integrity performance by using CBC-MAC (Cipher Block Chaining Message Authentication Code).

However, WPA2 is expensive for the already deployed networks due to AES, which requires new or additional equipments for the current deployed wireless networks. In addition, WPA2 is vulnerable to security risks because it completely trusts on secrecy session keys. Thus, WPA2 is not introduced to replace WPA. Indeed, WPA2 is considered to be a complex and secure way for the wireless networks from a security point of view, especially for the enterprise network.

12.2.2.2 WiMAX Security

The IEEE 802.16 (WiMAX) Working Group wants to avoid the well-known and documented security design issues with IEEE 802.11 by incorporating a preexisting standard into IEEE 802.16 [43]. However, as the standard evolved from 802.16, the requirements evolved from line-of-sight to mobile WiMAX. As a result, the security requirements and standards also evolved in order to address changing needs. The security features introduced in the initial IEEE 802.16 standard have been greatly enhanced in the IEEE 802.16e standard. Key new features include: (1) PKMv2 (Privacy Key Management version 2 protocol), (2) message authentication is performed using the HMAC/CMAC (Hash-Based Message Authentication Code or Cipher-Based Message Authentication Code) scheme, (3) device/user authentication is carried out using EAP, and (4) confidentiality is achieved using AES-based encryption.

Our analysis indicates that the industry's main challenge will be to balance security needs with the cost of implementation, performance, and interoperability. In addition, since WiMAX utilizes IP as its transport mechanism for handling control/signaling and management traffic, network operators will have to defend against general IP-related security threats as well.

12.2.2.2.1 Architecture Design

WiMAX, based on the IEEE 802.16 standard, provides strong support for authentication, key management, encryption and decryption, control and management of plaintext protection, and security protocol optimization.

1. Authentication and authorization. Generally, WiMAX supports three types of authentication, which are handled in its security sublayer: (1) RSA-based authentication, (2) EAP-based authentication, and (3) RSA-based authentication followed by EAP-based authentication.

   Prior to use for the first time by a user, WiMAX devices require X.509 digital certificates to be loaded on the device and also to be programmed in the home network's Authentication Authorization and Accounting (AAA) server. The X.509 certificate used in RSA-based authentication is issued by the subscriber station (SS) manufacturer and contains the SS's public key (PK) and MAC address. When requesting an authorization key (AK), the SS transmits its certificate to the BS, which validates it and then uses the PK to encrypt an AK and transmit it to the SS.

   In EAP authentication, the SS is authenticated by a X.509 certificate or by a unique operator-issued credential such as Subscriber Identity Module (SIM), Universal Subscriber Identity Module (USIM), or user ID and password. The WiMAX standard allows for use of any of the three EAP authentication schemes: EAP-AKA (Authentication and Key Agreement), EAP-TLS (Transport Layer Security), and EAP-TTLS MSCHAPv2 (Tunneled Transport Layer Security with Microsoft Challenge Handshake Authentication Protocol version 2). EAP-TTLS is used to support establishment of secure connections in a roaming environment and protection of user credentials.

   In general, for authentication and authorization, once the mobile station (MS) requests registration with the base station (BS), the BS communicates with the MS's home AAA server using EAP. On verifying the identity of the MS, the AAA server returns an intermediate key (master session key) to the authenticator in the MS visiting network. Ultimately, the master session key (MSK) is translated into another intermediate key—the pairwise master key (PMK). The AK is then generated from the PMK.

2. Key management. Once the identity of the SS/MS has been validated, traffic keys are then exchanged. IEEE 802.16e defines privacy key management (PKM) for secure key distribution between the MS and the BS. In addition to ensuring synchronization of keying data between the BS and SS/MS, the protocol is used by the BS to authorize SS/MS access to the network.

   The IEEE 802.16e standard supports two versions of the PKM protocol: (1) PKM version 1 (PKMv1), which provides a basic set of functionality, and (2) PKM version 2 (PKMv2), which incorporates a number of enhancements. Both versions of the PKM protocol facilitate two functions:

2. Access control: The protocol provides the means for the BS to authenticate an SS/MS. It also provides the BS with the capability to authorize the SS/MS access to the network and any subscribed services. In addition, the protocol enforces periodic re-authentication and reauthorization. b. Key management: The protocol enables the secure exchange of key information between the BS and the SS/MS.

Key management is implemented using a client-server model in PKM. The SS/MS (PKM client) requests keying material from the BS (PKM server). The client receives keying material for services that it is authorized to access. WiMAX communication is secured through the use of five kinds of keys: (1) AK, (2) key encryption key (KEK), (3) downlink HMAC key, (4) uplink HMAC key, and (5) traffic encryption key (TEK).

PKMv2 addresses the requirement for mutual authentication between the SS/MS and BS. It includes new security features such as support for (1) a new key hierarchy for AK derivation and (2) the EAP. Mutual authentication allows the BS to validate the identity of the SS/MS and also allows the SS/MS to validate the identity of the BS. PKMv2 supports both an RSA-based authentication process and an EAP-based process.

Similar to PKMv1, the RSA authentication process in PMKv2 involves a three-message exchange. However, the contents of the last two messages in the sequence have changed. In addition, instead of the AK being generated by the BS and sent encrypted to the SS/MS, the AK is now generated by the SS/MS using an encrypted pre-PAK (pre-primary authorization key). The SS/MS takes the pre-PAK and generates a PAK. The PAK is then used to generate the AK. The BS follows the same process to create the AK. The two generated AKs must match for subsequent communications between the SS/MS and BS to succeed.

The authentication process begins when the SS/MS sends an authentication information message containing the X.509 certificate of the SS/MS manufacturer to the BS, as in PKMv1. Immediately following this message, an authorization request message is sent by the SS/MS, as in PKMv1. However, the PMKv2 message has an additional random number generated by the SS/MS. Once the SS/MS identity has been validated by the BS, the BS replies with an authorization reply message containing the following: (1) the random number generated by the SS/MS and received in the authorization request message, (2) a random number generated by the BS (this number is used to verify the freshness of the current message), (3) the pre-PAK that is RSA encrypted using the SS/MS public key (the pre-PAK will be used by the SS/MS to generate the AK), (4) the key lifetime of the PAK, (5) the PAK sequence number, (6) the service set identifier (SSID) list, (7) the X.509 certificate for the BS, and (8) the RSA signature of the BS [46].

With the information contained in the authorization reply message, the SS/MS can verify the identity of the BS and generate the AK.

3. Encryption. Only after the successful exchange of keys can encrypted data be transmitted via the WiMAX connection. WiMAX security includes an encapsulation protocol for securing data across the wireless link. The BS uses the AK to generate the TEK. The TEK is utilized for secure encryption of data across the wireless link. Other keys are generated by the BS to facilitate a secure three-way handshake in transmitting the TEK to the SS/MS.

12.2.2.2.2 Attacks and Defenses in WiMAX

There are various sources of potential vulnerabilities in WiMAX 802.16e. Some of these sources include:

1. The fact that management MAC messages are never encrypted, providing adversaries an ability to listen to the traffic and potentially gain access to sensitive information.

2. The fact that some messages are not authenticated (no integrity protection). Typically, a HMAC is used as a digest. However, this is not used for broadcasts and a few other messages. Simple forgery can affect communication between an MS and BS.

3. Weakness in authentication and authorization procedures is an enabler for the BS or SS masquerading threat. It is not easy to get the security model correct in a mobile environment due to limited bandwidth and computation resources.

4. Issues with key management such as the size of the TEK identifier and TEK lifetime are considered potential sources of vulnerabilities for WiMAX security.

Since the security sublayer lies at the bottom of the MAC layer in order to protect data exchanged between the MAC layer and the PHY layer, security threats in WiMAX are mainly in the PHY and MAC layers. Below, we discuss attacks in WiMAX at the PHY and MAC layers, respectively.

1. PHY is vulnerability to attacks from wireless links such as jamming attacks. In addition to threats from jamming, 802.16 is also vulnerable to other attacks, such as water torture attack and forgery attack.

   a. Jamming attack. Jamming attack is achieved by introducing a source of noise strong enough to significantly reduce the capacity of the channel. It is not difficult to perform a jamming attack because necessary information and equipment are easy to acquire. We can prevent jamming attacks by increasing the power of signals or by increasing the bandwidth of signals using spreading techniques such as frequency-hopping spread spectrum (FHSS) or direct sequence spread spectrum (DSSS). Furthermore, since it is easy to detect jamming by using radio spectrum monitoring equipment and the sources of jamming are easy to be located by using radio direction finding tools, we can also ask for help from law enforcement to stop the jammers.

b. Water torture attack. This is also a typical attack in which an attacker forces an SS to drain its battery or consume computing resources by sending a series of bogus frames. This kind of attack is considered even more destructive than a typical DoS attack since the SS, which is usually a portable device, is likely to have limited resources. To prevent this kind of attack, a sophisticated mechanism is necessary to discard bogus frames, thus avoiding running out of battery or computational resources [44].

c. Other attacks. In addition to attacks like jamming, and water torture attacks, 802.16 is also vulnerable to other attacks such as forgery attacks in which an attacker with an adequate radio transmitter can write to a wireless channel. In mesh mode, 802.16 is also vulnerable to replay attacks in which an attacker resends valid frames that the attacker has intercepted in the middle of forwarding (relaying) process. WiMAX can defend these attacks by mutual authentication.

2. There are a lot of threats at the MAC layer in WiMAX. First, some threats arise from its authentication scheme, such as masquerading attacks on the authentication protocol of PKM. In addition, there are some other serious attacks in MAC layers, such as man-in-the-middle and DoS attacks. Below we present four categories of attacks at the MAC layer.

a. Threats to authentication. Many serious threats also arise from the WiMAX's authentication scheme in which masquerading and attacks on the authentication protocol of PKM are the most considerable.

A masquerade attack is a type of attack in which one system assumes the identity of another. WiMAX supports unilateral device-level authentication, which is a RSA/X.509 certificate-based authentication. The certificate can be programmed in a device by the manufacturer. Therefore, sniffing and spoofing can make a masquerade attack possible. Specifically, there are two techniques to perform this attack: identity theft and rogue BS attack. In identity theft attack, the attacker will reprogram a device with the hardware address of another device. The address can be stolen by interfering with the management messages. In a rogue BS attack, the SS can be compromised by a forged BS, which imitates a legitimate BS. The rogue BS makes the SSs believe that they are connected to the legitimate BS; thus, it can intercept SSs' whole information.

PKMv2 provides a three-way authentication with a confirmation message from SS to BS. There are two possible attacks, as follows. First, a replay attack can be performed if there is no signature by SS. Second, even with the signature from SS, an interleaving attack is still possible.

b. Man-in-the-middle attack. Man-in-the-middle attack is possible due to the vulnerabilities in initial network entry procedures in WiMAX. WiMAX standard does not provide any security mechanism for the SS/BC negotiation parameters [44]. Through intercepting and capturing a message in the SS/BC negotiation procedure, an attacker can imitate a legitimate SS and send a tamped SS/BC response message to the BS while interrupting the communication between them. The spoof message would inform the BS that the SS only supports low-security capabilities or has no security capability.

If the BS still accepts, then the communication between the SS and the BS will not have strong protection. Under these circumstances, the attacker is able to wiretap and tamper all the information transmitted. Tao Han et al. also proposed a solution to this kind of attack, which they called SINEP [43].

c. DOS attack. Denial of service (DoS) attacks are a concern for WiMAX networks. Some noticeable DoS attacks may include the following:

DoS attacks based on ranging request/response (RNG-REG/RNG-RSP) messages: An attacker can forge a RNG-RSP message to minimize the power level of SS to make SS hardly transmit to BS, thus triggering an initial ranging procedure repeatedly. An attacker can also perform a water torture DoS by maximizing the power level of SS, effectively draining the SS's battery.

DoS attacks based on mobile neighbor advertisement (MOB_NBR_ADV) message: A MOB_NBR_ADV message is sent from the serving BS to publicize the characteristics of neighbor base stations to SSs searching for possible handovers. This message is not authenticated. Thus, it can be forged by an attacker in order to prevent the SSs from efficient handovers, downgrading the performance or even denying the legitimate service.

DoS attacks based on fast power control (FPC) message: A FPC message is sent from the BS to ask a SS to adjust its transmission power. This is also one of the management messages that are not protected. An attacker can intercept and use a FPC message to prevent a SS from correctly adjusting transmission power and communicating with the BS. He can also use this message to perform a water torture DoS attack to drain the SS's battery.

DoS attacks based on authorization-invalid (Auth-invalid) message: The Auth-invalid is sent from a BS to a SS when a AK shared between a BS and a SS expires or a BS is unable to verify the HMAC/CMAC properly. This message is not protected by HMAC, and it has a PKM identifier equal to zero. Thus, it can be used as a DoS tool to invalidate a legitimate SS.

DoS attacks based on reset command (RES-CMD) message: This message is sent to request a SS to reinitialize its MAC state machine, allowing a BS to reset a nonresponsive or malfunctioning SS. This message is protected by HMAC, but still has potential to be used to perform a DoS attack.

In order to prevent DoS attacks, vulnerabilities in the initial network entry should be fixed first. Reference [44] suggests that the authentication mechanism should be extended to as many management frames as possible. It also suggests using digital signatures as an authentication method.

12.2.2.3 3GPP LTE Security

The Long Term Evolution (LTE) architecture design is greatly different from that used by the existing 3G network. That difference brings with LTE a need to adapt and improve the security functions. Research work in [64] surveys that there are four main requirements for security functions in LTE:

■ Provide at least the same level of security as the 3G network without affecting user convenience.

■ Provide security techniques to defend against current attacks from the Internet.

■ The security functions provided by LTE shall not affect the stepwise transition from 3G to LTE.

■ Allow continued used of the USIM card.

12.2.2.3.1 Architecture Design

LTE security requirements cover three levels: (1) which protects communication between the user equipment (UE) and Evolved Universal Terrestrial Radio Access Network (EUTRAN) or Mobility Management Entity (MME) (2) which provides protection between elements in the wireline network, and (3) which provides secure access to the mobile station. In general, when compared to 3G, LTE security has (1) an extended authentication and key agreement, (2) a more complex key hierarchy and (3) additional security for the eNB (evolved Node B), which will be introduced in the following [17].

1. Extended authentication and key agreement. Authentication, encryption, and integrity protection procedures in LTE focus on the following:

   a. Freshness: The authentication vector that is at the heart of the authentication procedure is guaranteed to be fresh, i.e., not previously utilized. This is achieved via the sequence numbers exchanged in the messages that serve as input to the ciphering and integrity algorithms.

   b. Security algorithms: The algorithms used in the home environment (HE) and USIM to compute the authentication vectors are mostly one-way mathematical functions, where the output is obtained with a given set of inputs, using a predefined algorithm. Thus, it is extremely complex for an attacker to try to obtain the inputs using the outputs.

   c. Use of Internet Protocol Security (IPSEC): The IPSEC protocol is utilized to ensure confidentiality of user traffic as it is transmitted between nodes in the LTE EPS (Evolved Packet System). In addition, IPSEC tunnels are utilized for communication between various nodes in the visited and home networks—for mobile nodes. This introduces a requirement to have serving gateway (SGW) and MME nodes with sufficient processing power to handle encryption and decryption at required speeds so that performance is not hampered.

Mutual authentication of the UE and network is a cornerstone of the LTE security framework. The AKA procedure is utilized to achieve this by ensuring that the serving network authenticates the user's identity and the UE validates the signature of the network.

2. Key hierarchy. For data encryption, LTE uses a stream encryption method in which data are encrypted by taking an exclusive OR (XOR) of the data and key stream in the same way as is done in 3G. However, in LTE, in order to achieve a higher data rate, a hierarchical key system is designed to allow key updating without executing AKA, which may take several hundreds of milliseconds for key computation.

In the hierarchical key system, LTE utilizes five different keys, each used for a specific purpose and valid only for certain duration. Different keys are used for wireless communication. We believe this approach greatly reduces the effect of any possible security compromise. All the keys are derived using the key derivation function [46]. The five critical security keys derive their basis from the K key with a number of intermediate keys utilized as well. K is the permanent key stored on the USIM on the UE. To minimize the harm that may result if one of the keys used for encryption or integrity protection becomes compromised, the same key isn't stored and used at multiple locations on the network in LTE key system.

3. Additional security for the eNB. As some of the Radio Network Controller (RNC) functions are integrated into the eNB in LTE, the 3G security architecture cannot be reused like for the Radio Access Network in LTE. Therefore, measures described below are specified to minimize the harm that may result when a key is stolen from an eNB.

   a. Separation of AS and NAS security functions. Because a nonaccess stratum (NAS) message is exchanged with idle mode UEs, NAS security associations are established between the UE and core network nodes, i.e., the MME. Different from NAS, access stratum (AS) security is applied to all communication between the UE and the eNB. The algorithm used for AS is negotiated independently from the algorithm used for NAS.

   b. Handover security. The model for key transmission at handover in LTE achieves forward security. Here, forward security means that, even with knowledge of the current key shared by UE and eNB, the attacker cannot infer the future key of this eNB. Forward security is used to achieve authorized access to eNB installed in an exposed location. In that case, the key for eNB is changed from time to time and even if the current key is leaked, the scope of harm is limited because future keys will be generated without using the current key.

12.2.2.3.2 Attacks and Defenses in LTE

Below we will review attacks in the MAC layer and higher layer for LTE.

1. Attacks at the MAC layer. Attacks at the MAC layer mainly include the following:

   a. User location tracking. Location tracking refers to tracking the UE presence in a particular cell or across multiple cells. Location tracking as such does not pose a direct security threat, but it is a security breach in the network and can be a potential threat. Location tracking is made possible by tracking a combination of the Cell Radio Network Temporary Identifier (C-RNTI) with handover signals or with packet sequence numbers as described below.

   The C-RNTI is a unique and temporary UE identifier (UEID) at the cell level. As the C-RNTI is transmitted in clear text, a passive attacker can determine whether the UE using the C-RNTI is still in the same cell or not. During handover, a new C-RNTI is assigned to the UE via the handover command message. A passive attacker can link the new C-RNTI from the handover command message and the old C-RNTI unless the allocation of C-RNTI itself is confidentiality protected. This allows tracking of the UE over multiple cells. If continuous packet sequence numbers are used for the user plane or control plane packets before and after a handover, then mapping between the old and new C-RNTIs is possible based on the continuity of packet sequence numbers.

   b. Bandwidth stealing. Bandwidth stealing could emerge as a security issue in LTE. For example, fake buffer status reports can be utilized. The buffer status report is used as input information for packet scheduling, load balancing, and admission control. Sending false buffer status reports on behalf of another normal UE can change the behavior of these algorithms. By changing the packet scheduling behavior at the eNB, it is possible to carry out a bandwidth stealing attack making the eNB believe that the UE does not have anything to transmit.

   c. Security issues due to open architecture. The 4G LTE network will be an IP network with a large number of devices that are highly mobile and dynamic with activity periods ranging from a few seconds to hours. Diversity in device types and security levels coupled with the open architecture of an IP-based LTE network will result in greater numbers of security threats than seen in 3G networks. At present, handheld mobile devices (mainly cellular phones) are the most widespread users of wireless networks. Such devices have typically been proprietary in their design and makeup. While there is initial evidence of malicious activity in cellular networks, large-scale infection of cellular smart phones has not yet occurred.

   d. DoS attack. In LTE networks, there may be two possible ways to carry out a DoS. The first type of DoS attack would be against a specific UE. A malicious radio listener can use the resource scheduling information along with the C-RNTI to send an uplink control signal at the scheduled time, thus causing a conflict at the eNB and service problems for the real UE.

Newly arriving UEs are susceptible to a second type of DoS attack. In LTE, the UE is allowed to stay in active mode, but turn off its radio transceiver to save power consumption. This is achieved via the discontinuous reception (DRX) period. During a long DRX period, the UE is still allowed to transmit packets because the UE may have urgent traffic to send. However, this can create a potential security hole.

A third type of DoS attack can be based on the buffer status reports used by an eNB for packet scheduling, load balancing, and admission control. Attackers can send reports impersonating a real UE. If the impersonator sends buffer status reports that report more data to send than are actually buffered by the real UE, this will cause a change in the behavior of admission control algorithms [55]. If the eNB sees many such fake buffer status reports from various UEs, it may believe that there is a heavy load in this cell. Consequently, the eNB may not accept newly arrived UEs.

2. Attacks at the higher layers. Apart from end user equipment posing traditional security risks, it is expected that new trends such as Spam over Internet Telephony (SPIT) will also become a security concern in 4G LTE. Other Voice over Internet Protocol (VoIP)-related security risks are also possible, such as Session Initiation Protocol (SIP) registration hijacking where the IP address of the hijacker is written into the packet header thus, overwriting the correct IP address [51].

12.2.3   Possible Threats on 4G

Possible security risks mostly arise from the open nature of 4G as summarized next. First, a large number of external connectivity points with peer operators, with third-party application providers, and with the public Internet, as well as numerous heterogeneous technologies accessing the infrastructure, serve as potential security holes if the security technologies do not fully interoperate. Moreover, multiple service providers share the core network infrastructure, meaning that compromise of a single provider may result in collapse of the entire network infrastructure. Finally, service theft and billing fraud can take place if there are third parties masquerading as legitimate ones [5]. New end user equipments can also become a source of malicious (e.g., DoS) attacks, viruses, worms, spam mails and calls, and so on. In particular, the SPIT and the new spam for VoIP, will become serious problems just like the email spam today [4]. For example, SPITs targeting VoIP gateways can consume available bandwidth, thereby severely degrading QoS and voice quality. Clearly, the open nature of VoIP makes it easy for the attackers to broadcast SPITs similarly to the case of spam emails. Other possible VoIP threats include: (1) spoofing that misdirects communications, modifies data, or even transfers cash from a stolen credit card numbers, (2) eavesdropping of private conversations that intercepts and crypt-analyzes IP packets, and (3) phishing attacks that steal user names, passwords, bank accounts, credit cards, and even social security numbers [45].

12.3   Security in Ad Hoc Networks

An ad hoc wireless network is a collection of wireless mobile nodes that self-configure to construct a network without the need for any established infrastructure or backbone. People and vehicles can thus be internetworked in areas without a preexisting communication infrastructure or when use of such infrastructure requires wireless extension. Due to the absence of any fixed infrastructure and its wireless nature, it becomes difficult to make use of the existing security techniques in ad hoc networks, and this poses a number of challenges in ensuring the security of the communication [61].

12.3.1   Vulnerabilities of the Ad Hoc Networks

Compared with the wired network, the ad hoc network will need more robust security schemes due to the following reasons [34]:

1. Lack of centralized machinery: For high survivability ad hoc networks should have a distributed architecture with no central entities. Centrality increases vulnerability.

2. Varying topology: An ad hoc network is dynamic due to frequent changes in topology. In particular, there is no such clear secure boundary because of the freedom for the nodes to join, leave, and move inside the network. Thus, some of the nodes may be compromised by the adversary and thus perform some malicious behaviors that are hard to detect. Even the trust relationships among individual nodes also change, especially when some nodes are found to be compromised. And continuously changing scale of the network has set a higher requirement to the scalability of the protocols and services in the ad hoc network [29].

3. Resource constraints: Many types of attacks are possible in the ad hoc networks, such as packet dropping attack, resource consumption attack, fabrication attack, DoS attack, route invasion attack, node isolation attack, flooding attack, spoofing attack, and impersonation attack.

4. Roaming in dangerous environment: Any malicious node can create a hostile attack or deprive all other nodes from providing any service.

12.3.2   Security Requirements to Ad Hoc Networks

Previous work [22] has outlined availability, confidentiality, integrity, authentication, and nonrepudiation as the main security goals to consider for ad hoc networks.

Availability means to ensure that the network is operative whenever it is needed. With focus on intentional faults (i.e., attackers), availability implies protection, detection, and recovery from attacks. Confidentiality implies protecting the network content (e.g., routing table updates) from unauthorized disclosure. In particular, for military, emergency, and crisis management operations, it may be vital not to disclose the network participants to the outside world. Integrity ensures that data have not been altered during transmission, either intentional or unintentional. Authentication allows a node to verify the identity of its peers, preventing nodes from acting on behalf of another. Nonrepudiation means that nodes cannot deny transmitting a message. Moreover, there are some other security criteria that are more specialized and application-oriented, which include location privacy, self-stabilization, and byzantine robustness, all of which are related to the routing protocol in the ad hoc network [55].

In wireless environments, we also consider the following security requirements. Availability ensures legitimate parties can access resources and services from ad hoc networks successfully. This requirement is very important, as a network is useless if it cannot provide services. To ensure availability, security solutions should offer resistance to DoS attacks, including memory-DoS, computation-DoS, and network bandwidth-DoS attacks. A new requirement of ad hoc networks is anonymity, which requires the identity of the mobile user to be protected from the network he gains access to. This requirement implies user location privacy and unlinkability between two communications, and protects the user's motion pattern from being disclosed. At the end, an important requirement on security schemes for ad hoc networks is efficiency. The security solution should be efficient in both computation and communications, as mobile devices are usually resource constrained and the bandwidth is limited in ad hoc networks.

12.3.3   Attacks to Ad Hoc Networks

Attacks in ad hoc networks can cause congestion, propagate incorrect routing information, prevent services from working properly, or shut them down completely. In particular, the security attacks in ad hoc networks can be roughly classified into two major categories: passive attacks and active attacks.

12.3.3.1   Passive Attacks

A passive attack does not disrupt the normal operation of the network, and the attackers snoop the data transmitted in the network without altering it. Typical passive attacks include eavesdropping and traffic analysis.

1. Eavesdropping. Eavesdropping attack aims to obtain some confidential information that should be kept secret during the communication. Attackers can stay in the network for a while or a long time to capture messages transmitted within their hearing range. Based on the obtained messages, attackers can get some private information, such as the identities, transmission relationship, and so on.

2. Traffic analysis. Through traffic analysis, attackers monitor packet transmissions and then infer important information, such as a source, destination, and source-destination pair.

12.3.3.2 Active Attacks

Active attacks an range from deleting messages to injecting error messages to impersonating a node and so on. Thus, active attacks violate availability, integrity, authentication and nonrepudiation. Nodes that carry out the active attacks are considered to be malicious, and are referred to as compromised, while nodes that just drop the packets, with the aim of saving battery life, are considered to be selfish. A selfish node does not participate in the routing protocols or in forwarding packets in the network. Hence, we need to consider malicious attacks not only from outside, but also from within the network from compromised nodes. According to different attack targets and schemes, we review existing active attacks as follows:

1. Information disclosure. In this attack, an attacker reveals information regarding the location of nodes or the structure of the network. It gathers the node location information, such as a route map, and then plans further attack scenarios. In traffic analysis attack, adversaries try to analyze traffic to learn the network traffic pattern, track changes in the traffic pattern, and hence figure out the identities of communication parties. The leakage of such information is devastating in security-sensitive scenarios.

2. Information modification. In a message modification attack, adversaries make some changes to the routing messages or information stored in nodes [2]. Some typical information modification attacks are discussed below:

    a. Wormhole attack. This kind of attack is based on the modification of the metric value for a route or by altering control message fields. This is the simplest way for a malicious node to disturb the operations of ad hoc networks. The only task the malicious node needs to perform is to announce better routes (to reach other nodes or just a specific one) than the ones presently existing. There information modification can be achieved by:

        ■ Modifying the route sequence number [41]: When choosing the optimum path to take through a network, the node always relies on a metric of values, such as hop delays, etc. The smaller the delays, the more optimum the path. Hence, a simple way to attack a network is to change this value with a smaller number than the last "better" value.

        ■ Altering the hop count [41]: This attack is more specific to some protocols wherein the optimum path is chosen by the hop count metric. A malicious node can disturb the network by announcing the smallest hop count value to reach the compromised node. In general, an attacker would use a value zero to ensure to the smallest hop count.

For example, an attacker records packets at one location in the network, tunnels them to another location, and retransmits them there into the network. This could potentially lead to a situation where, it would not be possible to find routes longer than one or two hops, probably disrupting communication.

■ Altering routing information [41]: This attack can be achieved by modifying the routing path information in compromised nodes. Any packet passing through a compromised node will be transmitted to an incorrect node that will not accept this packet. Another instance can be seen when considering a category of attacks called the black hole attacks. Here, a malicious node uses the routing protocol to advertise itself as having the shortest path to the node whose packets it wants to intercept. Once the malicious node has been able to insert itself between the communicating nodes, it can do anything with the packets passing between them. It can then choose to drop the packets, thereby creating a DoS attack [47].

b. Byzantine attack. A compromised intermediate node works alone, or a set of compromised intermediate nodes work in collusion and carry out attacks, such as creating routing loops, forwarding packets through nonoptimal paths, or selectively dropping packets, which results in disruption or degradation of the routing services.

c. State pollution attack. If a malicious node gives incorrect parameters in reply, it is called a state pollution attack [22]. For example, in best effort allocation, a malicious allocator can always give the new node an occupied address, which leads to lots of repeated broadcast of duplication address detection messages throughout the ad hoc networks and the rejection of the new node.

3. Information spoofing/fabrication. There are basically three subcategories for fabrication attacks:

a. Node spoofing. This is a special case of integrity attacks whereby a compromised node impersonates a legitimate one due to the lack of authentication in the current ad hoc routing protocols. If a malicious node impersonated some nonexistent nodes, it will appear as several malicious nodes conspiring together, which is called a sybil attack. This attack aims at network services when cooperation is necessary, and affects all the autoconfiguration schemes and secure allocation schemes based on the trust model as well. The main result of the node spoofing attack is the misrepresentation of the network topology that may cause network loops or partitioning.

b. Node information spoofing. This attack occurs when false information is injected into or transmitted by compromised nodes. We classify this attack into two categories as follows:

- Error messages sending: Whenever a node moves, the closest node sends an "error" message to the other nodes to inform them that a route is no longer accessible. Thus, an attacker can send fake error messages to nodes nearby and then isolate nodes from one another.

- Error information cache: This occurs when information stored in routing tables is injected with false information. A node overhearing any packet may add the routing information contained in that packet's header to its own route cache, even if that node is not on the path from source to destination. The vulnerability of this system is that an attacker could easily exploit this method of learning routes and poison route caches by broadcasting a message with a spoofed IP address to other nodes. When they receive this message, the nodes would add this new route to their cache and would now communicate using the route to reach the malicious node.

c. Message spoofing. Instead of modifying or interrupting the existing routing packets in the networks, malicious nodes also could fabricate their own packets to cause chaos in the network operations. They could fabricate message by injecting huge packets into the networks. However, message fabrication attacks are not only launch by the malicious nodes. Such attacks also might come from the internal misbehaving nodes. With a lack of integrity and authentication in routing protocols, fabrication attacks will result in erroneous and bogus routing messages.

4. Resource exhaustion. DoS is the classic resource exhaustion attack, where the attacker injects a large amount of junk packets into the network. These dummy packets spend a significant portion of network resources, and introduce wireless channel contention or even congestion in ad hoc networks. Here, we review three main types of DoS attacks in ad hoc networks:

a. Jamming attack. Jamming is a particular class of DoS attacks. The objective of a jammer is to interfere with legitimate wireless communications. A jammer can achieve this goal by preventing either a real traffic source from sending out a packet or the reception of legitimate packets [21].

b. Flooding attack. In flooding attack, the attacker exhausts the network resources, such as bandwidth, and consumes a node's resources, such as computational and battery power, or disrupts the routing operation, causing severe degradation in network performance [22]. For example, the attacker creates a large number of half-opened TCP connections with a victim node, but never completes the handshake to fully open the connection. As a result, all of the node battery power, as well as network bandwidth, will be consumed and could lead to DoS.

c. Sleep deprivation attack. The sleep deprivation attack aims to consume the batteries of a victim node. This attack is most specific to wireless ad hoc networks, but may be encountered in conventional WSNs or wired networks. The idea behind this attack is to request the services a certain node offers, over and over again, so it cannot go into an idle or power-preserving state, thus depriving it of its sleep. This can be very devastating to networks with nodes that have limited resources, for example battery power. It can also lead to constant business of the component, hindering other nodes to legitimate request services, data, or information from the targeted entity. An attacker or a compromised node can attempt to consume battery life by requesting excessive route discovery or by forwarding unnecessary packets to the victim node [22].

12.3.4 Security Solutions to Ad Hoc Networks

In order to provide solutions to the security issues involved in ad hoc networks, we must elaborate on two of the most commonly used approaches: prevention, and detection and reaction [47].

12.3.4.1 Prevention

Prevention mechanisms require encryption techniques to provide authentication, confidentiality, integrity, and nonrepudiation of information. Some existing preventive approaches use symmetric algorithms, some use asymmetric algorithms, while the others use one-way hashing, each having different trade-offs and goals. Among the existing preventive approaches, the most typical one is to secure routing protocols, which can defend against different attacks.

The general purpose of securing ad hoc routing protocols is to prevent attackers from modifying routing messages or injecting harmful routing messages. In addition, routing protocols for ad hoc networks must handle outdated routing information to accommodate dynamic changing topology. False routing information generated by compromised nodes can also be regarded as outdated routing information. So integrity and authenticity of routing messages should be guaranteed. Confidentiality can be ensured easily, e.g., by encryption, but it will increase overhead. Route establishment should be a fast process. If too many security mechanisms are built in, the efficiency of the routing protocol may be sacrificed [63]. So there is a trade-off between security and efficiency.

For data authentication, Zapata and Asokan proposed a secure AODV routing protocol (SAODV) [59]. SAODV authenticates mutable data in route request messages by hash chains. However, for nonmutable data, the protocol uses only digital signatures. A node requesting a route to the destination generates a random seed for the hash chain and computes the maximum hash chain value by repeated hashing of the seed until reaching the maximum hop count. The signature on all fields but the seed and hop count is appended to the message. Intermediate nodes verify the signature, and the maximum hash chain value is reached after hashing the received seed for max hop count times. If verification holds, the hop count is stepped and the seed is updated by hashing it. In order to allow intermediate nodes to respond whenever this node holds a valid route in its route cache, the double signature scheme is proposed. Route error messages do not use the hash chain mechanism, but instead are digitally signed. Since it is not considered relevant which node initially started the error message, the signature is replaced for each hop, rather than appended. The protocol provides authentication for end nodes, but not for intermediate ones, allowing adversaries on the path to forge their identity. The hash chain mechanism guarantees that malicious nodes cannot reduce the hop count value, but may increase it or omit updating it.

A different solution, trust-based routing [49] for ad hoc networks, was developed. The role of trust in ad hoc networks was identified in [49]. When a network entity establishes trust in other entities, it can predict the future behaviors of others and diagnose their security properties. Trust helps in assistance in decision making to improve security and robustness, adaptation to risk leading to flexible security solutions, misbehavior detection, and quantitative assessment of system-level security properties.

12.3.4.2 Detection and Reaction

Prevention mechanisms by themselves cannot ensure complete cooperation among nodes in the network. Detection, on the other hand, attempts to identify clues of any malicious activity in the network and take punitive actions against such nodes [60]. An effective way to identify when an attack occurs in an ad hoc network is the deployment of an intrusion detection system (IDS) [52]. The IDS is a sensoring mechanism that monitors network activity in order to detect malicious actions and, ultimately, an intruder. However, many IDS solutions have been proposed for wired networks, which are defined on strategic points such as switches, gateways, and routers; they cannot be implemented on the ad hoc network with many mobile nodes. Thus, the wired network IDS characteristics must be changed prior to being implemented in the ad hoc network. An IDS in ad hoc networks can be divided by (1) the architecture, which exemplifies the operational structure of the IDS, and (2) the detection engine, which is the mechanism used to detect malicious behavior(s), as can be seen in Figure 12.1.

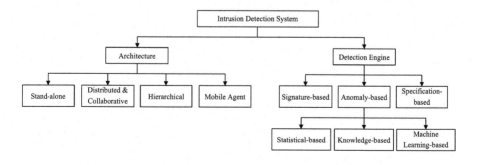

Figure 12.1: IDS classification.

12.3.4.2.1 Architectures

The optimal IDS architecture for the ad hoc networks may depend on the network infrastructure itself. The existing architectures for ad hoc networks fall under four basic categories: stand-alone IDS, distributed and collaborative IDS, hierarchical IDS, and mobile agent for IDS [38].

1. In the stand-alone architecture, an IDS is implemented on each node and does the operation of intrusion detection. Whatever decision is made by each node is based on the information of the node itself, and there is no cooperation among the nodes of the network. In this architecture, because there is no cooperation and interchange on information among the IDSs, as well as the extent of view of each node on the network being limited, the intrusion detection operation has a low accuracy. Because of this, this kind of architecture is rarely applied in ad hoc networks.

2. In the distributed and collaborative architecture every node in the ad hoc network must participate in intrusion detection and response by having an IDS agent running on it. The IDS agent is fully responsible for detecting and collecting local events and data to identify possible intrusions, as well as initiating a response independently [39].

3. The extended version of the distributed and collaborative IDS is the hierarchical architecture. This architecture suggests multilayered network infrastructures in which the network is divided into clusters. The architecture has cluster heads, in some sense, that act as control points, which are similar to switches, routers, or gateways in wired networks.

4. Mobile agents are applied in ad hoc networks as a concept in the same intrusion detection techniques. These agents can move easily throughout a major network, and each has a specific duty. Because one or more agents can be placed inside a node, the intrusion detection operation can be distributed throughout the network. There are several privileges in using the mobile agents. For example, some operations are not proportioned to every node. And because of this, there will be a decrease in consuming the resources. The mobile agents-based IDS can be considered a kind of distributed and cooperative IDS. Some techniques also use the combination of the mobile agents with hierarchical IDSs.

12.3.4.2.2 Detect Engine

We classify these employed intrusion detection engines into three main categories: (1) signature-based engines, which rely on a predefined set of patterns to identify attacks; (2) anomaly-based engines, which rely on particular models of nodes' behavior and mark nodes that deviate from these models as malicious; and (3) specification-based engines, which rely on a set of constrains (i.e., description of the correct operation of programs/protocols) and monitor the execution of programs/protocols with respect to these constraints [57].

1. In the signature-based method known patterns of attacks are kept. Then the behavior of the network and its nodes is controlled, and when any suspicious behavior is observed, it is compared with the existing patterns to detect the intrusion. If a behavior matches with existing patterns, it is considered an attack [27].

2. In the anomaly-based technique, normal behavior of the target system (network and nodes) is defined, and after that the normal behavior model is constructed. Based on the model, a threshold is defined to show the boundary of normal and abnormal behaviors. Then the nodes and the network are under control, and if any behavior unmatched with the normal behavior is observed, it is considered an attack. According to the processing type of behavior model of the target system, anomaly-based detection techniques can be classified into three types: statistical-based, knowledge based, and machine learning based [27].

   a. In statistical-based techniques, the traffic activities of the network are monitored and a profile is created based on some metrics, such as traffic rate, number of packets for each protocol, communication rate, and different IP addresses, to detect network abnormal behavior. During the process of detection, both data related to the currently observed profile and data related to the normal profile are considered. When an anomaly occurs in the network, the current profile is created and an anomaly privilege is calculated by comparing the current and normal profile. If the privilege exceeds the defined threshold, the IDS reports the anomaly. The statistical-based method doesn't require background knowledge on the target system's normal activity. In this technique, statistical methods provide exact reports on malicious activities that are done during a long period.

   b. Knowledge-based techniques are widely used in expert systems based on a set of specific rules for classifying the audit data. Commonly, they contain three steps: First, according to the data gathered during the process of training, different features and classes result. Then in the second stage, a set of necessary rules are extracted to classify the parameters or functions. Finally, in the third stage, the gathered data are classified according to these rules. If the specifications are complete enough, this model will be able to detect illegal behavior patterns; then, as a result, the false positive rate will decrease. Knowledge-based techniques are robust and flexible. However, it is difficult to get high-quality knowledge, which is necessary in knowledge-based techniques.

   c. Machine learning techniques are based on a created model that can categorize patterns of observed behaviors. Data are first labeled in order to learn the behavior model of the nodes and the whole network. The data labeling process demands high resources, such as energy and bandwidth. In many cases, the function of the principle of machine learning is simultaneous with statistical techniques. In these cases, the constructor of the model, in addition to use of the statistical techniques, uses the results of the previous stages to improve the performance of the intrusion detection system. So machine learning-based anomaly detection is able to change its strategy to obtain new information. The key problem related to these methods is the costly nature of them in terms of required resources. Several machine learning-based schemes have been used in the anomaly-based detection technique, such as Bayesian networks, Markov models, Neural networks, fuzzy logic techniques, genetic algorithms, and clustering detection.

3. In the specification-based technique, a series of constraints is predefined to describe the correct operation of a program or protocol. Then the system monitors the execution of the program or protocol by these defined constraints. If a behavior goes beyond these constraints, it is reported as an attack.

Below, we will discuss some typical IDSs proposed especially for the ad hoc network.

In ad hoc networks, a node is distributed, and hence cooperation is required with other nodes. Zhang et al. [62] proposed a distributed and collaborative design of IDS. In his proposed architecture model, each and every node is fully responsible for detecting signs of intrusion locally and independently, but neighboring nodes can collaboratively investigate in a broader range. Individual IDS agents are placed on each and every node. Each IDS agent runs independently and monitors local activities. The agent detects intrusion from local investigation and start response. If delicacy is detected in the local data, neighboring IDS agents will cooperatively participate in global intrusion detection actions. These individual IDS agents collectively form the IDS system to protect from attack to the wireless ad hoc network. Sterne et al. [48] further suggested a dynamic intrusion detection hierarchy that is potentially applied to large networks use the clustering technique. Nodes on the first level are cluster heads, while nodes on the second level are leaf nodes. In this model, every node has the task to monitor, log, analyze, respond, and alert or report to cluster heads.

12.3.5 Challenges in Ad Hoc Networks

A number of restriction and technical difficulties are faced by researchers, which are explained in existing research [22]. These general problems must be taken into account for further security research in ad hoc networks. Some of these are:

1. The basic idea from ad hoc networks is that each node in the network is mobile, and can move from one place to another within the coverage area, but still the mobility is limited. So securing mobility challenge is a hard problem.

2. The connectivity among nodes can be highly ephemeral, and maybe will not happen again, nodes traveling throw a coverage area and making connections with other nodes will be lost, as many nodes are mobile, and maybe will move in the opposite direction. Ad hoc networks are unstable and lack the relatively

long life context, so personal contact of a user's device to a hot spot will require a long life password, and this will be impractical for securing ad hoc.

3. The ad hoc network does not require any infrastructure, so it is very difficult to carry out any kind of centralized management and control.

4. Another challenge to ad hoc networks is the resource constraint. The wireless channel is bandwidth-constrained and shared among multiple networking entities. And the computational capabilities of mobile devices are powered by batteries with their inherent limitations.

12.4   Security in Wireless Sensor Networks

As an important part of the Internet of things, WSNs are becoming increasingly popular with applications ranging from habitat monitoring to the battlefield. Many aspects of the security in WSNs have been examined, including secure and efficient routing, secure data fusion, key management, secure localization, and so on. In addition to traditional security issues in WSNs, some emerging security techniques, such as privacy preservation, have been attracting more and more research. On the other hand, for the requirement of having been used in some special and important environments, new WSNs with an unattended nature or able to work underwater are proposed. Such emerging WSNs bring new threats and challenges. So in this part, we focus on the main emerging security techniques (privacy preservation) in WSNs and three emerging WSNs: USNs, UWSNs, and WMSNs.

12.4.1   Privacy Preservation in WSNs

Privacy is a serious concern for many wireless networks, whereas privacy protection for hardware-limited sensor networks is a complicate issue. Nowdays, the critical privacy concern on information being collected, transmitted, and analyzed in WSNs has received lots of attention. Such private information of concern may include payload data collected by sensors and transmitted through the network to a centralized data processing server. There are two main types of privacy concerns: data oriented and context oriented [33].

12.4.1.1   Data Oriented

Data-oriented privacy protection focuses on the privacy of data collected from or a query posted to a WSN. Here, *data* refers to not only sensed data collected within a WSN, but also queries posed to a WSN by users. Data -oriented privacy concerns may be violated by data analysis attacks. In the case of a data analysis attack, a malicious node of the WSN abuses its ability of decrypting data to compromise the payload being transmitted [33].

There are two types of adversaries that may compromise data-oriented privacy. One is an external adversary that eavesdrops on the data communication between sensor nodes in a WSN. The second type is an internal adversary that is also a participating node of the WSN, but has been captured and manipulated by malicious entities to compromise private information.

12.4.1.1.1 External Adversary

This type of adversary can be effectively defended against using the traditional techniques of cryptographic encryption and authentication. It is an important technique for energy saving in WSNs. Furthermore, data aggregation is also a common and effective method to preserve private data against an external adversary, as the process reduces the amount of data at the intermediate sensor nodes.

12.4.1.1.2 Internal Adversary

Compared with external adversaries, the internal adversaries can be more powerful because they consist of nodes/base stations captured and controlled by malicious entities, and therefore may have knowledge of encryption keys used in a WSN. Since a participating node is allowed to decrypt data legally, the traditional encryption and authentication techniques may no longer be effective. Thus, the main challenge for protecting data-oriented privacy is to prevent an internal adversary from compromising the private information, while maintaining the normal operation of the WSN. There are two types of data privacy we would like to protect against internal adversaries: the privacy of data being collected and the privacy of queries being posed to WSNs.

1. For data collection privacy, end-to-end encryption between the data source and the base station can be used to defend against internal adversaries. With this approach, no intermediate node, including the internal adversaries, can compromise the privacy of data being transmitted without knowing the key shared by only the two end nodes. Nonetheless, this approach also impedes the normal operation of a WSN. In particular, it renders data aggregation (i.e., node-by-node selection of transmitted data to reduce traffic volume) infeasible in intermediate nodes, due to their incapability of decrypting the transmitted data. This leads to a significant amount of additional traffic being transmitted, jeopardizing the energy consumption of a power-constrained WSN. Thus, a main challenge for privacy-preserving data aggregation is to defend against internal adversaries under hop-by-hop (i.e., link) encryption, where nodes that are directly communicating with each other share a private key for encryption and decryption [3]. If an aggregator is compromised, it may perform either passive or active attacks.

   a. As for passive attacks, the compromised aggregator can decrypt the transmitted data and then obtain the content privacy. The passive attack aims to compromise the confidentiality of private data.

   b. An aggregator may launch active attacks, including raw data tampering or bogus data injection. Different from the passive attacks, active attacks aim to destroy the integrity of collected data. Techniques are designed to defend against such active attacks by perturbed data values. Data mining techniques such as perturbation and randomization can be developed in order to work with these privacy protection aggregation schemes. Randomized data are used to mask the private values. In another technique, perturbation technique, certain distribution is added to the private data. Given the distribution of the random perturbation, the aggregated result can be recovered in the base station. However, data perturbation techniques have the drawback that they do not yield accurate aggregation results. There are two types of data perturbation. In additive perturbation, randomized noise is added to the data values. The overall data distributions can be recovered from the randomized values. Another is multiplicative perturbation, where the random projection or random rotation techniques are used in order to perturb the values. In tune of their argument, we can apply the second technique of masking the private data by some random numbers to form additive perturbation.

Ukil and Sen [53] consider a scenario where data aggregation needs to be done in a privacy-preserved way for a distributed computing platform. There are several data sources that collect or produce data. The data collected or produced by the sources are private, and the owner or the source does not like to reveal the content of the data. But the collected data from the source are to be aggregated by an aggregator, which may be a third party or part of the network, where the data sources belong. The data sources do not trust the aggregator. So the data need to be secure and privacy protected. The computation for the aggregation is based on the concept of SMC [40]. SMC allows parties with similar backgrounds to compute results upon their private data, minimizing the threat of disclosure. Consider a set of parties who trust neither each other nor the channels by which they communicate. Still, the parties wish to correctly compute some common function of their local inputs, while keeping their local data as private as possible. Generally, this problem can be seen as a computation of a function on a series of private inputs with different participants, where each participant knows only its input and no more information before the output is revealed to all participants in the computation.

2. Different from the protection of private data being collected in a WSN, the query issued to a WSN (to retrieve the collected data) is often also of critical privacy concerns. In the medical WSN, if an adversary learns that queries have been frequently issued to the part of the WSN that covers a patient's house, the adversary can certainly infer the health of the patient is getting more attention probably due to his or her health problems. For data query privacy, the base station usually not only collects the target queried data, but also adds some other data (e.g., perturbed data) available in WSNs to prevent adversaries from inferring the private query based on the data being accessed. Thus, a main challenge for protecting query privacy is to minimize the amount of dummy information while maintaining the privacy of queries being issued. From the

perspective of energy preservation, query processing should be restricted to as small a targeted range as possible. Nonetheless, narrowing down the range also increases the chance for the query to be inferred by adversaries. In the following, we will present the typical technique for protecting the privacy of data aggregated and queries posed. Private data query represents significant challenges to the design of a resource-constrained WSN. For energy saving, a target region transformation technique [6] was proposed to conceal the true target queried region. The transformation function is used to map one region into several different regions, so that the target region cannot be recognized from the other random chosen regions. In order to achieve a different security and energy saving level, three transformation functions, including uniform, randomized, and hybrid, were proposed. In the union transform (UT), the interesting region of each query is mapped into the set of all regions. In the randomized transform (RT), each query is transformed to randomly chosen regions involving the target one. The hybrid transform (HT) is a combination of UT and RT. In comparison with the data query problem in the domain of databases, these schemes are similar to the k-anonymity algorithms, which hide the target's real identity using other k similar objects so that it is impossible for the adversary to distinguish the target from the k other objects. But the cost of such an anonymity-based technique is high in a WSN, because query dissemination and data collection in the uninteresting regions consume a large amount of energy. The larger the region saturated with queries, the more effectively does safeguarding privacy work, but at the cost of more energy. Consequently, a proper trade-off between energy consumption and privacy protection is important.

12.4.1.2 Context Oriented

Context-oriented privacy protection mainly concentrates on contextual information. Similar to data-oriented privacy, context-oriented privacy may also be threatened by both external and internal adversaries. Nonetheless, existing research has mostly focused on defending against external adversaries, because such adversaries may be able to compromise context privacy easily by monitoring wireless communication. Within the category of external adversaries, one can further classify adversaries into two categories, local attackers and global attackers, based on the strength of attacks an adversary is capable of launching. Local attackers can only monitor a local area within the coverage area of a WSN, and therefore have to analyze traffic hop by hop to compromise traffic context information. On the other hand, a global attacker has the capability (e.g., a high-gain antenna) of monitoring the global traffic in a WSN. One can see that a global attacker is much stronger than a local one. The contextual information in WSNs mainly includes location and timing of traffic flows.

12.4.1.2.1   Location Privacy

Location privacy is a hot and important security issue in WSNs. An effective location privacy preservation protocol for WSN can prevent attackers from identifying (and then capturing) important nodes (such as source and base station) by hiding their locations. Local passive attackers can locate a node by localization techniques such as triangulation, angle of arrival, signal strength, and so on. Moreover, if an attacker knows the location of each node, he will be able to selectively compromise more important nodes, which will allow him to get much more information or cause more damage to the network.

A major challenge for location privacy protection is that an adversary may be able to compromise private information even without the ability of decrypting the transmitted data. In traffic analysis attacks, a third-party adversary does not have the ability to decrypt data payloads. Instead, he eavesdrops on the wirelessly transmitted data and tracks the traffic flow information hop by hop. In particular, he may derive the locations of the base station and data source by observing and analyzing the traffic patterns between different hops (to track down the base station or the data source).

As it is the traffic that exposes the privacy information, location of some critical nodes, various traffic pattern obfuscation techniques have been proposed to hide the real traffic pattern. A common idea of such techniques is to introduce random factors to packet routing, in order to increase the uncertainty of traffic patterns observed by the adversary and to counteract the adversarial traffic analysis attacks [33]. With this common idea, the cost associated with privacy protection is the delay of delivering data packets, the (possible) reduction of successful delivery probability, or even a huge amount of energy consumption. Consequently, a proper trade-off between energy consumption and privacy protection is important. In the following, we review the existing techniques for protecting the locations of data source and base station.

1. Data source location protection. The source location protection problem is known as the classic Panda Hunter Game model, which was first introduced in [26]. In the Panda Hunter Game, a large number of sensors are deployed to monitor pandas' habitat. When sensors detect a panda, they will generate event messages and transmit them toward the base station. In the meantime, a panda hunter also attempts to identify the location of the data source to find the panda. In this game, the objective of the defender is to properly obtain the panda's moving information in order to enable biological research. He also intends to hide the location information from being known by the panda hunters, which have the ability to eavesdrop the wireless communication between different sensor nodes, but do not have the key to decrypt the payload. The objective of the panda hunter is to infer the location of the data source (and thereby the panda) by analyzing the traffic flow in the WSN.

   For a local adversary, he is able to monitor the local traffic. There are two approaches for an adversary to start the attack, arbitrarily choosing a place to stay in the network to monitor traffic or staying around the base station with the prior knowledge about the location of the base station. In the following, we discuss some existing techniques against the disclosure of the location of the data source in WSNs.

   In [26], phantom routing is first presented to protect the source location from a local attacker. Different from the shortest path routing, phantom routing first transmits a message a few hops from the data source, and then, by employing a probabilistic flooding scheme, the message is transmitted toward the base station. The premise of this approach is that even if an adversary is able to track back along the routing path, it would only be able to figure out the terminal node of the random walk instead of the original data source.

Another simple solution is to generate some fake messages by intermediate nodes with some probability [58]. When receiving a fake packet, a sensor node just discards it. Although this approach perturbs the local traffic pattern observed by an adversary, it also has limitations on privacy protection. Specifically, to maintain the energy efficiency of the WSN, the length of each path along which fake data are forwarded is only one hop; therefore, an adversary is able to quickly identify fake paths and eliminate them from consideration.

As for global adversaries, they can observe the whole network traffic. Thus, to further protect the location of the data source, fake data packets can be introduced to perturb the traffic patterns observed by the adversary. In particular, a simple approach is to globally inject dummy data as well as keeping the transmission of real data the same as that of dummy data. However, this approach may introduce significant delay to the data transmission process. To cope with this concern, various techniques have been proposed in the literature. For example, a special distribution of data transmission interval was suggested in [34], such that as long as every sensor node follows such a predetermined distribution, the delay of real data can be minimized without allowing an adversary to identify the real traffic. Two other schemes, proxy-based filtering scheme (PFS) and tree-based filtering scheme (TFS), were proposed in [34] to filter partial dummy data without threatening the source privacy. The whole network is divided into cells. The proxies are responsible for relaying real data from cells around them and filtering dummy data. After filtering all dummy packages from cells and buffering real data, the proxy will send a data package, including buffered real data and new generated dummy data, at the same rate of transmission. Thanks to the filtering procedure, a large number of dummy data is removed so that energy consumption on overhead communication is greatly reduced.

2. Base station location protection. Due to the important role of base station in wireless sensor networks, the location of the base station is of critical importance. In a WSN, a base station is not only in charge of collecting and analyzing data, but also used as the gateway connecting the WSN with an outside wireless or wired network. Consequently, destroying or isolating the base station may lead to the malfunction of the entire network. In the following, we discuss some typical existing privacy-preserving techniques for defense against three major adversaries.

Existing base station location attacks include packet-tracing attack [1], rate monitoring attack [1], and zeroing-in attack [35]. In [12], Deng et al. present a few techniques to safeguard the base station against packet-rate monitoring and time correlation attacks. A protection method called DEFP with techniques of multipath routing and fake message injection is proposed. However, these measures would take a long time to find the base station as attackers concentrate on the traffic rates on different locations. In [9], Conner et al. propose a fake base station protocol for protecting the base station. The work creates a dummy base station away from the real base station. All the data are first forwarded to the dummy base station. Then the aggregated data are rerouted to the real base station. This scheme implicitly assumes that the fake base station is with powerful computation and storage ability. However, this may not be true in a homogenous network. Once the adversary destroys the fake base station and gets its private information, he can track the real base station easily. Jian et al. [23] go further to design a new location-privacy routing protocol with fake packet injection to provide path diversity and minimize the information that an adversary can deduce from the overheard packets about the direction toward the receiver. After that, Acharya and Younis [1] extend popular metrics for measuring anonymity to suit the unique characteristic of WSNs and presents two approaches for boosting the anonymity of the base station through packet retransmission (BAR) and by repositioning the base station (RIA). In BAR, the base station selectively transmits data packets that get forwarded through the network in order to confuse the adversary. Meanwhile, the RIA approach introduces the concept of dynamically relocating the base station in order to safeguard it. However, how to sense and measure the threat to the base station is not presented, which is very important and hard to address. In [35], Liu and Xu have investigated zeroing-in attacks that utilize hop counts and the packet time of arrival (ToA). A few adversaries observe the network metrics by eavesdropping on the local communication and collectively determine the sink location by solving the least squares problem over the observations. A zeroing-in attack cannot be launched to routing protocols that do not use hop count information.

12.4.1.2.2 Timing Privacy

Timing privacy, on the other hand, concerns the time when sensitive data are created at the data source, collected by a sensor node, and transmitted to the base station. When sensors detect the target, they generate event messages and forward them to the base station. If an adversary can identify a specific time when an event message is generated, the adversary and use the information to track the target, or even predict the target's next move. Here the attacker's confidence depends on the assumption that the delay time of data passing through each intermediate sensor along the routing path is the same. When the adversary eavesdrops on the message near the base station, it can first obtain the arrival time and then deduct from it the multiplication of average delay and the hop count. This type of privacy is also of primary importance, especially in the mobile target tracking application of WSNs, because an adversary with knowledge of such timing information may be able to pinpoint the nature and location of the tracked target without learning the data being transmitted in WSNs. Furthermore, the adversary may be able to predict the moving path of the mobile target in the future, violating the privacy of the target.

In order to defend such an attack, in [25], each intermediate sensor located along the routing path locally buffers the data for a random period of time, such that an adversary cannot accurately estimate the original generation time of the message. Apparently, the random delay leads to increasing costs of buffer space at the intermediate nodes. How to make a trade-off between the protection of timing privacy and the efficiency of buffer space is of primary concern. Furthermore, a rate-controlled adaptive delaying scheme is also proposed in [25] to adjust the delay distribution as a function of the incoming traffic rate and the available buffer space.

12.4.2 Security in Unattended Wireless Sensor Networks

In recent years, while many WSNs operate in general settings such as real-time data uploading with a constantly present base station, there are emerging WSN applications that fall outside the real-time data collection model. These networks can be deployed in hostile environment (e.g., military or law enforcement environments) and have been known as unattended wireless sensor networks (UWSNs). Different from traditional WSNs, a UWSN is left unattended for most of the time after deployment. A UWSN has a mobile base station (BS) that visits the network with some frequency. In UWSNs, the unattended feature makes nodes extremely vulnerable to attacks that happen between visits of the mobile BS.

In prior security research, it has been often assumed that attackers considered in UWSNs are featured as mobile and powerful hardware equipment [36]. They can compromise up to a certain number of sensors within a given time interval. Attackers move unpredictably, and thus they are untraceable. According to different attack purposes and strategies, they can be classified into three categories: curious attacks, search-and-erase attacks, and search-and-replace attacks. Here, we discuss these attacks and their defensive schemes one by one.

12.4.2.1 Curious Attackers

The attacker's goal is to learn as many nodes' privates as possible while keeping himself unobservable [36]. An attacker may compromise nodes one by one, obtain their secrets, and then leave the network without being noticed. Curious attackers may cause severe damage to the network. First, if the node secrecy information is leaked, any cryptographic protocol that depends on the secrecy (e.g., keys) would become useless. Second, the attacker may decrypt and obtain important sensing data.

Forward secrecy means that even if an attacker obtains the sensor's current secrets, he cannot decrypt (or forge authentication tags for) data collected and encrypted (or authenticated) before compromise. *Backward secrecy* means that an attacker who obtains the sensor's current secrets cannot decrypt (or forge authentication tags for) data after compromise [36].

Forward security is relatively easy to obtain by key evolution such as hash function, which doesn't help on backward security. Pietro et al. [13] provide both forward and backward secrecy by using public key cryptography. Unfortunately, public key cryptography is not suitable for WSNs due to the large computational overhead.

DISH [37] and POSH [64] achieve both forward and backward security by key evolution and node cooperation. In DISH, each node requests random data from randomly selected nodes and then updates its key based on the random data and its current key. Different from DISH, in POSH each node selects some nodes as the recipients and sends random data to each of them. POSH does not need to send an extra data request message, and hence achieves much lower communication cost than DISH.

However, both POSH and DISH don't consider node failures and message losses, which are common in real sensor applications. Furthermore, both POSH and DISH generate random data by a pseudorandom number generator (PRNG). If an attacker compromises a sensor, he can obtain the PRNG algorithm and compute all subsequent random values. Under this attack, a PRNG cannot provide backward security. Hence, with more rounds of attacks, in DISH/POSH the number of nodes that can provide secure random data decreases. After several rounds, no secure node exists in the network and no secure random data can be provided. As a result, the node self-healing capability decreases to zero. An alternative way to presensor PRNGs is to use a true random number generator (TRNG). Compared to PRNGs, TRNGs extract randomness from physical phenomena, and hence the random numbers are nondeterministic and cannot be precomputed. However, TRNG is only suitable for nodes equipped with extra hardware. TRNG is not suitable for small sensors.

12.4.2.2 Search-and-Erase Attackers

Attackers aim to find out sensitive data and then erase them before they reach the sink. Thus, without data storage protection, the accumulated data will be lost under this attack. To achieve data survivability, cryptographic or noncryptographic approaches have been considered. Cryptographic approaches are suitable for high-end sensors that are equipped with a pseudorandom number generator and can rely on considerable computational power. In [13], an adversary that can control a fixed portion of the network deployment area, and that can compromise all sensors that move within this attack, is considered. The proposed scheme is based on public key cryptography, but it uses an evolution mechanism based on nodes collaboration to generate one-time symmetric random keys. Almost all these works require sensors with some cryptographic ability. However, public cryptography-based approaches need computationally intensive calculations. In contrast, noncryptographic approaches are more suitable for low-cost sensors. One simple and direct approach is data replication: rather than having each sensor storing one datum, each datum is replicated and data copies are stored in different sensors. Existing data replication techniques to grant data survival are classified by different storage node selection schemes: pure random selection, selection based on location, selection based on hops, and so on [15].

12.4.2.3 Search-and-Replace Attackers

Different from attackers considered before, the search-and-replace attackers' goal is to alter the sensors' data so that the falsified data can mislead the sink. In spite of the paramount importance of this authentication problem, only a few solutions are proposed [14, 56]. In [14], two novel authentication schemes, CoMAC and ExCo, were proposed to secure collected data. Unfortunately, the simple collaboration among the sensors in CoMAC and ExCo incurs more attacks, such as path-based DoS (PDoS) and false endorsement DoS (FEDoS) attacks [14]. Wood & Stankovic [56] work on the data antifraud problem in unattended tiered sensor networks. They employ the bucketing technique to ensure data confidentiality and also query result authenticity verification while ensuring efficient multidimensional queries processing.

Besides data security, mobile sink compromise and user privacy have been considered in the context of UWSNs. UWSNs are assumed to employ mobile sinks for periodic data collection and network maintenance. Compromise of a mobile sink is obviously quite dangerous. To this end, privilege restriction schemes were proposed to grant a mobile sink the least privilege while not impeding its ability to carry out its intended tasks. Also, a UWSN might be accessed by multiple clients (subscribers). To prevent malicious network operators from linking users with their data access patterns, some privacy-preserving schemes have been proposed to hide either client identity or client search interest and query patterns.

12.4.3   Security in Underwater Wireless Networks

Recently, for the development of ocean exploration and the needs of military combat, more research has been focused on underwater acoustic communication. Underwater sensor networks are a new field that combines the wireless network with acoustic communication technology, and it has been widely applied to oceanographic data collection, pollution monitoring, disaster prevention, and offshore exploration. Security is the basis of safe operation of UWNs. Security is vital to the acceptance and use of underwater sensor networks for many applications. Below, we analyze the major threats and attacks that strongly influence the security of UWNs.

12.4.3.1   Vunerablilities and Attacks in UWNs

The unique characteristics of the underwater condition make UWNs vulnerable to malicious attacks due to the high bit error rates, large and variable propagation delays, and low bandwidth of acoustic channels. Specifically, high bit error rates cause packet errors. Consequently, critical security packets can be lost. Wireless underwater channels can be eavesdropped on. Attackers may intercept the information transmitted and attempt to modify or drop packets. Malicious nodes can create out-of-band connections via fast radio (above the water surface) and wired links, which are referred to as wormholes. Since sensors are mobile, their relative distances vary with time. The dynamic topology of the UWNs not only facilitates the creation of wormholes, but also complicates their detection [54]. Since power consumption in underwater communications is higher than in terrestrial radio communications, and underwater sensors are sparsely deployed, energy exhaustion attacks to drain the batteries of nodes pose a serious threat for the network lifetime.

There are a large number of threats and attacks to which UWSNs are susceptible. They can be classified as data security attacks, DoS (denial of service) attacks, impersonation attacks, replication attacks, and physical attacks. Among these attacks, due to the special features of UWSNs, DoS attack is more destructive than others. Even if UWSN is well protected by an encryption algorithm, it is still threatened with DoS attack. A DoS attack can disrupt communication and cooperation between nodes and decrease availability of the whole network, wasting precious power. A DoS attack is low cost, deadly and hard to detect. A malicious adversary can cause great damage with very low cost. Malicious adversary impersonates a legal node to deceive neighbor nodes. In the power exhausttion attack, an attacker imposes a particularly complex task to a sensor node in order to shorten its battery life.

12.4.3.2   Security Techniques

Existing secure research in UWNs mainly focuses on some of the DoS attacks and defenses.

To reduce the jamming attack in UWNs, some solutions are already proposed for traditional wireless networks such as spread spectrum [56]. Frequency-hopping spread spectrum (FHSS) and direct-sequence spread spectrum (DSSS) in underwater communications are drawing optimum attention for their performance considering the noise and multipath interference. Another defense against jamming is to switch nodes to a lower duty cycle. Due to the limited energy, UWSNs mostly apply a sleep and wake-up scheme. Only when the nodes have to send or receive data, do the nodes wake up; other times; the nodes sleep to save energy.

The distributed visualization of wormhole (Dis-VoW) is proposed to detect wormhole attacks in three-dimensional underwater sensor networks [54]. In Dis-VoW, the distance between every sensor is collected, and then broadcast to its neighbors; every node is able to construct the local network topology, such as virtual outline within two hops using multidimensional scaling. Some nodes far away may appear as neighbors due to wormhole threats. The virtual outline at certain regions and the result of disagreement can be detected visualizing the virtual outline.

Bidirectional link verification can help protect against a hello flood attack, although it is not accurate due to node mobility and the high propagation delays of UWNs.

Authentication is also a possible defense. Authentication can be used to counter the selective forwarding attack, sybil attack, and sinkhole attack.

12.4.3.3   Open Questions

Although UWNs are attracting more and more research interest, many security issues are yet to be addressed.

12.4.3.3.1   Secure Routing

Secure routing is specially challenging in UWNs due to the large propagation delays, the low bandwidth, the difficulty of battery refills of underwater sensors, and the dynamic topologies [16]. Therefore, routing protocols should be designed to be energy aware, robust, scalable, and adaptive. Many routing protocols have been proposed for underwater wireless sensor networks. However, none of them have been designed with security as a goal. Although the attacks against routing in UWNs are mostly the same as in ground-based sensor networks, the same countermeasures are not directly applicable to UWNs due to their differences in characteristics. Some open issues in routing need to be addressed:

■ Develop new techniques against sinkholes and wormholes, and improve existing ones.

■ Quick and powerful encryption and authentication mechanisms against outside intruders should be devised for UWNs because the time required for intruder detection [24] is high due to the long and variable propagation delays, and routing paths containing undetected malicious nodes can be selected in the meantime for packet forwarding.

■ Sophisticated mechanisms should be developed against insider attacks such as selective forwarding, sybil attacks, hello flood attacks, and acknowledgment spoofing.

■ There is a need to develop reputation-based schemes that analyze the behavior of neighbors and reject routing paths containing selfish nodes that do not cooperate in routing. The proper functioning of these schemes is challenging because they do not work well in mobile environments, the time required to detect compromised nodes increases substantially in UWCNs due to the long propagation delays, and they must be adapted to tolerate short-term disruptions.

12.4.3.3.2 Secure Localization

Secure localization approaches proposed for ground-based sensor networks do not work well underwater because long propagation delays, multipath, and fading cause variations in the acoustic channel. Bandwidth limitations, node mobility, and sparse deployment of underwater nodes also affect localization estimation [16]. Localization schemes can be classified into range-based schemes and range-free schemes [42]. However, none of these localization schemes were designed with security. Some localization-specific attacks (replay attack, sybil attack, wormhole attack) are severe threats to UWNs. Therefore, effective cryptographic primitives against injecting false localization information in UWNs need to be developed. Moreover, it is necessary to design resilient algorithms able to determine the location of sensors even in the presence of sybil and wormhole attacks. In addition, techniques to identify malicious or compromised anchor nodes and to avoid false detection of these nodes are required.

12.4.3.3.3 Secure Time Synchronization

Secure time synchronization is essential in many underwater applications, such as coordinated sensing tasks. Also, scheduling algorithms such as time division multiple access (TDMA) require precise timing between nodes to adjust their sleep-wake-up schedules for power saving. Achieving precise time synchronization is especially difficult in underwater environments due to the characteristics of UWNs. For this reason, the time synchronization mechanisms proposed for ground-based sensor networks cannot be applied, and new mechanisms are required.

Time synchronization disruption due to masquerade, replay, and message manipulation attacks can be addressed using cryptographic techniques. However, countering other possible attacks, such as delays (deliberately delaying the transmission of time synchronization messages) and DoS attacks, requires the use of other strategies. A correlation-based security model for water quality monitoring systems has been proposed in [16] to detect outlier timestamps due to insider attacks. The authors prove that the acoustic propagation delays between two sensors in neighboring depth levels fit an approximately normal distribution, which means that the timestamps between them should correlate. However, this correlation is lost if a captured inside node is sending falsified timestamps. With proper design of a timestamp sliding window scheme, insider attacks are detected. However, identifying a neighbor node as malicious is difficult, because sometimes timestamps can be corrupted due to propagation delay variations caused by the channel rather than deliberately.

In all, because of the high and variable propagation delays of UWNs, the time required to synchronize nodes should be investigated. Efficient and secure time synchronization schemes with small computation and communication costs need to be designed to defend against delay and wormhole attacks.

12.4.3.3.4 Cross-Layer Security

Due to the underwater complex environment (e.g., water current, ambient noise, aquatic organism) and UWN constraints (e.g., node failure, new node joint to the network), network topology changed. Hence, the security system of UWN should be adaptive and selective to fit the demand of the network. Unfortunately, existing layered security designs are nonadaptive [19]. In practice, a malicious adversary would apply blended attacks to disrupt the network, and a layered security protocol cannot protect the network against cross-layer attacks. Hence, cross-layer security design is necessary. In addition, energy efficiency is an issue a cross all the network layers. It cannot be considered and solved in just one layer. The cost of security service is energy consumption. Cross-layer design can provide more safety services. Moreover, it can minimize the energy consumption by offering appropriate security service.

Thus, cross-layer security architecture for UWNs is absolutely necessary. The architecture includes: a cross-layer energy management scheme, cross-layer key management scheme, cross-layer authentication scheme, cross-layer intrusion detection scheme and cross-layer trust model.

1. Cross-layer energy management. Through cross-layer design, we can make a compromise between energy and safety to prolong the network lifetime.

2. Cross-layer key management. Due to limited energy, computation, and storage, the encryption should be simple and stable.

3. Cross-layer intrusion detection. Existing intrusion detection scheme are layered based. Most of them only consider intrusion detections in the MAC layer.

These schemes ignore physical layer attacks, which are hard to detect and fatal. If the channel is interfered, with the schemes based on the MAC layer cannot discover the problem. Hence, it is necessary to use a cross-layer scheme to detect intrusion.

4. Cross-layer authentication. Existing authentication schemes in UWNs are cross-layer techniques to defend various attacks. Hence, a well-designed cross-layer authentication can protect networks effectively.

5. Cross-layer trust model. A cross-layer trust model for UWNs can improve the system security and performance of UWNs, and efficiently defend against the attacks of malicious nodes.

12.4.4   Security in Wireless Multimedia Sensor Networks

A WMSN is composed of numerous multimedia sensors that exchange sensed data with the base station using a wireless channel. WMSNs are used in many application domains, such as surveillance systems, telemedicine, and so on. In order to ensure a broad deployment of such innovative services, strict requirements on security, privacy, and distributed processing of multimedia contents should be satisfied, also taking into account the limited technological resources (in terms of energy, computation, bandwidth, and storage) of sensor nodes. Thus, with respect to classic WSNs, the achievement of these goals is more challenging due to the presence of multimedia data, which usually requires complex compression and aggregation algorithms.

Many works in the literature concern the processing of images (compression, extraction, analysis, and aggregation), some works concern the problem to transmit multimedia contents by means of wireless networks, and other works are focused on real-time applications [7]. Notice that although privacy and security policies play a fundamental role in the context of WMSN applications, WMSN security is indeed a very young research field. According to Kundur et al. [30], so far researchers have focused on the problem of privacy in WMSN. Besides this, there are only a few more prior studies on the issue, mainly considering authentication, secure node localization, trust management, and data. We will introduce the study of these issues, respectively. Furthermore, the future direction of research for secure wireless multimedia sensor networks is then discussed.

12.4.4.1   Security Challenges in WMSNs

Though some of the previous security solutions for traditional WSNs can be easily adapted for WMSN, WMSN also has some novel features that stem from the fact that some of the sensor nodes will have video cameras and higher computation capabilities. This brings new security challenges as well as new protection opportunities. Here, we summarize the main security opportunities and challenges in WMSNs:

1. Opportunities:

   a. Sensor nodes in WMSN should be powerful enough to satisfy the computational and communication demands required to manage multimedia data. From the security point of view, more powerful nodes mean we could have the capability to employ more advanced solutions. In particular, we believe that the use of public key cryptographic solutions is feasible for most WMSN platforms [20].

   b. Sensor nodes (at least an important percentage of the sensor nodes in a WMSN) will have bigger storage capabilities than normal sensor nodes in a WMSN. That extra storage capability can be exploited to use security schemes that were not possible in WSN, e.g., the storage of other nodes' public keys and many session keys, signed information, nodes' reputation statistics, etc.

   c. Video sensing capability brings the opportunity to detect and identify the attacker in WMSN applications. Therefore, while specifying the threat model for WMSN, we should take this new capability into account.

2. Challenges:

   a. In addition to data delivery modes typically found in WSN, WMSN should support snapshot and streaming multimedia transmission [18]. Snapshot multimedia means that event-triggered observations are transmitted to the base station in a short time period. Snapshot delivery mode is arguably more vulnerable to DoS attacks, because if the snapshot information is blocked, it might not be detected by the base station. Digital streams have properties distinguishing them from regular messages.

   b. Privacy issues are of concern in WSNs, if the collected data are private and sensitive. Video, image, and audio data are typically more sensitive than scalar data, such as temperature. Hence, privacy enhancing techniques, such as source location, hiding, and distributed visual secret sharing may be crucial for WMSNs.

   c. Bandwidth demand to carry uncompressed multimedia data is excessive for most scenarios. Compressing data with existing encoders uses complex algorithms and entails high energy consumption, which is not feasible on most power-constrained sensor nodes. For WMSNs, reducing the amount of data using multimedia in-network processing techniques (distributed compression, distributed itering, etc.) is possible. Nevertheless, this requires a different trust model and new security architecture; i.e., all sensor nodes participating in distributed and collaborative processing should be trustworthy.

d. High-speed requirements for carrying multimedia data draw an interest in free-space optical (FSO) communication. There are security aspects of networking issues unique to FSO WMSNs [31].

12.4.4.2 Privacy in WMSNs

Privacy is a key requirement for numerous application scenarios of WMSNs. As an example, consider the systems of telemedicine or military surveillance. In both cases, data are sensitive and are required to be adequately protected. In other words, in such contexts it is fundamental to guarantee the confidentiality of the communication among the nodes within the network and between the nodes and the sink.

There are a lot of privacy-related problems, however, a few of which have been considered in WMSNs. Czarlinska [1] proposed a novel paradigm for securing privacy and confidentiality in a distributed manner. The suggested paradigm is based on the control of dynamical systems, and it requires low complexity for the processing and communication. Czarlinska et al. [10] then presented attacks that affect the data privacy in visual sensor networks and proposed privacy-promoting security solutions established upon a detected adversary using a game theoretic analysis and keyless encryption. Czarlinska and Kundur [11] investigated the event acquisition properties of wireless image sensor networks, which include different techniques at the camera nodes to recognize the differences between event and nonevent frames in risky environments that are prone to security attacks.

Another area that is receiving increasing interest from the research community is free-space optical (FSO) sensors [25]. Although FSO communications are very interesting for WMSNs due to their high bandwidth support, their application in realistic scenarios is challenging. Since they require LoS (line-of-sight) communications, uneven terrain and obstacles found in real-life deployments could reduce the number of links. Weather conditions, such as fog, rain, heavy snow, and the sun, might render a whole network useless.

Since the use of FSO completely changes the nature of the routing protocols, new security solutions are needed. OPSENET is a novel secure routing protocol for FSO WMSNs that prevents outsider attacks via lightweight cryptographic mechanisms [31]. In OPSENET, global picture of the network is established at the base station, which can be employed as a watchdog. This provides an effective means to identify some of the most devastating insider attacks [31]. Another interesting new concept is distributed visual secret sharing, introduced in [31], in which the images collected by the sensor nodes are used to generate a large number of copies with a large amount of noise and distributed to different nodes of the network. Attackers with only a few copies will not be able to generate the original image. But, the base station will be able to generate it if it receives a sufficient number of copies.

At present, the existing solutions that guarantee the privacy of data in the context of WMSNs are still in a primitive state, and many open problems still exist and are yet to be discovered; hence, further research work is required in this field.

12.4.4.3 Authentification in WMSNs

At present, no solution to image data authentication is specifically deployed for WM-SNs. Therefore, new approaches exploiting the characteristics of multimedia nodes should be developed. In particular, it is now possible to think of new solutions based on public key digital signature schemes. The capabilities of multimedia nodes can be used to simplify the authentication procedures and to realize more effective key management approaches based on asymmetric cryptography. This research direction could be supported by nanotechnology-based solutions, which promise to greatly enhance the capabilities of sensor nodes. Some schemes have been proposed in the field of traditional WSNs, but they cannot be effectively applied to WMSNs due to the memory and processing limitations of sensors. On the contrary, WMSN sensors should be able to support new solutions based on multimedia content (e.g., schemes based on watermarking).

12.4.4.4 Secure Node Localization in WMSNs

According to the proposed vision of an integrated design approach, the problem of the authentication is strictly related to the secure node localization issue, given that authentication can be used to ensure reliable information. Due to the distributed nature of WMSNs, in several application scenarios the localization of the multimedia sensors is required to assure the supply of the services. Therefore, the integrity and confidentiality of localization information are fundamental, and it is necessary to define countermeasures versus possible malicious attacks. Authentication mechanisms can improve the security of the localization information, but they are not enough to guarantee the complete reliability of the contained information. Moreover, although further approaches were defined to address such an issue, they do not adequately satisfy some requirements, such as the real-time scheduling constraints imposed by some multimedia applications.

12.4.4.5 Trust Management in WMSNs

Issues of authentication and secure localization information fall in the general problem of trust management. In a distributed and collaborative environment like a WMSN, trust management becomes a real challenging aspect. The analysis of the trust relationships among the components of a network drives one to choose ad hoc security-oriented countermeasures that aim at guaranteeing the protection of data, the secure routing, the exchange of localization information, and so on. However, the definition of an effective model of trust becomes a complex task in a highly distributed environment characterized by strict performance requirements. Each node should be equipped with an autonomous evaluation and analysis capabilities that aim at measuring the trust relationships with the other members of the network; notice that such relationships depend on the communication and cooperation needs of the nodes. In other words, it is required to move from the classic centralized and static approach proposed for the most widely used trust management solutions, to a fully distributed and dynamic approach that assumes that no trust relationship is defined a priori among the nodes of the network. At present, no trust management schemes are proposed for WMSNs due to the relevant computational effort required by the multimedia traffic and to the real-time constraints that are not suited to the limited power resources of current sensor nodes.

12.4.4.6 Security in Data Aggregation for WMSNs

In WMSNs, data include multiple compressed frames coming from different video sensors. In fact, due to the distributed elaboration of the multimedia contents and the limited bandwidth and power resources of WMSNs, it is necessary to introduce secure aggregation algorithms that decrease the total amount of information to elaborate, transmit, and protect at the same time the quality of the multimedia message. Lazos & Poovendran [32] propose a survey on the most important solutions, but they can be hardly applied to multimedia data. For instance, the encryption of images is a highly power consuming task; hence, the most innovative solutions come to the arrangement to adopt selective encryption schemes for the multimedia contents.

12.4.4.7 Security Challenges for Future WMSNs

Visual surveillance is an application where the industry (both civilian and military) will be willing to invest. The future of security in WMSNs is going to be application and environment dependent. In the case of video surveillance and monitoring, most of the time it can be assumed that all the nodes are trusted initially but later on can be compromised. In this scenarios, DoS is going to be one of the most common attacks. Other problems, like resilience to traffic analysis and compromised nodes, are also going to be of utmost importance.

1. DOS attacks. DoS will be the main attack to worry about in most of the WMSN scenarios. One advantage of WMSN nodes is that they are going to be more powerful than scalar nodes, which translates into more computational power for thwarting DoS attacks. Nevertheless, attackers can always bring even more powerful computing resources to perform their attacks. DoS attacks can be performed in many different ways, and against any of the different communication layers (i.e., physical, link, network, transport, and application layers). In addition, protecting only against a subset of them is obviously useless.

   In an attempt to detect DoS attacks—including currently unidentified ones—IDS monitors have been designed [24]. These monitors try to find out (1) nodes that are not behaving according to certain parameters that characterize what is considered as normal behavior and (2) nodes that are performing a well-known attack. Unfortunately, it is complicated to design them in such a way that the number of false alarms is low enough and identification of new attacks comes in a timely manner.

2. Traffic analysis attacks. A very common type of attack (just second to DoS) is the traffic analysis attacks. Moreover, traffic analysis attacks are not limited

to trying to eavesdrop on the information the sensor network is transmitting (which can be easily prevented by the use of encryption). They can actually go much further. Two main traffic analysis countermeasures against the attacks are concealing the geographical position of the sink and achieving event source unobservability [20].

3. Compromised nodes attacks. Compromised nodes in MWSNs are trying to disrupt the normal network operation. Nevertheless, among the most serious damage that a compromised node can do in a WMSN: is eavesdrop on what the sensor network is detecting. It is just like tapping into the video surveillance security camera system of a building you are trying to rob. In order to prevent an attacker (by compromising a very small number of intelligently selected sensor nodes) of a WMSN from getting a fairly clear picture of what the network is sensing, schemes that ensure that each of the nodes only has very partial information would prove very valuable. This is a very complex and exciting research area, and it will receive a considerable amount of attention in the near future.

12.5   Conclusion

In this chapter we studied the security issues in wireless networks including the 4G system, ad hoc networks, and wireless sensor networks. These issues highlight some of the latest research results in this area and present some novel and innovative security and privacy techniques especially used in wireless networks. These issues summarize the current state-of-the-art research and provide valuable insights into future directions and challenges in the field.

References

1. U. Acharya and M. Younis. Increasing base-station anonymity in wireless sensor networks. *Journal of Ad Hoc Networks*, 8(8):791–809, 2010.

2. A. Alomari and M. L. Mihailescu. Improvement authentication of routing protocols for mobile ad hoc networks. *Journal of Advances in Computer Networks and Its Security*, 2(1):83–88, 2012.

3. R. Bista and J. W. Chang. Privacy-preserving data aggregation protocols for wireless sensor networks: a survey. *Journal of Sensors*, 10(5):4577–4601, 2010.

4. H. K. Bokharaei, A. Sahraei, Y. Ganjali, R. Keralapura, and A. Nucci. You can spit, but you can't hide: spammer identification in telephony networks. *Proceedings of IEEE INFOCOM*, 2011, pp. 41–45.

5. E. Bou Harb. *A distributed architecture for spam mitigation on 4G mobile networks.* PhD thesis, Concordia University, 2011.

6. B. Carbunar, Y. Yu, W. D. Shi, et al. Query privacy in wireless sensor networks. *Journal of ACM Transactions on Sensor Networks (TOSN),* 6(2):14, 2010.

7. M. Chen, V. C. M. Leung, L. Shu, and H. Chao. On multipath balancing and expanding for wireless multimedia sensor networks. *Journal of Ad Hoc and Ubiquitous Computing,* 9(2):95–103, 2012.

8. S. Chen and K. Zeng. Hearing is believing: detecting mobile primary user emulation attack in white space. *Proceedings of IEEE INFOCOM,* 2011, pp. 36–40.

9. W. Conner, T. Abdelzaher, and K. Nahrstedt. Using data aggregation to prevent traffic analysis in wireless sensor networks. *Proceedings of the International Conference on Distributed Computing in Sensor Systems,* 2006, pp. 202–217.

10. A. Czarlinska, W. Huh, and D. Kundur. On privacy and security in distributed visual sensor networks. *Proceedings of Image Processing,* 2008, pp. 1692–1695.

11. A. Czarlinska and D. Kundur. Wireless image sensor networks: event acquisition in attack-prone and uncertain environments. *Journal of Multidimensional Systems and Signal Processing,* 20(2):135–164, 2009.

12. J. Deng, R. Han, and S. Mishra. Countermeasures against traffic analysis attacks in wireless sensor networks. *Proceedings of Security and Privacy for Emerging Areas in Communications Networks,* 2005, pp. 113–126.

13. R. Di Pietro, G. Oligeri, C. Soriente, and G. Tsudik. Intrusion-resilience in mobile unattended WSNS. *Proceedings of IEEE INFOCOM,* 2010, pp. 1–9.

14. R. Di Pietro, C. Soriente, A. Spognardi, and G. Tsudik. Collaborative authentication in unattended wsns. *Proceedings of the Second ACM Conference on Wireless Network Security,* 2009, pp. 237–244.

15. R. Di Pietro and N. V. Verde. Epidemic data survivability in unattended wireless sensor networks. *Proceedings of the Fourth ACM Conference on Wireless Network Security,* 2011, pp. 11–22.

16. M. Domingo. Securing underwater wireless communication networks. *Journal of Wireless Communications,* 18(1):22–28, 2011.

17. D. Forsberg and W. D. Moeller. *LTE security.* Wiley-Interscience Press, 2011.

18. R. Gennaro and P. Rohatgi. How to sign digital streams. *Journal of Information and Computation,* 165(1):100–116, 2001.

19. A. Gkikopouli, G. Nikolakopoulos, and S. Manesis. A survey on underwater wireless sensor networks and applications. *Proceedings of Control and Automation (MED),* 2012, pp. 1147–1154.

20. M. Guerrero-Zapata, R. Zilan, K. Bicakci, et al. The future of security in wireless multimedia sensor networks. *Journal of Telecommunication Systems*, 45(1):77–91, 2010.

21. A. Hamieh and J. Ben-Othman. Detection of jamming attacks in wireless ad hoc. *Proceedings of IEEE ICC*, 2009.

22. P. M. Jawandhiya, M. M. Ghonge, M. S. Ali, and J. S. Deshpande. A survey of mobile ad hoc network attacks. *Journal of Engineering Science and Technology*, 2(9):4063–4071, 2010.

23. Y. Jian, S. G. Chen, Z. Zhang, and L. Zhang. Protecting receiver-location privacy in wireless sensor networks. *Proceedings of IEEE INFOCOM*, 2007, pp. 1955–1963.

24. A. K. Jones and R. S. Sielken. Computer system intrusion detection: a survey. *Technique report. University of Virginia Computer Science Department*, 2000.

25. P. Kamat, W. Y. Xu, W. Trappe, and Y. Y. Zhang. Temporal privacy in wireless sensor networks. *Proceedings of Distributed Computing Systems*, 2007, pp. 23–23.

26. P. Kamat, Y. Y. Zhang, W. Trappe, et al. Enhancing source-location privacy in sensor network routing. *Proceedings of Distributed Computing Systems*, 2005, pp. 599–608.

27. D. Kheyri and M. Karami. A comprehensive survey on anomaly-based intrusion detection in MANET. *Journal of Computer and Information Science*, 5(4):132–139, 2012.

28. M. Knysz and H. Xin. Open WiFi networks: lethal weapons for botnets. *Proceedings of IEEE INFOCOM*, 2012, pp. 2631–2635.

29. S. Kumari and M. Shrivastava. A study on the security and routing protocols for ad-hoc network. *Journal of Advanced Trends in Computer Science and Engineering*, 1(3):2278–3091, 2012.

30. D. Kundur, W. Luh, U. N. Okorafor, and T. Zourntos. Security and privacy for distributed multimedia sensor networks. *Proceedings of the ICDCS*, 2008, pp. 112–130.

31. D. Kundur, W. Luh, U. N. Okorafor, and T. Zourntos. Security and privacy for distributed multimedia sensor networks. *Journal of Proceedings of the IEEE*, 96(1):112–130, 2008.

32. L. Lazos and R. Poovendran. Serloc: robust localization for wireless sensor networks. *Journal of ACM Transactions on Sensor Networks (TOSN)*, 1(1):73–100, 2005.

33. N. Li, N. Zhang, S. K. Das, and B. Thuraisingham. Privacy preservation in wireless sensor networks: a state-of-the-art survey. *Journal of Ad Hoc Networks*, 7(8):1501–1514, 2009.

34. W. J. Li and A. Joshi. Security issues in mobile ad hoc networks—a survey. *The 17th White House Papers, Graduate Research in Informatics at Sussex*, 2008, pp. 1–23.

35. Z. H. Liu and W. Y. Xu. Zeroing-in on network metric minima for sink location determination. *Proceedings of the Third ACM Conference on Wireless Network Security*, 2010, pp. 99–104.

36. D. Ma, C. Soriente, and G. Tsudik. New adversary and new threats: security in unattended sensor networks. *Journal of Network*, 23(2):43–48, 2009.

37. D. Ma and G. Tsudik. DISH: Distributed self-healing. *Proceedings of International Symposium on Stabilization, Safety, and Security of Distributed Systems*, 2008, pp. 47–62.

38. S. Madhavi. An intrusion detection system in mobile ad hoc networks. *Proceedings of Information Security and Assurance*, 2008, pp. 7–14.

39. S. Menaria, S. Valiveti, and K. Kotecha. Comparative study of distributed intrusion detection in ad-hoc networks. *Journal of Computer Applications*, 8(9):11–16, 2010.

40. D. Mishra, D. Kumar, N. Koria, et al. A secure multi-party computation protocol for malicious computation prevention for preserving privacy during data mining. *Journal of Computer Science and Information Security*, vol. 3, no. 1, pp. 0908–0913, 2009.

41. R. Molva and P. Michiardi. Security in ad hoc networks. *Journal of Personal Wireless Communications*, vol. 4, no. 2, 756–775, 2003.

42. M. Moradi, J. Rezazadeh, and A. S. Ismail. A reverse localization scheme for underwater acoustic sensor networks. *Journal of Sensors*, 12(4):4352–4380, 2012.

43. T. Han, N. Zhang and K. M. Liu. Analysis of Mobile WiMAX Security: Vulnerabilities and Solutions. *Proceedings of IEEE MASS*, 2008.

44. T. Nguyen. *A survey of WiMax security threats*. White paper. *Computer Science Department, Washington University*, 2009.

45. Y. Park and T. Park. A survey of security threats on 4G networks. *Proceedings of GLOBECOM Workshops*, 2007, pp. 1–6.

46. N. Seddigh, B. Nandy, R. Makkar, and J. F. Beaumont. Security advances and challenges in 4g wireless networks. *Proceedings of Privacy Security and Trust (PST)*, 2010, pp. 62–71.

47. C. Sreedhar, S. M. Verma, and N. Kasiviswanath. A survey on security issues in wireless ad hoc network routing protocols. *Journal on Computer Science and Engineering*, 2(2):224–232, 2010.

48. D. Sterne, P. Balasubramanyam, D. Carman, et al. A general cooperative intrusion detection architecture for MANETS. *Proceedings of Information Assurance*, 2005, pp. 57–70.

49. Y. Sun, Z. Han, and K. J. R. Liu. Defense of trust management vulnerabilities in distributed networks. *Journal of Communications Magazine*, 46(2):112–119, 2008.

50. K. Tan, G. H. Yan, J. W. Yeo, and D. Kotz. Privacy analysis of user association logs in a large-scale wireless LAN. *Proceedings of IEEE INFOCOM*, 2011, pp. 31–35.

51. J. Tang, Y. Cheng, and Y. Hao. Detection and prevention of SIP flooding attacks in voice over IP networks. *Proceedings of IEEE INFOCOM*, 2012, pp. 1161–1169.

52. J. Tang, Y. Cheng, and W. H. Zhuang. An analytical approach to real-time misbehavior detection in IEEE 802.11 based wireless networks. *Proceedings of IEEE INFOCOM*, 2011, pp. 1638–1646.

53. A. Ukil. Privacy preserving data aggregation in wireless sensor networks. *Proceedings of Wireless and Mobile Communications (ICWMC)*, 2010, pp. 435–440.

54. W. C. Wang, J. J. Kong, B. Bhargava, and M. Gerla. Visualisation of wormholes in underwater sensor networks: a distributed approach. *Journal of Security and Networks*, 3(1):10–23, 2008.

55. W. Wei, F. Y. Xu, and Q. Li. Mobishare: flexible privacy-preserving location sharing in mobile online social networks. *Proceedings of IEEE INFOCOM*, 2012, pp. 2616–2620.

56. A. D. Wood and J. A. Stankovic. A taxonomy for denial-of-service attacks in wireless sensor networks. *Handbook of sensor networks: compact wireless and wired sensing systems*, CRC Press, 2004, pp. 739–763.

57. C. Xenakis, C. Panos, and L. Stavrakakis. A comparative evaluation of intrusion detection architectures for mobile ad hoc networks. *Journal of Computers and Security*, 30(1):63–80, 2011.

58. Y. Yang, M. Shao, S. C. Zhu, et al. Towards event source unobservability with minimum network traffic. *Proceedings of Wireless Network Security (WiSec)*, 2008.

59. M. G. Zapata and N. Asokan. Securing ad hoc routing protocols. *Proceedings of the 1st ACM Workshop on Wireless Security*, 2002, pp. 1–10 .

60. K. Zeng, K. Govindan, D. Wu, and P. Mohapatra. Identity-based attack detection in mobile wireless networks. *Proceedings of IEEE INFOCOM*, 2011, pp. 1880–1888.

61. J. Zhang, L.Y. Fu, and X. B. Wang. Impact of secrecy on capacity in large-scale wireless networks. *Proceedings of IEEE INFOCOM*, 2012, pp. 3051–3055.

62. Y. G. Zhang, W. K. Lee, and Y. A. Huang. Intrusion detection techniques for mobile wireless networks. *Journal of Wireless Networks*, 9(5):545–556, 2003.

63. L. D. Zhou and Z. J. Haas. Securing ad hoc networks. *Journal of Network*, 13(6):24–30, 1999.

64. A. Zugenmaier and H. Aono. Security technology for SAE/LTE. *Journal of NIT DOCOMO*, 11(3):27–30, 2010.

Chapter 13

Content Dissemination and Security in Device-to-Device (D2D) Communication

Aiqing Zhang, Jianxin Chen, and Liang Zhou

*Key Lab of Broadband Wireless Communication and Sensor Network Technology
(Nanjing University of Posts and Telecommunications), Ministry of Education, China
A. Zhang is also with the Anhui Normal University, China*

CONTENTS

Device-to-device (D2D) communication has been proposed as a promising data offloading solution in future networks. The key technologies, including peer discovery, mode selection, resource allocation, power control, and interference management, have been widely investigated in the research community, while data dissemination and security solutions are rarely directly considered. In this chapter, we address the above two issues by reviewing the current contributions in other wireless networks, which may shed a light on constructing the corresponding mechanisms in a D2D communication system. Furthermore, we put forward the challenges and open issues for future research.

13.1   Introduction

Generally, D2D communication refers to the technologies that enable devices to communicate directly without an infrastructure of access points or base stations and the involvement of wireless operators [1]. Especially, and in most applications, D2D refers to device-to-device communication under the control of a cellular network. Due to its inherent characteristics, e.g., improving resource utilization, enlarging users' throughput, extending battery lifetime, etc. [2], it has been proposed as a new data offloading solution and spectrum efficiency enhancement method, thus gaining substantial attention recently in the research community. Different from traditional device-to-device communication techniques, such as Bluetooth or Wi-Fi Direct, the D2D communication works on a licensed band, which promises to provide a planned environment instead of an uncoordinated one, resulting in better user experience. Thereafter, it is prospective to implement content dissemination over device-to-device (D2D) communication under the assistance of cellular networks [3–6].

However, with the particular characteristics and heterogeneity of the D2D systems as well as limited computational capacity and energy of the devices, data dissemination is never an easy task in D2D communication. Up to this end, in the wireless communication literature, content dissemination in D2D systems has not been directly considered, while in their counterparts, such as mobile social networks (MSNs) [7], vehicle ad hoc networks (VANETs) [8], and other wireless networks [9–11], there are abundant studies, which may draw experiences for D2D communication. On the other hand, as the connections happen directly between the proximity devices, the D2D communication could be subject to many security threats, such as modification and fabrication of the data, violation of the user's privacy, and so on. Evidently, any malicious behavior of users may cause serious consequences and lead to deteriorating user experience. Furthermore, availability must be achieved in the sense that users might be frustrated if the services are intermittent or they suffer from long waiting times for sharing the content. Therefore, it is ultimately important to develop some elaborate and carefully designed security mechanisms to achieve security and availability objectives in D2D communication before its practical implementation. In order to address the two issues above, this chapter focuses on data dissemination and security mechanisms in D2D communication. The contributions can be concluded as threefold.

First, we give the network architecture of D2D communication and identify the main components. Moreover, the technical specifications are also addressed.

Second, in order to take a step forward for designing content dissemination protocols in the D2D system, we propose a system architecture and then draw references from the data dissemination strategies in other wireless networks. Furthermore, the challenges and open issues are put forward.

Last but not least, security solutions are shaped by clarifying security requirements, threat models, and proposing possible approaches with reviewing security schemes in WBANs and VANETs.

13.2 D2D Communication Technical Specifications

This section focuses on the network architecture and key technologies for LTE-A-based D2D communication.

13.2.1 Network Architecture

D2D communication as an underlay to the LTE-Advanced (LTE-A) network is a technology in which adjacent UEs under a cellular communication network set up a D2D link using a cellular interface. This is done by directly exchanging data through the D2D link. Extensively, a D2D communication system includes three types of entities: gateway (GW) and eNB of a cellular network and user equipment (UE) of cellular users. Their relationship is presented in Figure 13.1, which shows an example of traffic traversing in the LTE-A system [6].

GW: GW, which serves as the gate from the local subsystem to the core network, is able to route IP packets from and to the internet, as well as detecting the potential D2D traffic by proximity service control function (PSCF). The PSCF earmarks the traffic flows and finds out pairs of D2D-enabled devices.

eNB: The eNB is a very important element in E-UTRA (Evolved Universal Terrestrial Radio Access) of the LTE-A network, as it is responsible for resource allocation of the cellular network as well as coordination of D2D devices by ensuring two peers meet in space, time, and frequency. The eNB can also control the transmit power of the cellular users to limit the interference and implement user authentication in a cellular network.

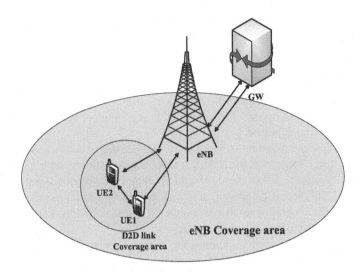

Figure 13.1: System model for D2D communication.

UE: UEs[1] are the peer entities of the D2D connections. The data dissemination protocols are mainly designed to share information among users by D2D communications without increasing additional traffic load to the cellular network.

In conventional cellular communications, UEs communicate via radio area uplink, core network uplink, core network downlink, and radio area downlink despite the proximity to each other. D2D-enabled UEs, however, can communicate directly with other D2D-enabled UEs using the D2D link. D2D communication can handle data traffic without the core network load and the additional radio network load [6].

13.2.2   Key Technologies of D2D Communication

In order to realize the potential gains of D2D communication, it is important to address the key technologies, including peer discovery, mode selection, resource allocation, power control, and interference management.

> *Peer discovery.* Before the D2D connection is established, the devices in proximity must find each other, which is called peer and service discovery. Generally, there are two kinds of approaches for peer discovery: centralized and distributed methods. In the centralized approach, a certain entity in the cellular network, e.g., Packet Data Network (PDN) gateway or Mobility Management Entity (MME), detects that it may be better for two communicating UEs to set up a D2D connection. This entity then informs the eNB to

---

[1]In most D2D communication systems, UEs refer to mobile phones, which are ubiquitous handheld devices with functionalities varying from communication of voice data to transmission of video streaming.

request measurements from the UE to check if the D2D communication offers higher throughput. If so, the eNB decides that the two UEs can communicate in D2D mode [12]. The distributed approach does not need the involvement of the base station while it is typically time- and energy-consuming since the UE broadcasts identity periodically so that other UEs may be aware of its existence and decide whether they shall start a D2D communication with it [13].

Mode selection. Mode selection refers to the selection between the traditional infrastructure path and D2D path. Usually, mode selection is combined with resources allocation [14]. Specifically, the eNB collects up-to-date information on channels, buffer status, and traffic load, selects the mode, and allocates resources (e.g., schedules and power for UL/DL) on a timescale that is similar to LTE-A's scheduling and transmission time interval. It is necessary to note that the automatic switching between cellular and D2D connection should be reliable and seamless to guarantee user satisfaction.

Resource allocation. Resource allocation in D2D communication can be considered an optimization mechanism aiming to improve the performance of energy consumption, throughput, and interference by methods of scheduling, mode selection, and power control. Basically, resource allocation of D2D connections can be either distributively determined by the UEs themselves or centrally performed by the eNB in LTE-A systems. Specifically, distributed resource allocation schedules a channel state-aware maximal independent set at any given time slot based on the current traffic and channel condition, while in centralized resource allocation, the eNB has full control over the resources allocated to each D2D connection and needs to inform the D2D UEs of the scheduled resources for data transmission via physical downlink control channel (PDCC) control signaling [1, 15].

Power control and interference management. Interference management in a cellular environment with D2D links is a critical issue, as the interference from D2D links can be expected to reduce the cellular capacity and efficiency. Due to the strong near-far effect, power control plays a key role in mitigating intracell interference [16]. Particularly, D2D links may find short time intervals and frequency proportions, or the eNB may assign dedicated physical resource blocks for D2D connections [17].

Especially, mode selection, resource allocation, and power control algorithms are usually considered jointly in order to achieve the optimal performance. Additionally, interference may be used in physical layer security mechanisms to improve secrecy capability, and access control may be implemented during the process of peer discovery.

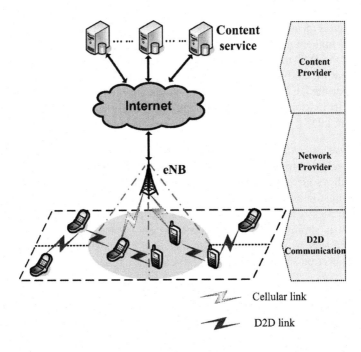

Figure 13.2: System architecture of data dissemination for D2D communication.

13.3 Content Dissemination

As mentioned above, content dissemination is an advantageous application in D2D communication systems. This section describes a typical system architecture of content dissemination in D2D communication and analyzes content dissemination strategies in wireless networks, which sheds a light on the mechanism formation in D2D.

13.3.1 System Architecture of Content Dissemination

Motivated by [3], a system architecture of content dissemination is proposed, displayed in Figure 13.2. Generally, the architecture is composed of three layers: content provider, network provider, and D2D communication layer. The content provider layer possesses various materials that are intended to be disseminated among the devices. In the network provider layer, eNB works as an internet gateway to link devices and content providers. In the D2D communication layer, the devices may communicate directly with each other for data sharing or search the eNB for getting the content from the Internet.

13.3.2   Content Dissemination Strategies

Since there are few content dissemination strategies directly considered in D2D systems, some protocols can be obtained by extending current algorithms [3, 7, 18, 20–22] that are known to address data dissemination problems in wireless networks. Specifically, the authors in [3] study the heterogeneous media provision in peer-to-peer (P2P)-based vehicular networks and develop fully dynamic service schemes with the goals of maximizing the total user satisfaction and achieving a certain amount of fairness. Unlike prior works that target bandwidth or demand fairness, they propose a media-aware satisfaction fairness strategy, which is aware of the characteristics of user satisfaction and media content and ensures max-min satisfaction fairness sharing among multiple vehicles. Importantly, both schemes are designed in a distributed manner, so it is helpful to shed insight on protocol design and provide fundamental guidelines on establishing an efficient data dissemination scheme in D2D communication systems.

In [20], the authors evaluate the benefits of a hybrid system that combines a peer-to-peer and a centralized client-server approach against each method acting alone. In particular, they employ a word-of-mouth demand evolution model due to Bass [23] to represent the evolution of interest in a piece of content. This paper also studies the relative performance of peer-to-peer and centralized client-server schemes, as well as a hybrid of the two—from the point of view of both consumers and the content distributor. What is most interesting is that the work models demand for a file as a function of time using the Bass diffusion model, representing the temporal evolution of popular files. Consequently, the results provide a guidance to understand how server provisioning affects performance of content distribution in D2D communication systems. Moreover, the contribution, which calculates the average per user delay in each setting and explicitly characterizes the extent to which the hybrid approach reduces the capacity required to achieve a target delay, is heuristic for the mechanism design of data dissemination in D2D communication systems as a cellular network has central control over the D2D link.

In order to study the dissemination of dynamic content over a mobile social network, [7] defines a global fairness objective to determine how the service provider can allocate its bandwidth optimally to make the content at users as fresh as possible. Additionally, the authors specify a condition under which the system is highly scalable; that is, even if the total bandwidth dedicated by the service provider remains fixed, the expected content age at each user grows slowly with the number of users. It is expected that D2D communication applications could be explored with the model, as it essentially outlines how to compute a user's sensitivity to the injection rates of other users, in a distributed manner. In addition, as dynamic content sharing is also an important application in D2D communication, this research provides us with a good direction.

In [21] and [22], the authors address the data dissemination problem in vehicle ad hoc networks. Specifically, [21] focuses on push-based data dissemination, where the data can be efficiently delivered from moving vehicles or fixed stations to other vehicles, while [22] studies the pull-based data dissemination/access, where a

Table 13.1 Data Dissemination Strategies and Useful Experiences for D2D

Literature	Context in USE	Contributions	Experiences for D2D
[3]	P2P-based VANETs	Dynamic schemes, considering user satisfaction and fairness	Distributed manner, dynamic content dissemination
[20]	Internet	Evaluating the benefits of a hybrid system combining peer-to-peer and centralized approaches	Hybrid approach, content demand model
[7]	MSNs	Dynamic content, considering fairness and scalability	Distributed manner, dynamic content dissemination
[21, 22]	VANETs	Considering data-delivery delay and dissemination capacity	Distributed manner, dynamic content dissemination

vehicle is enabled to query information about specific targets. Generally, the push-based approach is used to disseminate data that are useful for many people, whereas the pull-based approach is used to query data that are specific for some user. Furthermore, the authors provide analytical models to explore the dissemination capacity (DC) of the proposed schemes, which provide guidelines on choosing the system parameters to maximize the DC under different delivery ratio requirements. In practice, a hybrid of push and pull can be used to improve the system performance.

Table 13.1 summarizes the contributions of the literature listed above and provides useful experiences or references for data dissemination mechanism design in D2D.

13.3.3 Challenges and Open Issues

Although data dissemination has been studied in the database community and the network community, addressing the content dissemination problem in a D2D communication system is challenging for a variety of reasons.

Device (user) mobility. The fact that the users follow various mobility directions and velocities, results in different contact opportunities for content transmission. Even worse, the mobility parameters are a priori unknown. In this sense, any practical content dissemination policy cannot obtain this information directly, which further complicates this problem [24]. Therefore, it is vital to investigate the impacts of network dynamics on content dissemination.

Content diversity and storage capacity. Users may not necessarily be interested in the same content; that is, different users usually have diverse content demands. Moreover, even for this same content, its priority level for different users may be different as well. Meanwhile, users may frequently transmit the popular contents and store the contents that are likely to be of use by either themselves or other users in proximity, while the devices typically have limited storage capacity. Accordingly, determining the content transmission and store fashions becomes a very difficult part of content dissemination in a D2D communication system.

Distributed implementation. Although in a D2D communication system eNB has full control over the connections, practically, a D2D link is established for freeing up the cellular network from heavy data loading. If the data dissemination policy is centrally implemented, the advantages of D2D communication will disappear. Alternatively, similar to traditional peer-to-peer systems, mobile users expect to transmit and store contents without the control of any central authority for security purposes. Consequently, distributed implementation of the content dissemination is another significant challenge.

User experience. Quality of experience (QoE), defined by the International Telecommunication Union (ITU) as "overall acceptability of an application or particular service, as perceived subjectively by the end users," has been widely accepted by researchers and industry for its appealing achievements [25]. However, the parameters that affect QoE are complicated and sometimes unmeasurable, which causes a harsh task to construct the estimation model. In D2D systems, things are worse as the network is heterogeneous, the demand of the users is diversity, and the operators usually pose incomplete information for the customers. As a result, it remains a big challenge to design advantageous data dissemination protocols with an accurate QoE prediction model.

13.4 Security in D2D Communication

Security is a basic requirement for D2D communication before putting it into practical applications. In this section we first identify the potential threats and security requirements in D2D communication. Then the possible security solutions in D2D communication are presented and open issues are illustrated.

13.4.1 Threat Models

Generally, D2D communication is wireless, which may introduce a number of security vulnerabilities. The possible attacks are listed as follows:

Eavesdropping. Due to the open nature of the D2D link, information is sent via a broadcast that anyone can receive. Thus, the unauthorized users may eavesdrop on the in-transmit message.

Data deceit attack. As data fabrication or alternation can be performed by impersonation, the correctness of application data is a major vulnerability. The adversary, modifying the source data or forging the material, is supposed to share the fake information with other users. Alternatively, the contaminated devices infect their neighbors without conscious, which causes large portions of the peers to be occupied by the epidemic.

Free-riding attack. Some UEs, which receive data from their pairs, may not be willing to share the material with others since the data transmission process is energy consumption. Such selfish behavior is referred to as a free-riding attack, which causes a serious threat and reduces the system availability in D2D communication.

Privacy violation. Some privacy-sensitive information, such as the user's name, SIM card number, and position, could be intentionally derived so that personal privacy is jeopardized.

Denial of service (DoS) attack. The adversary sends a bulk of irrelevant messages to take up the channel and consume the computational resources of the other users. It is characterized by an explicit attempt by attackers to prevent legitimate users of a service from using that service.

13.4.2 Security Requirements

For mitigation of the potential threats to the greatest possible extent, D2D communication should meet the following security requirements:

Data confidentiality. Confidentiality ensures that the data are well protected and not revealed to unauthorized users. Even though the message may be eavesdropped by the unintended receiver, the confidentiality objective, which is achieved by end-to-end encryption, is expected to thwart the thief from getting any useful information.

Data authority and correctness. The original authority message should not be altered in the delivery and can be detected by the receiver if a modification incident happens, which means any false data fabricated by the adversary should be easily found out upon reception.

Entity authentication. Entity authentication techniques verify the identity of the devices in communications and distinguish legitimate users from unauthorized ones.

Non-repudiation. Non-repudiation is necessary to prevent legitimate users from denying transmission or reception of their messages. To achieve this security objective, a digital signature is usually adopted, which is efficient in transmission non-repudiation, while failing to deal with reception non-repudiation. Therefore, a carefully designed protocol is expected to resist reception non-repudiation.

Privacy preservation. Privacy preservation is a critical security requirement for a D2D communication system to be considered to be of practical implementation and commercialization. In particular, it is necessary to prevent users (both transmitters and receivers) from obtaining others' privacy information, such as identity, SIM card number, and position, during the processes of D2D communication. However, the privacy protection in D2D communication should be conditional, where senders and receivers are anonymous to each other while traceable by the trust authority (TA). With traceability, the TA is supposed to reveal the source ID of false messages.

Availability. Users may be frustrated if services become temporarily unavailable due to attacks such as free riding or DoS. Moreover, the system availability features are also influenced by the time period that the users wait for services in D2D communication.

13.4.3 Security Solutions

Despite its importance, very limited efforts have been made on security issues in D2D communication. In [19], instead of interference mitigation and avoidance, the authors novelly introduce D2D communication into the problem of secrecy capacity, such that a D2D pair could achieve its own transmission while the secrecy requirement of cellular communication is still satisfied. Moreover, they utilize secrecy outage probability to better depict the imperfect channel state information at the eavesdropper and derive the optimal transmission power capacity of cellular users. Evidently, the contribution is an effective way to enhance information-theoretic secrecy capacity while it does not consider the security requirements of D2D pairs.

The authors in [26] introduce the physical-social graphs to capture the physical constraints for feasible D2D cooperation and the social relationships among devices for effective cooperation. The authors propose a coalitional game theoretic approach to find the efficient D2D cooperation strategy and develop a network-assisted relay selection mechanism for implementing the coalitional game solution, which is a promising direction for treating security problems in D2D communication. However, this work does not consider and analyze the potential threats, such as eavesdropping or fabricating.

The authors in [27] propose an adaptation of a single cooperative jammer to increase the ambiguity at all malicious users by distracting them with artificial interference. Meanwhile, they derive the optimal joint transmission scheme by beamforming solution to maximize the secrecy rate of the intended user with the help of the cooperative jammer. The work improves the security of down link cellular networks without considering D2D communication scenarios. Yet they provide guidance for security in D2D communication.

Cryptography approaches for D2D communication security. The cryptography technique, which is widely used for information security in wireless communication, is expected to work for D2D communication. As little research has been contributed for D2D communication in this area, technologies may be inherited from existing work, such as the security mechanisms in wireless body area networks (WBANs) and VANETs, because they have network features similar to those of the D2D communication system, in spite of some differences. Next we review the main security approaches in WBANs and VANETs for a comprehensive understanding of the security strategies in different wireless communication environments, which cultivates an insight into D2D communication security.

In WBANs, to protect the patient-related data against malicious modification and to ensure privacy preservation, much research has focused on designing data security and privacy mechanisms. Considering the stringent energy and memory constraints, symmetric cryptography, which seems to be inherently well suited for low-end devices due to their relatively low overhead, is widely studied in WBANs. Previous works mainly concentrate on security issues in terms of key management [28], data transmission protocol [29, 30], and access control [31]. However, increasingly interest has been raised in exploring the feasibility of implementation of asymmetric encryption in wireless sensor networks, aiming at improving the security level by taking advantage of public key cryptography [32].

As the nodes in WBAN communicate with each other directly or in an ad hoc manner, which has the same characteristics as D2D communication, it is possible to apply the security approaches of WBAN in D2D systems with some modifications according to the demand of the specific D2D system.

In VANETs, the security issues become more challenging due to the unique features, such as high-speed mobility of the network, and the extremely large amount of network entities. Generally, the energy constraint is not as vital as that in WBANs, while the time efficiency is considered to be a critical constraint. As public key infrastructure satisfies most VANET security requirements, it is the most viable mechanism in terms of secure data transmission [33, 34], conditional privacy [35, 36], and authentication [37]. Meanwhile, to achieve the objectives of non-repudiation and certificate revocation [38], a digital signature is implemented via asymmetric cryptography, by which the sybil attack [39] is proposed as well.

Table 13.2 lists the comparisons of security issues in D2D, WBANs, and VANETs. It is notable that the network architecture of the VANETs, including trust authority (TA), roadside units (RSUs), and onboard units (OBUs), is similar to that D2D of the communication system composed by GW, eNB, and UEs. The differences between them lie in D2D links between UEs being an integral part of the normal cellular network, while in VANETs only dedicated short-range communication (DSRC) happens among the entities. Therefore, the security approaches in D2D communication environments are distinguished from those in VANET, yet some common grounds exist.

Cooperation stimulation for D2D communication. As D2D communication happens among the devices, the availability of the system largely depends on the cooperation degree of the devices. Accordingly, cooperation stimulation is a big challenge in D2D communication when applying to large-scale application. Currently, there is abundant research on cooperation stimulation in the context of mobile ad hoc networks, delay-tolerant networks, mobile social networks, VANETs, and so on. In general, the strategies fall into the following three categories:

Table 13.2 Security Approaches and Objectives Comparison

Systems		Security Approaches	Security Goals
D2D	[19]	—	Secrecy capacity and eavesdropping resistance
	[26]	—	Cooperation promotion
	[27]	—	Secrecy rate
WBANs	[28]	Symmetric encryption	Resilience to adversary intervention and network configurations
	[29]	Symmetric encryption	Privacy preservation and access control
	[30]	Symmetric encryption	Symmetric key generation based on physiological signals
	[31]	Symmetric encryption	Privacy preservation and authentication
	[32]	Hybrid	Content-oriented and contextual privacy, forging attack resistance
VANETs	[33]	PKI	Privacy preservation and cooperative data forwarding
	[34]	Symmetric encryption	Collisions and hidden terminal avoidance
	[35]	PKI	Sybil attack mitigation and privacy preservation
	[36]	PKI	Privacy preservation, traceability, and replication
	[37]	Hybrid	Anonymous authentication and traceability

'—' non cryptography approach is employed; 'PKI' public key infrastructure-based cryptography; 'Hybrid,' both symmetric and asymmetric encryption approaches are employed.

■ Reputation based. In reputation-based mechanisms, nodes monitor each other's behavior, evaluate each node's trustworthiness, cooperate with those who maintain good reputation, and detect misbehavior [40, 41].

■ Credit based. Credit-based systems treat packet-forwarding or data-sharing services as transactions so the communication nodes pay credit (or virtual currency) to the service providers [42, 43]

■ Tit for tat based. The basic goal of tit for tat-(or punishment-) based mechanisms is to cooperate with nodes that are cooperative and punish those that are selfish [44, 45].

Besides the above incentive approaches with one incentive measurement, there are also studies on integrating two of the strategies for fairness and effectiveness of cooperation stimulation [46, 47]. Moreover, the game theory approach is extensively used to realize cooperation stimulation [43, 47, 48].

Because D2D communication systems have similar characteristics with the wireless networks listed above, it is reasonable to assume that the cooperation stimulation mechanisms above are applicable in D2D communication with small changes considering the particular features of users with the devices, e.g., user experience, social relationships, and types of UEs.

13.4.4 Challenges and Open Issues

The security solutions of current wireless networks may give some guidance for security strategies of D2D communication systems. But many unique characteristics of the D2D systems bring out new research challenges and open issues in this area.

Cryptography schemes. Encryption may satisfy D2D communication security requirements in terms of data confidentiality and integrity and entity authentication. However, it still faces challenges including key management, privacy preservation, and access control. Moreover, how to integrate symmetric and asymmetric encryption into a D2D communication system remains an open issue.

Game theory-based cooperation stimulation. Game theory is a prospective method likely to contribute to cooperation stimulation in D2D communication. The main obstacles lie in forming a proper game and searching the optimal equilibrium, which causes the users to be cooperative.

Physical layer security. Physical layer security for wireless communications is currently a wide open research area, particularly due to the fact that physical layer security depends heavily on extensive knowledge of security theory and on physical layer architecture expertise [49]. Current proposals for addressing physical layer security can be of interest in developing future security mechanisms combining with resource allocation and interference management for D2D wireless communications, which is a significant direction worthy of investigation.

Mobile social network-based D2D communication security. A specially interesting fashion that may be devoted to D2D communication security is mobile social networks (MSNs), which are attracting increasing attention in the research community and industry [50]. The main advantage for exploring mobile social network-based D2D communication security is the members of these social networks themselves, as they are able not only to control the devices but also to verify their peers from social judgment. But the perfect outcome poses obstructions, as it is a harsh task to construct social trust and social ties with privacy preservation, which remain open issues.

13.5 Conclusions

We have addressed content dissemination and security schemes for the D2D communication system in this chapter. We propose possible solutions by reviewing and analyzing the current contributions on these issues in wireless networks, comparing the characteristics of the corresponding network with those of the D2D system and drawing from the experiences of mature strategies. In particular, based on the research results on VANETs, MSNs, and P2P networks, we propose the system architecture of content dissemination for the D2D system and propose that devices mobility, content diversity, distributed implementation and QoE are the main challenges in content dissemination for D2D system. Meanwhile, we identify threat models and security requirements in D2D communication, and propose that cryptography approaches, game theory-based cooperation stimulation, physical layer security, and mobile social network-based D2D communication security may dominate the possible security strategies in D2D communication.

References

1. Lei, L., Zhong, Z., Lin, C., and Shen, X. (2012). Operator controlled device-to-device communications in LTE-Advanced networks. *IEEE Wireless Communications*, 19(3), 96–104.

2. Doppler, K., Rinne, M., Wijting, C., Ribeiro, C., and Hugl, K. (2009). Device-to-device communication as an underlay to LTE-Advanced networks. *IEEE Communications Magazine*, 47(12), 42–49.

3. Zhou, L., Zhang, Y., Song, K., Jing, W., and Vasilakos, A. (2011). Distributed media-service scheme for P2P-based vehicular networks. *IEEE Transactions on Vehicular Technology*, 60(2), 692–703.

4. Golrezaei, N., Dimakis, A., and Molisch, A. (2012). Wireless device-to-device communications with distributed caching. IEEE International Symposium on Information Theory, Cambridge, MA, USA.

5. Zhou, L., Wang, X., Tu, W., Mutean, G., and Geller, B. (2010). Distributed scheduling scheme for video streaming over multi-channel multi-radio multi-hop wireless networks. *IEEE Journal on Selected Areas in Communications*, 28(3), 409–419.

6. Yang, M., Lim, S. Y., Park, H. J., and Park, N. H. (2013). Solving the data overload: device-to-device bearer control architecture for cellular data offloading. *IEEE Vehicular Technology Magazine*, 8(1), 31–39.

7. Ioannidis, S., Chaintreau, A., and Massoulie, L. (2009). Optimal and scalable distribution of content updates over a mobile social network. IEEE International Conference on Computer Communications, Rio de Janeiro.

8. Qian, Y. and Moayeri, N. (2008). Design of secure and application-oriented VANETs. IEEE Vehicular Technology Conference, Singapore.

9. Lu, K., Qian, Y., and Chen, H. (2007). A secure and service-oriented network control framework for WiMAX networks. *IEEE Communications*, 45(5), 124–130.

10. Lu, K., Qian, Y., Guizani, M., and Chen, H. (2008). A framework for a distributed key management scheme in heterogeneous wireless sensor networks. *IEEE Transactions on Wireless Communications*, 7(2), 639–647.

11. Zhou, L., Chao, H., and Vasilakos, A. (2011). Joint forensics-scheduling strategy for delay-sensitive multimedia applications over heterogeneous networks. *IEEE Journal on Selected Areas in Communications*, 29(7), 1358–1367.

12. Fodor, G., Dahlman, E., Mildh, G., Parkvall, S., Reider, N., and Miklós, G. (2012). Design aspects of network assisted device-to-device communications. *IEEE Communications Magazine*, 50(3), 170–177.

13. Chao, S. L., Lee, H. Y., Chou, C. C., and Wei, H. Y. (2013). Bio-inspired proximity discovery and synchronization for D2D communications. *IEEE Communications Letters*, 17(12), 2300–2303.

14. Wen, S., Zhu, X., Zhang, X., and Yang, D. (2013). QoS-aware mode selection and resource allocation scheme for device-to-device (D2D) communication in cellular networks. IEEE Internet Conference on Communications, Budapest, Hungary.

15. Yu, C. H., Klaus, D., Cássio B. R., and Olav, T. (2011). Resource sharing optimization for device-to-device communication underlaying cellular networks. *IEEE Transactions on Wireless Communications*, 10(8), 2752–2763.

16. Erturk, M. C., Mukherjee, S., Ishii, H., and Huseyin, A. (2013). Distributions of transmit power and SINR in device-to-device networks. *IEEE Communications Letters*, 17(2), 273–276.

17. Fodor, G., Belleschi, D. D. P. M., Johansson, M., and Abrardo, A. (2013). A comparative study of power control approaches for device-to-device communications. IEEE Internet Conference on Communications, Budapest, Hungary.

18. Zhou, L., and Chao, H. (2011). Multimedia traffic security architecture for Internet of things. *IEEE Network*, 25(3), 35–40.

19. Yue, J., Ma, C., Yu, H., and Zhou, W. (2013). Secrecy-based access control for device-to-device communication underlaying cellular networks. *IEEE Communication Letters*, 17(11), 2068–2071.

20. Shakkottai, S. and Johari, R. (2010). Demand-aware content distribution on the Internet. *IEEE/ACM Transactions on Networking*, 18(2), 476–489.

21. Zhao, J., Zhang, Y., and Cao, G. (2007). Data pouring and buffering on the road: a new data dissemination paradigm for vehicular ad hoc networks. *IEEE Transaction on Vehicular Technology*, 56(6), 3266–3277.

22. Zhao, J. and Cao, G. (2008). VADD: vehicle-assisted data delivery in vehicular ad hoc networks. *IEEE Transacion on Vehicular Technology*, 57(3), 1910–1922.

23. Bass, F. M. (1969). A new product growth model for consumer durables. *Management Science*, 15, 215–227.

24. Zhou, L., Wang, H., and Guizani, M. (2012). How mobility impacts video streaming over multi-hop wireless networks. *IEEE Transactions on Communications*, 60(7), 2017–2028.

25. ITU-T Recommendation J.144R1. (2007). *Objective perceptual video quality measurement techniques for digital cable television in the presence of a full reference.*

26. Chen, X., Proulx, B., Gong, X., and Zhang, J. (2013). Social trust and social reciprocity based cooperative D2D communications. ACM International Symposium on Mobile Ad Hoc Networking and Computing (MOBIHOC), Bangalore, India.

27. Jeong, S., Lee, K., Huh, H., and Kang, J. (2013). Secure transmission in downlink cellular network with a cooperative jammer. *IEEE Wireless Communications Letters*, 2(4), 463–466.

28. Yu, C., Lu, C., and Kuo, S. (2010). Noninteractive pairwise key establishment for sensor networks. *IEEE Transactions on Information Forensics and Security*, 5(3), 556–569.

29. Lu, R., Lin, X., and Shen, X. (2012). SPOC: a secure and privacy-preserving opportunistic computing framework for mobile-healthcare emergency. *IEEE Transactions on Parallel and Distributed Systems*, 24(3), 614–624.

30. Mana, M., Feham, M., and Bensaber, B. (2011). Trust key management scheme for wireless body area networks. *International Journal of Network Security*, 12(2), 75–83.

31. He, D., Bu, J., Zhu, S., Chan, S., and Chen, C. Distributed access control with privacy support in wireless sensor networks. *IEEE Transactions on Wireless Communications*, 10(10), 3472–3481.

32. Lin, X., Lu, R., Shen, X., Nemoto, Y., and Kato, N. (2009). SAGE: a strong privacy-preserving scheme against global eavesdropping for eHealth systems. *IEEE Journal on Selected Areas in Communications*, 27(4), 365–377.

33. Liang, X., Li, X., Lu, R., Lin, X., and Shen, X. (2012). Morality-driven data forwarding with privacy preservation in mobile social network. *IEEE Transaction on Vehicle Technology*, 61(7), 3209–3222.

34. Hao, Y., Tang, J., and Cheng, Y. (2013). Secure cooperative data downloading in vehicular ad hoc networks. *IEEE Journal on Selected Areas in Communications*, 31(9), 523–537.

35. Lin, X. (2013). LSR: mitigating zero-day sybil vulnerability in privacy-preserving vehicular peer-to-peer network. *IEEE Journal on Selected Areas in Communications*, 31(9), 237–246.

36. Lin, X., Sun, X., Ho, P., and Shen, X. (2007). GSIS: a secure and privacy preserving protocol for vehicular communications. *IEEE transactions on Vehicle Technology*, 56(6), 3442–3456.

37. Park, Y., Sur, C., Jung, C., and Rhee, K. (2010). An efficient anonymous authentication protocol for secure vehicular communications. *Journal of Information Science and Engineering*, 26, 785–800.

38. Zhang, Y., Liu, W., Lou, W., and Fang, Y. (2006). Securing mobile ad hoc networks with certificateless public keys. *IEEE Transactions on Dependable Secure Computing*, 3(4), 386–399.

39. Douceur, J. R. (2002). The sybil attack. *IPTPS*, 251–260.

40. Yu, W. and Liu K. J. R. (2007). Game theoretic analysis of cooperation stimulation and security in autonomous mobile ad hoc networks. *IEEE Transacrions on Mobile Computing*, 6(5), 507–521.

41. Zhang, G., Yang, K., Liu, P., Yang, X., and Ding, E. (2012). Resource-exchange based cooperation stimulating mechanism for wireless ad hoc networks. IEEE Internet Conference on Communications, Ottawa, Ontario, Canada.

42. Mahmoud, M. E. A. and Shen, X. (2012). FESCIM: fair, efficient, and secure cooperation incentive mechanism for multihop cellular networks. *IEEE Transactions on Mobile Computing*, 11(5), 753–766.

43. Kang, X. and Wu, Y. (2013). A game-theoretic approach for cooperation stimulation in peer-to-peer streaming networks. IEEE Internet Conference on Communications, Budapest, Hungary.

44. Zhou, H., Chen, J., Fan, J., Du, Y., and Das, S. K. (2013). ConSub: incentive-based content subscribing in selfish opportunistic mobile networks. *IEEE Journal on Selected Areas in Communications*, 31(9), 669–679.

45. Niu, B., Zhao, H. V., and Jiang, H. (2011). A cooperation stimulation strategy in wireless multicast networks. *IEEE Transactions on Signal Processing*, 59(5), 2355–2369.

46. Lu, R., Lin, X., Zhu, H., and Shen, X. (2010). Pi: a practical incentive protocol for delay tolerant networks. *IEEE Transacrions on Wireless Communications*, 9(4), 1483–1493.

47. Li, Z. and Shen, H. (2012). Game-theoretic analysis of cooperation incentive strategies in mobile ad Hoc networks. *IEEE Transacrions on Mobile Computing*, 11(8), 1287–1303.

48. Chen, T., Zhu, L., Wu, F., and Zhong, S. (2011). Stimulating cooperation in vehicular ad hoc networks: a coalitional game theoretic approach. *IEEE Transactions on Vehicular Technology*, 60(2), 566–579.

49. Granjal, J., Monteiro, E., and Silva, J. (2013). Security issues and approaches on wireless M2M systems. *Wireless Network and Security*, 133–164.

50. Vastardis, N. and Yang, K. (2013). Mobile social networks: architectures, social properties, and key research challenges. *IEEE Communications Surveys and Tutorials*, 15(3), 1355–1371.

Index